WATER

A COMPREHENSIVE TREATISE

Volume 6

Recent Advances

WATER

A COMPREHENSIVE TREATISE

Edited by Felix Franks

WATER
A COMPREHENSIVE TREATISE

Edited by Felix Franks

Department of Botany
University of Cambridge
Cambridge, England

Volume 6
Recent Advances

PLENUM PRESS • NEW YORK AND LONDON

Library of Congress Cataloging in Publication Data

Franks, Felix.
  Water: a comprehensive treatise.

  Includes bibliographies.
  CONTENTS: v. 1. The physics and physical chemistry of water.
—v. 2. Water in crystalline hydrates; aqueous solutions of simple nonelectrolytes.—[etc.]—v. 6. Recent advances.
  1. Water. I. Title.
QD169.W3F7   546'.22   78-165694
ISBN 0-306-40139-8 (vol. 6)

©1979 Plenum Press, New York
A Division of Plenum Publishing Corporation
227 West 17th Street, New York, N.Y. 10011

Printed in the United States of America

Preface

Since the publication of the previous volumes many new aspects of the physical and life sciences have been developed in which the properties of water play a dominant role.

Although, according to its preface, Volume 5 was to be the last one of the treatise, these recent developments have led to a revision of that statement. The present volume and its companion, still in preparation, deal with topics that were already mentioned in the preface to Volume 5 as gaining in importance.

The recent development of X-ray and, more particularly, neutron scattering techniques have led to studies of "structure" in aqueous solutions of electrolytes on the one hand, and to the role of water in protein structure and function on the other. Both these topics have reached a stage where reviews of the present state of knowledge are useful.

The application of *ab initio* methods to calculations of hydration and conformation of small molecules has a longer history, but here again a critical summary is timely.

The role of solvent effects in reaction kinetics and mechanisms should have had a place in Volume 2 of this treatise, but, as sometimes happens, the author who had taken on this task failed to live up to his promise. However, since 1972 the physical chemistry of mixed aqueous solvents has made considerable strides, so that the belated discussion of this topic (by a new author) is built on evidence that was not available at the time of publication of Volume 2.

Hand in hand with the development of more penetrating experimental techniques there have been developments in theoretical approaches and computer simulation methods to probe the nature of the liquid state in general, and of water in particular. This volume therefore includes contributions on both these topics. At the time of writing it seems that the "hydrophobic" effect is still a topic of great importance; it crops up in several of the chapters constituting this volume, and it is at last being considered worthy of study by the theoreticians.

Of necessity, this volume, as distinct from its predecessors, has the character of a "Recent Advances" publication, the connecting thread being water. It becomes clear that we are still far from appreciating fully the extent to which the eccentric properties of water shape many physicochemical and most biochemical processes that are normally studied without due regard to solvent effects.

Once again it is a pleasure to thank Joyce Johnson for her willing and cheerful cooperation in the editorial work. I also thank the contributing authors, some of whom produced manuscripts at quite short notice.

My main debt of gratitude goes to my wife Hedy for her constant support, practical and spiritual, during the preparation of this volume, which coincided with a very difficult and upsetting period of my professional life.

Felix Franks

Department of Botany
University of Cambridge

Contents

Chapter 1

X-Ray and Neutron Scattering by Aqueous Solutions of Electrolytes

J. E. Enderby and G. W. Neilson

Chapter 4

Mixed Aqueous Solvent Effects on Kinetics and Mechanisms of Organic Reactions

J. B. F. N. Engberts

Chapter 5

Solvent Structure and Hydrophobic Solutions

D. Y. C. Chan, D. J. Mitchell, B. W. Ninham, and B. A. Pailthorpe

Chapter 6

Computer Simulation of Water and Aqueous Solutions

D. W. Wood

Contents of Earlier Volumes

X-Ray and Neutron Scattering by Aqueous Solutions of Electrolytes

J. E. Enderby and G. W. Neilson

H. H. Wills Physics Laboratory
Royal Fort
Tyndall Avenue
Bristol BS8 1TL

1. THE BASIC STRUCTURAL PROBLEM

1.1. Introduction

The composition of aqueous solutions* can be thought of as either (a) salt and water, or (b) anions, cations, and water, or (c) anions, cations, oxygen atoms, and hydrogen atoms. Thus, depending on circumstances, aqueous solutions can be thought of as either two-component, three-component, or four-component liquids. Macroscopic physical chemistry (e.g., measurements of solubility) uses the description implicit in (a), while in the microscopic Debye–Hückel theory, in its primitive form, the solution is regarded as a three-component system. When, however, we consider the scattering of radiation, solutions must be thought of in terms of description (c). Each "entity" scatters incident radiation in a characteristic way. In neutron scattering, for example, the primary scattering mechanism is the interaction of neutrons with each of the *nuclei* of the four components. In X-ray scattering, the situation is not quite so clear cut (Section 2), but for practical purposes we can ascribe to each chemical

---

* We are limiting our discussion to strong electrolytes of the form MX_n dissolved in D_2O or H_2O.

1

element a distribution of electrons that interacts *as a whole* with the primary beam. Formally, then, aqueous solutions must be thought of as four-component systems, each component of which has a characteristic scattering factor, which depends on the nature of the radiation used. For convenience we will call each component an "ion type"; thus, the four ion types in a solution of a salt MX_n dissolved in water H_2O are M, X, H, and O.

In order to set up a theory for the scattering of radiation from liquids that contain several ion types, let us label each ion type by the dummy suffices α and β, which may take values $1, 2, \ldots, j$ for a liquid containing j ion types. The atomic concentration of the α ion type is denoted by c_α and is subject to the sum rule

$$\sum_{\alpha=1}^{j} c_\alpha = 1$$

If any type of radiation is incident on a mixed assembly of scatterers a measure of the amplitude of the scattered waves is given by

$$\sum_{\alpha} f_\alpha \sum_{i(\alpha)} \exp(-i\mathbf{k} \cdot \mathbf{r}_{i(\alpha)})$$

where f_α is an appropriate scattering factor and $\mathbf{r}_i(\alpha)$ denotes the position of the ith ion of α type. The quantity $\hbar\mathbf{k}$ is known as the momentum transfer and is sometimes called $\hbar\mathbf{s}$ (in the X-ray literature), $\hbar\mathbf{Q}$ (in the neutron scattering literature), or $\hbar\mathbf{K}$ (in the electron transport theory). We use $\mathbf{k}$ in this article so as to conform with the notation employed by Hansen and MacDonald[381] in their excellent book on the liquid state. The mean absolute intensity, which we denote $d\sigma/d\Omega$, becomes

$$\sum_{\alpha} \sum_{\beta} f_\alpha f_\beta^* \overline{\sum_{i(\alpha)} \sum_{j(\beta)} \exp[i\mathbf{k} \cdot (\mathbf{r}_j(\beta) - \mathbf{r}_i(\alpha)]}$$
$$= \sum_{\alpha} \sum_{\beta} f_\alpha f_\beta^* N c_\alpha \left(\delta_{\alpha\beta} + \sum_{i \neq j} \cos \mathbf{k} \cdot \mathbf{r}_{ji} \right) \tag{1}$$

where N is the total number of ions in the sample.

Let us now introduce *partial structure factors* $S_{\alpha\beta}$ defined by

$$S_{\alpha\beta}(k) = 1 + \frac{4\pi N}{kV} \int_0^\infty dr [g_{\alpha\beta}(r) - 1] r \sin kr \tag{2}$$

where V is the volume of the sample. In (2), $g_{\alpha\beta}(r)$ is the *pair distribution function* which measures the average distribution of type β ion observed from an α ion at the origin. $g_{\alpha\beta}(r)$ tends, like the analogous quantity for a one-component system, to unity at large values of r. If there is an α ion at $r = 0$, the probability of finding a β ion at the same instant with its center in a small element of volume $d\mathbf{r}$ located at $\mathbf{r}$ is, for an isotropic liquid,

$$4\pi\varrho c_\beta g_{\alpha\beta}(r) r^2 \, dr$$

where $\varrho = N/V$, the total number density. In terms of $S_{\alpha\beta}$, we can rewrite (1) as

$$\frac{d\sigma}{d\Omega} = \sum_\alpha N c_\alpha f_\alpha f_\alpha^* + \sum_\alpha \sum_\beta N c_\alpha c_\beta f_\alpha f_\beta^* (S_{\alpha\beta} - 1)$$

and for those cases where f is real, more simply as

$$\frac{d\sigma}{d\Omega} = N\left[\sum_\alpha c_\alpha f_\alpha^2 + F_T(k) \right]$$

where

$$F_T(k) = \sum_\alpha \sum_\beta c_\alpha c_\beta f_\alpha f_\beta (S_{\alpha\beta} - 1) \tag{3}$$

It will be useful at this point to write down an explicit expression for $F_T(k)$ for the system we are considering, i.e., MX_n dissolved in H_2O. If c represents the atomic concentration of M, we find by a straightforward expansion of eqn. (3) that

$$
\begin{aligned}
F_T(k) = {}& \tfrac{1}{9}(1 - c - nc)^2 f_O{}^2 (S_{OO} - 1) + \tfrac{4}{9}(1 - c - nc)^2 f_H{}^2 (S_{HH} - 1) \\
& + \tfrac{4}{9}(1 - c - nc)^2 f_O f_H (S_{OH} - 1) + c^2 f_M{}^2 (S_{MM} - 1) \\
& + n^2 c^2 f_X{}^2 (S_{XX} - 1) + 2nc^2 f_X f_M (S_{XM} - 1) \\
& + \tfrac{2}{3}c(1 - c - nc) f_M f_O (S_{MO} - 1) + \tfrac{4}{3}c(1 - c - nc) f_M f_H (S_{MX} - 1) \\
& + \tfrac{4}{3}nc(1 - c - nc) f_X f_H (S_{XH} - 1) \\
& + \tfrac{2}{3}nc(1 - c - nc) f_X f_O (S_{XO} - 1)
\end{aligned}
\tag{4}
$$

In this article we shall refer to $F(k)$ as the "total scattering function." Since it depends explicitly on f_α we must distinguish between the total scattering functions from X-ray and neutron beams; to avoid a notation that is too clumsy we shall denote by $F_T(k)$ the total scattering using an unspecified radiation; X-ray scattering will be represented by $F_x(k)$ and neutron scattering by, simply, $F(k)$.

If we perform scattering experiments on systems containing more than one ion type, the quantity that can be extracted—and even then not

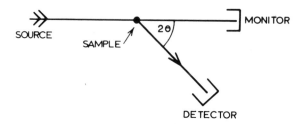

Fig. 1. Conventional layout for diffraction studies.

directly—is $F_T(k)$ and not $S_{\alpha\beta}$. In practice, we observe an intensity I of either neutrons, X-rays, or electrons as a function of a scattering angle 2θ in the geometry shown in Fig. 1.

Quite generally,

$$I(\theta) = \alpha(\theta)\left[\frac{d\sigma}{d\Omega} + \delta(\theta)\right] \tag{5}$$

and the challenge that faces experimentalists is the determination by theory, experiment, or both, the *calibration parameters*, $\alpha(\theta)$ and $\delta(\theta)$. An important series of internal consistency checks can be made on $\alpha(\theta)$ and $\delta(\theta)$ provided we know some general properties of $S_{\alpha\beta}(k)$. We therefore summarize these properties in Section 1.2.

1.2. The Properties of $S_{\alpha\beta}(k)$

1.2.1. Behavior at High k

The relationship between $g_{\alpha\beta}$ and $S_{\alpha\beta}$ [eqn. (2)] can be rewritten in the form

$$g_{\alpha\beta}(r) = 1 + \frac{1}{2\pi^2\varrho r} \int_0^\infty dk\,[S_{\alpha\beta}(k) - 1]k \sin kr \tag{6}$$

and for the integral on the right-hand side to be convergent, $S_{\alpha\beta}(k)$ must tend to unity at large k. We can therefore set

$$F_T(k)_{k\to\infty} = 0$$

which implies [from eqn. (3)] that

$$\left.\frac{d\sigma}{d\Omega}\right|_{k\to\infty} = N\sum c_\alpha f_\alpha^2 \tag{7}$$

1.2.2. Behavior at Low k

At low k, and at finite temperatures, the existence of both concentration and density fluctuations ensure that $S_{\alpha\beta}(0) \neq 0$. Each of the ten partial structure factors can be calculated, in the small-k (long-wavelength) limit from a knowledge of the thermodynamic parameters.[66]

Consider an open system of volume V containing $\mathscr{N}_a$ molecules of H_2O (the solvent) and $\mathscr{N}_b$ molecules of MX_n (the solute). Let $\langle \mathscr{N}_a \rangle$ and $\langle \mathscr{N}_b \rangle$ represent the ensemble average of $\mathscr{N}_a$ and $\mathscr{N}_b$ so that the deviations of particle numbers from their mean values may be expressed as

$$\Delta \mathscr{N}_a = \mathscr{N}_a - \langle \mathscr{N}_a \rangle \quad \text{and} \quad \Delta \mathscr{N}_b = \mathscr{N}_b - \langle \mathscr{N}_b \rangle$$

In terms of mean-square fluctuations, we can write

$$\langle (\Delta \mathscr{N}_a)^2 \rangle = \langle \mathscr{N}_a^2 \rangle - \langle \mathscr{N}_a \rangle^2$$

$$\langle (\Delta \mathscr{N}_b)^2 \rangle = \langle \mathscr{N}_b^2 \rangle - \langle \mathscr{N}_b \rangle^2$$

$$\langle \Delta \mathscr{N}_a \Delta \mathscr{N}_b \rangle = \langle \mathscr{N}_a \mathscr{N}_b \rangle - \langle \mathscr{N}_a \rangle \langle \mathscr{N}_b \rangle$$

Beeby[66] shows how $S_{\alpha\beta}(0)$ can be evaluated from a knowledge of these fluctuations. Generalizing the results to our system, we find that

$$S_{MM} = 1 + \frac{N}{\langle \mathscr{N}_b \rangle} \left(\frac{\langle (\Delta \mathscr{N}_b)^2 \rangle}{\langle \mathscr{N}_b \rangle} - 1 \right) \tag{8a}$$

$$S_{XX} = 1 + \frac{N}{\langle \mathscr{N}_b \rangle} \left(\frac{\langle (\Delta \mathscr{N}_b)^2 \rangle}{\langle \mathscr{N}_b \rangle} - \frac{1}{n} \right) \tag{8b}$$

$$S_{MX} = 1 + \frac{N}{\langle \mathscr{N}_b \rangle} \left(\frac{\langle (\Delta \mathscr{N}_b)^2 \rangle}{\langle \mathscr{N}_b \rangle} \right) \tag{8c}$$

$$S_{MO} = S_{XO} = S_{MH} = S_{XH} = 1 + N \frac{\langle \Delta \mathscr{N}_a \Delta \mathscr{N}_b \rangle}{\langle \mathscr{N}_a \rangle \langle \mathscr{N}_b \rangle} \tag{8d}$$

$$S_{OO} = 1 + \frac{N}{\langle \mathscr{N}_a \rangle} \left(\frac{\langle (\Delta \mathscr{N}_a)^2 \rangle}{\langle \mathscr{N}_a \rangle} - 1 \right) \tag{8e}$$

$$S_{HH} = 1 + \frac{N}{\langle \mathscr{N}_a \rangle} \left(\frac{\langle (\Delta \mathscr{N}_a)^2 \rangle}{\langle \mathscr{N}_a \rangle} - \frac{1}{2} \right) \tag{8f}$$

$$S_{OH} = 1 + \frac{N}{\langle \mathscr{N}_a \rangle} \left(\frac{\langle (\Delta \mathscr{N}_a)^2 \rangle}{\langle \mathscr{N}_a \rangle} \right) \tag{8g}$$

where N, as before, is the total number of ions in the sample.

The fluctuations are then related to the compressibility, x_T, the partial molar volumes, v_a and v_b, and the rate of change of osmotic pressure, p, with concentration x ($= \langle \mathcal{N}_b \rangle / \langle \mathcal{N}_a + \mathcal{N}_b \rangle$). This can be readily achieved by the methods of statistical mechanics with the following results:

$$\frac{\langle (\Delta \mathcal{N}_a)^2 \rangle}{\langle \mathcal{N}_a \rangle} = k_B T \frac{\langle \mathcal{N}_a + \mathcal{N}_b \rangle^2}{V} \left[\frac{(1-x) x_T}{\langle \mathcal{N}_a + \mathcal{N}_b \rangle} + \frac{x}{1-x} \frac{v_b^2}{V N_0 v_a p'} \right] \tag{9a}$$

$$\frac{\langle (\Delta \mathcal{N}_b)^2 \rangle}{\langle \mathcal{N}_b \rangle} = k_B T \frac{\langle \mathcal{N}_a + \mathcal{N}_b \rangle^2}{V} \left[\frac{x x_T}{\langle \mathcal{N}_a + \mathcal{N}_b \rangle} + \frac{v_a^2}{V N_0 v_b p'} \right] \tag{9b}$$

$$\frac{\langle \Delta \mathcal{N}_a \Delta \mathcal{N}_b \rangle}{\langle \mathcal{N}_b \rangle} = k_B T \frac{\langle \mathcal{N}_a + \mathcal{N}_b \rangle^2}{V} \left[\frac{(1-x) x_T}{\langle \mathcal{N}_a + \mathcal{N}_b \rangle} + \frac{v_b}{V N_0 p'} \right] \tag{9c}$$

In eqn. (9), N_0 is Avogadro's number, the partial molar volumes v_a and v_b are defined through

$$v_i = \frac{1}{N_0} \left(\frac{\partial v_i}{\partial \mathcal{N}_i} \right)_{P, \mathcal{N}_k} \qquad (i, k = a \text{ or } b)$$

and $p' = dp/dx$.

The final step is to evaluate v_i and p'. The partial molar volumes can be derived from a knowledge of the concentration dependence of the density. The osmotic pressure is in general difficult to measure directly but can be derived from the osmotic coefficient ϕ through

$$p = \frac{(n+1) N_0 k_B T}{v_a} \left(\frac{x}{1-x} \right) \phi$$

1.2.3. Intermediate Behavior

Two important results are worth mentioning as these are useful in deciding the validity or otherwise of proposed $S_{\alpha\beta}(k)$.[279] The quantity $d\sigma/d\Omega$ is positive definite, so that

$$c_\alpha + c_\alpha^2 [S_{\alpha\alpha}(k) - 1] > 0$$

$$c_\beta + c_\beta^2 [S_{\beta\beta}(k) - 1] - \frac{c_\alpha^2 c_\beta^2 [S_{\alpha\beta}(k) - 1]}{c_\alpha + c_\alpha^2 (S_{\alpha\alpha} - 1)} > 0$$

The fact that $g_{\alpha\beta}(0) \equiv 0$ means that each $S_{\alpha\beta}$ must satisfy a sum rule given by

$$\int_0^\infty dk \, [S_{\alpha\beta}(k) - 1] k^2 = -2\pi^2 \varrho \tag{10}$$

This latter condition is particularly useful for checking the accuracy of normalization procedures and will be exploited particularly in Sections 2 and 3. It follows immediately from eqn. (6) by considering the limit as $r \rightarrow 0$.

1.3. Termination Errors

In evaluating integrals like those in eqn. (6) a numerical problem arises from the fact that there is an upper limit for k beyond which data are inaccessible. With 0.5 Å neutrons from reactor sources for example, the practical limit is ~ 20 Å^{-1}. This premature termination introduces two sources of error; the first of these appears as high-frequency ripples in $g_{\alpha\beta}(r)$, which are particularly troublesome around the principal peak. The safest way to test for real structure is to change the maximum value of k used in evaluating the integral in eqn. (6) and to calculate $g_{\alpha\beta}(r)$ with and without a "window" function. Unfortunately not all experimenters have, in the past, applied these tests to their data so that some care in relating minor features in $g_{\alpha\beta}(r)$ to real structure is necessary.

The second problem arises from the fact that very sharp features in $g_{\alpha\beta}(r)$ produce oscillations in $S_{\alpha\beta}(k)$ out to high k values. Premature termination will lead to a serious loss in resolution and will broaden out features that are in fact well defined. This is a problem to which we shall return in Sections 2 and 3.

2. X-RAY STUDIES

2.1. Introduction

X-ray diffraction arises from the scattering of X rays by all the electrons in a liquid. It can be considered as an essentially elastic process and unlike neutron diffraction does not suffer from kinematic distortions (Section 3).

The principal contributions to the calibration parameters $\alpha(\theta)$ and $\delta(\theta)$ are those due to Compton scattering, polarization, absorption, and sample geometry. A full account of the methods that are used to deduce these quantities has been given by Narten and Levy.[645]

Once the corrections have been made, the data can be put on an absolute scale by making the scattering patterns at large k go to a sum over individual atomic scattering factors as shown in eqn. (7). It must be emphasized that the X-ray "form factor" $f(k)$ depends strongly on k and, in particular, falls to very low values at high k. All else being equal, form factors are proportional to the atomic number Z since X-ray scattering arises from electron–photon interactions. Hence the intensity of scattering (being proportional to the product of two form factors) will be that much greater for heavy elements. In practice it is impossible with laboratory X-ray sources to identify hydrogen atom positions directly; indirect methods use sophisticated data refinement techniques which assume an *a priori* knowledge of the position of some of the other elements in the sample. It is not possible to obtain hydrogen positions in aqueous solutions by X-ray methods alone.

If we consider the data expressed in r space, a total $G_x(r)$ can be obtained through

$$G_x(r) = \frac{1}{2\pi^2 \varrho r} \int dk \, F_x(k) k \sin kr \tag{11}$$

Thus because $F_x(k)$ contains both $S_{\alpha\beta}(k)$ and $f_\alpha(k)$, $G_x(r)$ is *not* a linear combination of $g_{\alpha\beta}(r)$. For this reason, some authors prefer to multiply $F(k)$ by

$$M(k) = \left[\sum_\alpha c_\alpha f_\alpha(k) \right]^{-2}$$

on the grounds that $f_\alpha(k)f_\beta(k)M(k)$ will be more or less independent of k. The modified distribution function $G_x^M(r)$ obtained in this way is related to $g_{\alpha\beta}(r)$ through a convolution function thus:

$$G_x^M(r) = \sum_\alpha \sum_\beta G_{\alpha\beta}(r)$$

with

$$G_{\alpha\beta} = \frac{1}{r} \int_{-\infty}^{\infty} du \, u g_{\alpha\beta}(u) T_{\alpha\beta}(u - r)$$

where

$$T_{\alpha\beta}(r) = \frac{1}{\pi} \int_0^\infty dk \, c_\alpha f_\alpha c_\beta f_\beta M(k) \cos kr$$

Once $G_x(r)$ [or $G_x^M(r)$] has been calculated in a self-consistent manner, features in it must be related to the ten radial distribution functions that characterize the solution. In the more systematic investigations this is usually achieved in two steps:

(a) In order to identify the peaks in $G_x(r)$, X-ray patterns are taken at a variety of concentrations. Changes in amplitude are monitored, so that a given peak can be related to a particular radial distribution function.

Reference is often made to data derived from studies of the hydrated solid. Calculations are then undertaken to estimate possible nearest-neighbor coordination numbers for particular ions.

(b) A model is proposed making use of the above information and attempts made to reconstruct $G_x(r)$. Acceptable fits can usually be obtained between this model $G_x(r)$ and that derived from the diffraction experiment. However, this can be misleading, as many features that appear at low k may become washed out in the Fourier inversion.

Thus even if a "perfect" experiment were carried out, X-ray methods *inherently* yield data with low resolution so far as structural studies of aqueous solutions are concerned. We conclude the section by stating explicitly why this is the case.

(a) $F_x(k)$ contains contributions from ten structure factors and is therefore an exceedingly complex quantity.

(b) $G_x(r)$ is not a linear combination of $g_{\alpha\beta}(r)$.

(c) The k dependence of $f(k)$ does not allow high-quality data to be obtained at high k, thereby reducing the resolution in real space.

(d) The Z dependence of $f(k)$ does not allow the positions of light elements to be determined.

In spite of these fundamental problems, X-ray data have, over the years, provided real insight into the nature of solutions, and in the next section we review some of the outstanding work carried out in this field.

2.2. A Survey of X-Ray Studies

X-ray diffraction experiments on aqueous solutions began in the 1930's. Although the data were not, by today's standards, of high quality, the scientific insight gained was often quite remarkable. In the classic papers by Bernal[96,99] and Fowler, a variety of modeling techniques to interpret diffraction data were employed. Their ideas concerning structural temperature and ion–water interaction remain part of the vocabulary of the subject today. Between 1940 and 1957, relatively few X-ray investigations were made, and structural studies did not begin again until the work of Brady in 1957.[139] Several excellent reviews of the early work have appeared in the literature. The general review by Kruh[523] covers investigations up to 1961 and describes the structural information that has been obtained for a variety of liquids. A more recent article by Karnicky and Pings[467] reviews the current status of X-ray diffraction studies for liquids and describes in detail the different approaches used in handling experimental data.

Table I summarizes basic structural properties derived from X-ray diffraction. Where errors have been estimated they are listed in parentheses, otherwise they are omitted. As far as we are aware there have been no systematic attempts to determine the experimental resolution of the data and how it effects the apparent structure. At low k, knowledge of the resolution functions is important since it is in this region that any long-range correlations manifest themselves.

The pioneering work of Brady[139] set out to investigate whether it would be possible to measure quantitatively the ion–ion and ion–water distances and the degree of ionic hydration. He also discussed the diffraction patterns qualitatively in terms of "order-producing" and "order-destroying" ions. In this terminology, order-producing ions are those whose electrostatic fields are sufficiently strong to create a well-defined structure around the ion. The order-destroying ions are those whose electrostatic fields are too weak to create a local order but strong enough to modify the water structure and change the effective interionic forces. Studies on solutions of KOH, KCl, and LiCl suggested hydration numbers of 4.6 for K^+, 4.6 for OH^-, and 8–9 for Cl^-.

Kruh, Wertz, and co-workers[72,524,538,821,905–908] initiated a systematic investigation of transition-metal complexing in aqueous solutions. Studies on concentrated $ZnCl_2$ in pure water and water–hydrochloric-acid solutions, for example, pointed to a pseudotetrahedral configuration around the Zn^{2+} ions. At molalities in the range 8.5–27.5 there appeared to be evidence for species of the form $(ZnCl_3)$. Below 8.5 m, the data indicated complexing of the form $Zn(Cl_2(H_2O)_2)$ or $Zn(Cl_3H_2O)$.

A complementary study of concentrated $CoCl_2$[907] solutions showed clear evidence for sixfold hydration around Co with a Co–O distance of 2.1 Å. In copper chloride solutions,[72] however, Cu^{2+} had six neighbors with extensive sharing of the chloride ions. The species in this case were inferred to be $Cu_3Cl_6(H_2O)_8$ at 3.18 M and $Cu_5Cl_{10}(H_2O)_{12}$ at 4.35 M.

Besides the work carried out in the USA, several structural investigations were undertaken in the Soviet Union.[258,259,728,775,799,800,801,857,858,926,927] Most of these studies are published in *Zhurnal Strukturnoi Khemii* (appearing in translation in *Journal Structural Chemistry*) and represent a new and useful source of additional information. The most interesting study concerned the measurement and identification of low-angle maxima in some transition-metal halide solutions. Dorosh and Skryshevskii[259] identified these peaks with cation–cation correlations and calculated corresponding interionic separations. The general interest, however, seemed, as in the US work, to be in determining ion–water distances and effective hydration numbers, as can be seen from Table I.

A thorough and systematic study is currently being undertaken by the X-ray diffraction group based in Cagliari, Sardinia.[165–167,556–559] A large number of systems have been studied including solutions of alkali halides, $NiCl_2$, $CaCl_2$, $CrCl_3$, and $Cr(NO_3)_3$. Particular emphasis was placed on modeling techniques; the authors pointed out that for the aqueous solutions, $G_x(r)$ is dominated by contributions from the strongly hydrated species so that ion–ion contributions cannot be readily identified. In the case of $Cr(NO_3)_3$ solutions, for example, the Cr^{3+} is octahedrally hydrated to six water molecules and the $(NO_3)^-$ appears to be unhydrated.

Another contribution came from Marques and Marques,[30] who studied concentrated aqueous solutions of $MgCl_2$, $MgBr_2$, $AlCl_3$, $InCl_3$ and $CaCl_2$. To explain certain low-angle features in the data, long-range ordering of the polyvalent hydrated cations was postulated. They claimed that the monovalent ions do not show such a superarrangement but, by an indirect argument, proposed that the distribution of polyvalent *anions* may be ordered.

The work summarized above is conventional in the sense that the authors were working with $F_x(k)$. Two attempts to improve the resolution of this function by novel methods are worthy of special note. One of the most interesting experiments yet reported in this subject was a study of divalent cation hydration in a group of nitrate solutions using the techniques of *isomorphic* replacement.[127] By differencing intensity data between 1 molar solutions which have a similar crystalline hydrate form [for example $Co(NO_3)_2$ and $Ca(NO_3)_2$] an estimate was made of an average cation-oxygen distance and the hydration number. In this way a more precise value can be obtained for hydration numbers and $\overrightarrow{OM}$ than from a single diffraction pattern. Besides being able to identify the first hydration shell of about six water molecules around the cation, the results suggest a second shell of 12 ± 2 water molecules at about 4.2 Å. This type of study relies on the questionable assumption that the only effect of changing the ion is to change the f value in $F_x(k)$; nevertheless, it confirms much of the earlier speculation about the nature of ionic hydration.

Perhaps the most distinguished attempt to improve the resolution is due to Narten and co-workers,[646,871] who combined the two techniques of X-ray and neutron diffraction on a range of LiCl solutions.[646] Heavy water was used as the solvent to avoid the substantial incoherent neutron scattering associated with protons (Section 3). The results were sufficiently

TABLE I. Hydration of Ions in Aqueous Solution Obtained from X-Ray Diffraction Studies

Ion	Solute	Reference	Concentration range[a]	Ion–oxygen distance[b] (Å)	Hydration number (h)[c]
Li$^+$	Li$_2$SO$_4$	800	2.22 m	2.08	4
	LiBF$_4$	728	10.40 m	2.14	6
	LiCl	136	6.86 m	—	4
	LiCl	646	18.5 m	2.25	4 ($\pm$1)
			6.9 m	1.95	4 ($\pm$1)
	LiCl	556	7.30 N	2.1	—
	LiBr	558	5.6 N	2.16*	4–6
			2.1 N	2.25*	4–6
Na$^+$	Na$_2$SO$_4$	800	2.22 m	2.38	4
	NaBF$_4$	775	9.05 m	2.4	6
K$^+$	KOH	136, 139	3.48 m	2.9	4
			2.02 m	2.9	4
	KI	926	5 M	2.9	2
			0.5 M	2.9	3
	K$_2$SO$_4$	800	0.55 m	2.8	4
Cs$^+$	CsF	103	23.92 m	3.13	
	CsCl	538	10 m	3.15	2–3
			5 m	3.15	2–3
			2.5 m	3.15	3–4
	CsBr	538	5 m	3.15	3–6
			2.5 m	3.15–3.22	5–6
	CsI	538	2.5 m	3.02–3.15	2–3
NH$_4$$^+$	NH$_4$NF$_4$	728	2.78 m	3.00	4
	NH$_4$F	640	15.56 m	2.88 ($\pm$0.01)	4
	NH$_4$Cl	640	6.51 m	2.80 ($\pm$0.01)	4
	NH$_4$Br	640	7.31 m	2.82 ($\pm$0.02)	4
	NH$_4$I	640	6.67 m	2.91	4
Mg^{2+}	MgCl$_2$	258	2.22 m	2.0	6
	MgCl$_2$	259	3.15 m	—	6
			2.22 m	—	6
			1.387 m	—	6
	MgCl$_2$	20	5.65 m	2.1	8
			4.27 m	2.1	8
	MgCl$_2$	30	5.72 m	2.1	6
			2.78 m	2.1	6

TABLE I (*continued*)

Ion	Solute	Reference	Concentration range[a]	Ion–oxygen distance[b] (Å)	Hydration number (h)[c]
	$MgBr_2$	30	5.72 m	2.1	6
			2.52 m	2.1	6
	$Mg(NO_3)_2$	30	4.75 m	2.1	6
			2.13 m	2.1	6
	$Mg(NO_3)_2$	127	1 M	2.1	6
Ca^{2+}	$CaCl_2$	259	2.22 m, 1.38 m	—	6
	$CaCl_2$	20	5.22 m	2.4	8
			3.26 m	2.4	8
	$CaCl_2$	559	4 M	2.42	6
			2 M	2.41	6
			1 M	2.42	6
	$CaBr_2$	557	2 N	2.44	6–8
			1.2 N	2.40	6–8
	$Ca(NO_3)_2$	127	1 M	2.26*	6
Sr^{2+}	$SrCl_2$	20	2.57 m	2.61	6–8
Ba^{2+}	$BaCl_2$	20	1.54 m	2.9	8
Co^{2+}	$CoCl_2$	259	3.83 m, 2.22 m, 1.39 m	—	6
	$CoCl_2$	907	3.75 m	2.1	6
	$Co(NO_3)_2$	127	1 M	2.11*	6
Ni^{2+}	$NiCl_2$	258	2.22 m	2.1	6
	$NiCl_2$	259	3.15 m, 2.22 m	—	6
	$NiCl_2$	167	4 M	2.06*	6
			2 M	2.05*	6
	$Ni(NO_3)_2$	127	1 M	2.07*	6
	$Ni(NO_3)_2$	779	0.1 M	2.05 (±0.10)	—
Cu^{2+}	$CuCl_2$	259	4.51 m	As in crystal hydrate	C
			2.22 m		C
			1.39 m		6
	$CuCl_2$	72	5.1 m	1.93	C
			3.55 m	1.93	C
	$Cu(SO_4)$	927	0.74 m	2.3	4–6
			0.45 m	2.3	4–6
	$CuBr_2$	275	0.056 m	1.97 ± 0.08	—

continued overleaf

TABLE I (*continued*)

Ion	Solute	Reference	Concentration range[a]	Ion–oxygen distance[b] (Å)	Hydration number (h)[c]
Zn^{2+}	$ZnCl_2$	524	27.5 m, 8.5 m, 5m	2.05	C
	$ZnCl_2$	905	3.5 m	2.05	C
	$ZnBr_2$	908	17.7 m, 8.9 m, 4.4 m	2.1	
	$Zn(NO_3)_2$	127	1 M	2.11*	6
Cd^{2+}	$CdCl_2$	30	2.22 m, 1.39 m	As in crystal hydrate	C
	$Cd(NO_3)_2$	30	1 M	2.26*	6
Al^{3+}	$AlCl_3$	30	9.2 N, 4.7 N	1.9	6
	$AlBr_3$	30	7.6 N	—	6
	$Al(NO_3)_3/HNO_3$	128	0.5 N	1.90	6
In^{3+}	$InCl_3$	30	15 N, 7.5 N	2.35	6
Cr^{3+}	$CrCl_3/HCl$	165	1 M	1.90	6
	$Cr(NO_3)_3/HNO_3$	128	0.5 N	1.98	6
Er^{3+}	$ErCl_3$	138	3.05 m, 1.46 m, 0.95 m	2.3	6
	ErI_3	138	1.26 m		
Fe^{3+}	$FeCl_3$	137	5.1 m–1.8 m	—	C
	$FeCl_3/HCl$	821	6.41 m–2.77 m	—	C
OH^-	KOH	72	2.01 m, 4.48 m	2.9	—
	KOH	136	17.5 m	2.9	6
Cl^-	LiCl	136	6.86 m	3.24	8–9
	LiCl	538	10 m	3.20	6–7
			5 m	3.20	5–9
			2.5 m	3.15	6–11
	LiCl	646	18.5 m	3.19*	6 (±1)
			6.9 m	3.10	6 (±1)
	LiCl	556	7.30 N	3.25	9–8
			2.15 N	3.25	8–8
	KCl	72	5.51 m	3.16	5–7
	HCl	871	13.9 m–0.58 m	3.13	4
	NH_4Cl	640	6.51 m	3.20	C
	$MgCl_2$	258	2.22 m	3.35	6

TABLE I (*continued*)

Ion	Solute	Reference	Concentration range[a]	Ion–oxygen distance[b] (Å)	Hydration number (h)[c]
	$MgCl_2$	20	5.65 m	3.2	8
			4.27 m	3.2	9
	$CaCl_2$	20	5.22 m, 3.26 m	3.2	8
	$CaCl_2$	559	4.52 m, 2.09 m, 1.00 m	3.14	6
	$CoCl_2$	907	3.75 m	3.1	—
	$NiCl_2$	258	2.22 m	3.35	6
	$NiCl_2$	167	4 M	3.14*	6
			2 M	3.13*	6
	$CrCl_3/HCl$	165	1 M	3.13*	—
Br^-	LiBr	538	10 m	3.43	7–9
			5 m	3.37	7–8
			2.5 m	3.40	7
	LiBr	556	4.26 m	3.40	7–10
	LiBr	558		3.29	6
	NH_4Br	799	7.3 m	3.36	C
	$CuBr_2$	275	0.056 m	3.14 ($\pm$0.1)	—
	NaBr	275	0.056 m	3.14 ($\pm$0.1)	
I^-	LiI	538	10 m	3.76	6–9
			5 m	3.65	7–8
			2.5 m	3.69	8–9
	LiI	926	5.56 m	3.7	10
			0.43 m	3.7	4
	KI	926	6.61 m	3.7	10
			0.51 m	3.7	4
	NH_4I	640	6.77 m	3.64	6
BF_4^-	NH_4BF_4	728	2.78 m	3.85 (B $\cdots$ O)	—
SO_4^{2-}	H_2SO_4	799	10 m–0.57 m	2.5	7

[a] *M*: Molarity; *N*: normality; *m*: molality.

[b] The ion–oxygen distances are not quoted to more than three significant figures; an asterisk indicates that the authors claim higher accuracy.

[c] Because of the inherent ambiguity referred to in the text, we have rounded *h* to the nearest whole number. Where a range of *h* values are quoted, this reflects the difficulty of estimating *h* from $G_x(r)$. In those cases where the existence of a chemical complex involving the other ion is suggested by the data, the concept of a well-defined value of *h* is not applicable. Such cases are indicated by C.

definitive to distinguish between various possible ion–solvent configurations. At the high concentrations, the data suggested that the chlorine ion is octahedrally coordinated with six water molecules orientated so that a single deuterium atom points along the O–D bond at the Cl ion. The Li ion apparently shows tetrahedral coordination with four water molecules placed symmetrically about the Li^+. The oxygen–oxygen configuration is clearly tetrahedral at the lower concentrations but becomes highly distorted with increasing LiCl concentration. A similar study on DCl–D_2O solutions[871] gave evidence for the existence of D_3O^+ ions with $\overrightarrow{OD} = 1.07$ Å. It was also found that the Cl^- ions are coordinated to oxygen atoms, but since the Cl–D distance was not established, the nature of the hydration sphere could not be determined.

Thus in spite of the fundamental problems that make X-ray diffraction a rather blunt tool for structural studies of solution, two themes strongly emerge from the work carried out over the last 50 years. First, high-valence positive ions do seem capable of coordinating water molecules in a way different from that expected on a purely statistical model. Because of the lack of a third dimension (a direct consequence of the low value of the X-ray scattering factor for hydrogen) the details of the cation–water conformation cannot be established from $F_x(k)$ data *alone*; by plausible arguments based on the crystallography of the hydrated solid, some authors claim to show that a particular arrangement is consistent with the observed $F_x(k)$. The situation for the negative ion is even less clear, although again some degree of hydration does seem generally accepted. Only in one case has a serious attempt been made to determine the anion–water conformation.[646]

The second theme that has emerged is the possibility that polyvalent ions in concentrated solutions exhibit long-range order. The origin of this order is far from clear and indeed its spatial extent is a matter of controversy.

Clearly, if real progress is to be made, a much sharper experimental tool is required. We believe that the technique pioneered in our laboratory (neutron diffraction combined with isotopic enrichment) represents such a tool. Before describing the technique in detail, we conclude by considering what developments are likely to occur in the field of X-ray studies over the next few years.

2.3. Future Prospects

The use of high-intensity synchrotron radiation offers several prospects for future structural studies. One possibility is to use synchrotron radiation to study EXAFS (extended X-ray absorption fine structure). It has already been demonstrated that with a conventional X-ray source EXAFS can be used to measure nearest-neighbor distances in amorphous materials.[275,779] As the name implies, EXAFS is inherently element selective, since the fine structure of an absorption edge is, in a complete theory, related to the atomic arrangement around the absorptive element. In a recent EXAFS study, significant measurements have been made of cation–oxygen distances to ± 0.1 Å in relatively dilute (1000 H_2O:1 MX_n) solutions of copper bromide and nickel nitrate (Table I).

Another possible use of synchrotron radiation is in anomalous scattering experiments. Here, the f factor depends on the incident wavelength provided one is working near an absorption edge. Until now the discrete nature of conventional X-ray energy sources has limited the technique to one or two carefully selected ions in solid state investigations. With a broad band of wavelengths of high intensity available from the synchrotron, however, it becomes feasible to study anomalous scattering by solutions and thereby extract at least some of the individual $S_{\alpha\beta}(k)$. In some senses the technique is equivalent to the isotopic method to be described in Section 3, although the possibility of *double* differencing to obtain terms like $S_{MX}(k)$ is not possible.

3. NEUTRON STUDIES

3.1. Introduction

Neutron diffraction arises from the scattering of neutrons by atomic nuclei, and for a general review of its application to the study of liquids the reader is referred to the article by Page.[679] The f factor (usually referred to as the scattering length), though dependent on the isotopic state of the nucleus (Table II), does not depend on k so that $F(k)$ ought to be better defined at high k than $F_x(k)$ and $G(r)$ is, in this case, a linear combination of $g_{\alpha\beta}(r)$. Thus in principle neutron diffraction ought to yield r-space data of resolution superior to that possible with X-ray methods. In practice there are factors specific to aqueous solutions that mitigate against this rather simplistic view, and these we will now consider.

It is necessary first of all to distinguish between *coherent* and *incoherent* scattering, the latter being a concept not relevant to X-ray diffraction. If all the nuclei in the sample are identical and have zero spin the scattering is said to be coherent. If the scattering length varies in a random way

TABLE II. Examples of Coherent Scattering Lengths (10^{-12} cm)

Element or isotope	f	Element or isotope	f
H	−0.372	Fe	0.951
D	0.670	^{54}Fe	0.42
^{6}Li	0.18	^{56}Fe	1.01
^{7}Li	−0.21	^{57}Fe	0.23
^{35}Cl	1.18	Ni	1.03
^{37}Cl	0.26	^{58}Ni	1.44
Ca	0.49	^{60}Ni	0.282
^{40}Ca	0.49	^{62}Ni	−0.87
^{44}Ca	0.18	^{64}Ni	−0.037

from nucleus to nucleus, there will be a structure-independent or incoherent contribution to the total scattering. This fluctuation in scattering length can either be isotopic in origin or arise from nuclear spin.

For simplicity we consider an element in which there are two stable isotopes. Let x_1 and x_2 be the concentrations of the isotopes with scattering lengths f_1 and f_2. The coherent scattering length is $f_{coh} = x_1 f_1 + x_2 f_2$ and defines a coherent cross section through

$$\sigma_{coh} = 4\pi(x_1 f_1 + x_2 f_2)^2 \tag{12}$$

The total cross section is $4\pi[x_1 f_1^2 + x_2 f_2^2]$ so that the incoherent cross section σ_{inc} is simply

$$\sigma_{inc} = 4\pi[x_1 f_1^2 + x_2 f_2^2] - 4\pi(x_1 f_1 + x_2 f_2)^2$$
$$= 4\pi x_1 x_2 (f_1 - f_2)^2 \tag{13}$$

If the nucleus has a spin I, the interaction with a neutron may give rise to a compound nucleus ($I + \frac{1}{2}$), each with a characteristic scattering length f_+ and f_-. By direct analogy with (12) and (13) we obtain σ_{coh} and σ_{inc} by noting that

$$x_1 \rightarrow (I + 1)/(2I + 1)$$
$$x_2 \rightarrow I/(2I + 1)$$
$$f_1 \rightarrow f_+$$
$$f_2 \rightarrow f_-$$

For our purpose incoherent scattering is significant in two ways. First, σ_{inc} for protons is extremely large so that heavy water must be used as solvent until such time as spin polarization techniques have improved, thereby allowing a separation of coherent and incoherent scattering. Secondly, incoherent scattering appears, in the general form of the scattering from real samples [eqn. (5)] as a contribution to $\delta(\theta)$.

A further complication characteristic of neutron work is the effect of multiple scattering. This introduces a substantial contribution to $\delta(\theta)$, which fortunately, under most experimental conditions, is isotropic and can be evaluated directly.[660]

The final distinction between X-ray work and neutron work involves the "inelastic" or "Placzek" corrections.[709,718] These arise because when a neutron interacts with a nucleus, a transfer of *both* momentum *and* energy takes place. Thus under normal experimental conditions in diffraction studies, an effective rather than a true cross section is measured. The crucial parameter that determines the importance of these corrections is the ratio of the neutron mass to the nuclear mass. For systems where the nuclear mass is ~ 30 neutron masses, the correction terms are reasonably well understood. However, for aqueous solutions the presence of D (or H) gives rise to correction terms, which so far have proved to be incalculable. Their general effect is to cause $F(k)$ to "droop" at high k (Fig. 2), thereby rendering invalid the asymptotic results for $F(k)$ and the calculation of $G(r)$. In terms of eqn. (5), Placzek corrections make a large contribution to $\delta(\theta)$ and are strongly θ dependent. They affect both the coherent and the

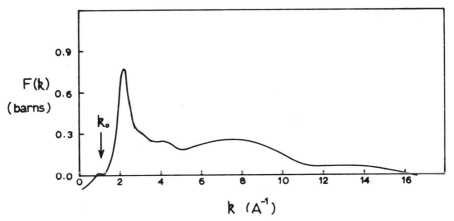

Fig. 2. $F(k)$ for a 4.41 m solution of $NiCl_2$ in D_2O. The "droop" at high k is associated with inelastic effects (the Placzek corrections). The vertical arrow at k_0 indicates the prepeak referred to in the text.

incoherent parts of the scattering. To summarize then, we have the following:

(a) The total scattering pattern for an aqueous solution arises from ten partial structure factors each weighted by appropriate scattering and concentration factors. Even for very concentrated solutions the ion–water contributions to the total scattering pattern are typically 10% and it is therefore exceedingly difficult to disentangle the ion–water terms unambiguously.

(b) Although in principle the analysis of neutron data is more direct, the necessity for large Placzek corrections for aqueous solutions proves to be a formidable obstacle. We now propose a solution[818] to both these problems, which exploits the fact that the f factor is isotope dependent (Table II).

3.2. First-Order Difference Scattering Functions

Let $\Delta_M^0(k)$ represent the algebraic difference between the corrected neutron scattering cross section in absolute units of two solutions that are identical in all respects except that the isotopic state of M has been changed. It follows[818] from eqn. (4) that

$$\Delta_M^0(k) = \Delta_M(k) + \text{correction terms} \tag{14}$$

where

$$
\begin{aligned}
\Delta_M(k) = &\tfrac{2}{3}c(1 - c - nc)f_O(f_M - f_M')[S_{MO}(k) - 1] \\
&+ \tfrac{4}{3}c(1 - c - nc)f_D(f_M - f_M')[S_{MD}(k) - 1] \\
&+ 2nc^2 f_X(f_M - f_M')[S_{MX}(k) - 1] + c^2[f_M^2 - (f_M')^2][S_{MM}(k) - 1] \\
= &A_1[S_{MO}(k) - 1] + B_1[S_{MD}(k) - 1] + C_1[S_{MX}(k) - 1] \\
&+ D_1[S_{MM}(k) - 1]
\end{aligned}
\tag{15}
$$

and the dominant contribution to the correction terms arises from Placzek effects. The change in incoherent scattering and multiple scattering are sufficiently small to be evaluated by elementary methods.

In (15) f_M and f_M' are the coherent scattering lengths for the two isotopic states of M used in the experiment.

Equation (15) differs from eqn. (4) in two significant ways. First, the relevant corrections are sufficiently small to allow $\Delta_M(k)$ to be derived from $\Delta_M^0(k)$ (see Section 3.3). Secondly, the water terms, which dominate the total scattering and therefore mask the solute–water interactions, no longer appear in (15). Thus the principal obstacles to progress have been overcome. The fact that the "droop" disappears from $\Delta_M(k)$ is of particular significance when considering the errors in normalizing the experimental data and in the evaluation of $G(r)$. There is, however, a third advantage of working with $\Delta_M(k)$. For a wide range of solutions C_1 and D_1 are much smaller than A_1 and B_1; hence, for practical purposes, $\Delta_M(k)$ is determined by S_{MO} and S_{MD}, that is to say, the short-range correlation between the water and M. This is precisely the quantity we wish to investigate.

We conclude this section by writing down an expression for $\Delta_X(k)$, the first-order difference obtained by isotopically substituting the anion X. By direct analogy with (15) it follows that

$$\Delta_X(k) = A_2[S_{XO}(k) - 1] + B_2[S_{XD}(k) - 1] + C_2[S_{XM}(k) - 1]$$
$$+ D_2[S_{XX}(k) - 1] \tag{16}$$

where

$$A_2 = \tfrac{2}{3}nc(1 - c - nc)f_O(f_X - f_X')$$
$$B_2 = \tfrac{4}{3}nc(1 - c - nc)f_D(f_X - f_X')$$
$$C_2 = 2nc^2 f_M(f_X - f_X')$$
$$D_2 = n^2c^2[f_X{}^2 - (f_X')^2]$$

The same remarks about the usefulness of eqn. (16) and in particular, the dominance of A_2 and B_2 once again apply. In this case, the anion–water conformation becomes directly accessible to the experimenter.

3.3. Placzek Corrections to the First-Order Difference $\Delta_{M}{}^0(k)$

We now show that the correction terms appearing in eqn. (14) can be handled by methods developed for the study of simple liquids.[660]

It is convenient at this point to express the effective differential scattering cross section for a multicomponent system[718] as the sum of two terms, an interference term and a self term, namely,

$$\left.\frac{d\sigma}{d\Omega}\right|_{\text{effective}} = \left.\frac{d\sigma}{d\Omega}\right|_{\text{effective}}^{\text{int}} + \left.\frac{d\sigma}{d\Omega}\right|_{\text{effective}}^{\text{self}} \tag{17}$$

If the isotopic state of the nucleus M is changed, the difference in $d\sigma/d\Omega\,|_{\text{effective}}$ is $\Delta_{M}{}^0(k)$ and may be written as

$$\Delta_{M}{}^0(k) = \Delta_M(k) + P_1(k) + \Delta_{M}{}^0(k)\,|^{\text{self}} \tag{18}$$

where $P_1(k)$ is the Placzek correction to the interference terms. We therefore wish to calculate $P_1(k)$ and $\Delta_M(k)\,|^{\text{self}}$, the dominant contributions to the correction terms.

The Placzek corrections to the self term in eqn. (17) can be obtained in terms of the moments of the energy transfer provided that the efficiency of the detector is known as a function of incident neutron energy. One frequent assumption[718] made is that the detector is black (i.e., constant efficiency of 100%), in which case it can be shown that

$$\frac{d\sigma}{d\Omega}\Big|^{\text{self}}_{\text{effective}} = \sum_j c_j \langle f_j^2\rangle \left(1 + \frac{m}{\mu_j}\right)^{-2}\left\{1 + \left(\frac{m}{\mu_j}\right)\left[2\cos\theta + \frac{1}{3}\left(\frac{\bar{K}_j}{E_0}\right)\right]\right.$$

$$+ \frac{1}{2}\left(\frac{m}{\mu_j}\right)^2 (3\cos^2\theta - 1)\left[1 + \frac{2}{3}\left(\frac{\bar{K}_j}{E_0}\right)\right]$$

$$\left. + \text{ terms in } \left(\frac{m}{\mu_j}\right)^3 \text{ and higher} \cdots\right\} \tag{19}$$

where $\bar{K}_j$ and μ_j are the mean kinetic energy and mass of the jth nucleus, m is the neutron mass of energy E_0, and θ is the scattering angle.

For a single isotopic substitution of the nucleus M, resulting in a change of scattering length f_M to f_M' and a change of mass μ_M to μ_M', the corresponding change in $d\sigma/d\Omega\,|^{\text{self}} = \Delta_M(k)\,|^{\text{self}}$ arises from those terms in (19) that contain the subscript $j = M$. Thus,

$$\Delta_M^0(k)\,|^{\text{self}} = c(\langle f_M^2\rangle - \langle f_M'^2\rangle)\left[1 - \frac{2m}{\mu_M^2}\mu_M'(1 - \cos\theta)\right] \tag{20}$$

where $\bar{K}/E_0$ and terms in $(m/\mu_M)^2$ are neglected. Equation (20) demonstrates that the correction to the "bound" cross sections is of order (m/μ_M), i.e., 3% for the case of nickel substitutions, and is typical of that for simple liquids.

The Placzek corrections for the interference term in eqn. (17) are not so straightforward and involve the momentum correlation between the two nuclei involved in any particular pair component of the interference term. Analogous to eqn. (19) we write

$$\frac{d\sigma}{d\Omega}\Big|^{\text{int}}_{\text{effective}} = \sum_{j,k}' c_j c_k f_j f_k \left\{\langle\exp(i\mathbf{k}\cdot\mathbf{r}_{jk})\rangle + \frac{\hbar^2}{\mu_j\mu_k}\left\langle\exp(i\mathbf{k}\cdot\mathbf{r}_{jk})\left[\frac{\hbar^2 k^4}{4E_0^2}\right.\right.\right.$$

$$\left.\left.\left. + \text{ terms involving momentum correlation}\right]\right\rangle + \cdots\right\} \tag{21}$$

$$= \sum_{j,k} c_j c_k f_j f_k \left\{[S_{jk}(k) - 1] + \frac{m^2}{\mu_j\mu_k}\hat{F}_{jk}(k) + \cdots\right\} \tag{22}$$

where $\hat{F}_{jk}(k)$ is, in general, a function of $S_{jk}(k)$ and its derivatives.

From (22), $P_I(k)$ is a sum of terms arising from every pair of nuclei involving M. The largest contribution to P_I is

$$-\frac{4}{3} c(1 - c - nc) f_D (f_M - f_M') \frac{m^2}{\mu_D \mu_M} \hat{F}_{MD}(k) \qquad (23)$$

and there are no contributions from the troublesome terms involving $\hat{F}_{DD}(k)$ and $\hat{F}_{DO}(k)$. Equation (23) shows that the Placzek correction to the interference term is order $m^2/\mu_D \mu_M$ and is, therefore even smaller than that for the self term.

3.4. Experimental Procedures

High-quality diffraction data must first be obtained over a wide range of k (e.g., by using the D4 spectrometer at the ILL, Grenoble). The results are put on an absolute scale by the vanadium technique.[660] Since $\Delta(k)$ does not "droop" at high k, further normalization, which allows for inaccuracies in the absorption corrections and the published values of the scattering lengths as well as the systematic error due to the slight variations in the light-water content of the samples, can be achieved by reference to the high- and low-k limits of $\Delta(k)$.

Great care must be taken with the preparation of the samples as the technique relies on the exact cancellation of the water terms. Soper et al.,[818] ensured that the difference in light-water content of each pair of samples was held constant to within $\pm 0.1\%$ by preparative techniques that involve the use of an infrared spectrometer. The samples were made so that the concentration scale was reproducible to within $\pm 1\%$. Mass spectrographic techniques were employed to determine to within 0.1% the abundances of the various isotopes characteristic of the raw materials used in the preparation of the samples.

In order to study ion-hydration effects in real space it is necessary to construct the Fourier transform of $\Delta(k)$. The quantity

$$\bar{G}(r) = \frac{1}{2\pi^2 \varrho r} \int dk\, \Delta(k) k \sin kr \qquad (24)$$

where the bar indicates that we are working with a difference, can be obtained by standard numerical quadrature.[660] It follows from the definition of $\Delta(k)$ that

$$\bar{G}_M(r) = A_1 g_{MO} + B_1 g_{MD} + C_1 g_{MX} + D_1 g_{MM} + E_1 \qquad (25)$$

and

$$\bar{G}_X(r) = A_2 g_{XO} + B_2 g_{XD} + C_2 g_{XM} + D_2 g_{XX} + E_2 \qquad (26)$$

where

$$E_1 = -(A_1 + B_1 + C_1 + D_1) \text{ and } E_2 = -(A_2 + B_2 + C_2 + D_2)$$

Since A and B are much greater than C and D, we have now accomplished what we set out to do, namely, devise a high-resolution technique to yield an appropriate combination of g_{MO}, g_{MD}, and g_{XO} and g_{XD}.

3.5. Second-Order Difference Scattering Functions

The second theme that we identified in our survey of the X-ray work was the possibility that, in certain circumstances, ions in concentrated solutions may be ordered. The concept of "order" in a liquid is not a precise one but in this context probably involves *at least* one of the following: (a) a phase relationship between the oscillations in g_{MM}, g_{MX}, and g_{XX} (charge ordering); (b) a tendency of the peak positions g_{MM} (or g_{XX}) to be commensurate; (c) the closest distance of approach and the mean separation of the ions to be roughly the same and to scale together with dilution.

In any event the test for ordering in ionic systems is unquestionably to measure $S_{MM}(k)$, $S_{XX}(k)$, and $S_{MX}(k)$ as functions of c, and we now show how that is possible by the technique of isotopic substitution.

Let us consider again a salt MX_n dissolved in D_2O and let $\Delta_{M_1}(k)$ and Δ_{M_2} represent two first-order differences for three solutions with M in the isotopic state M, $'M$, and $''M$. Similarly let $\Delta_{X_1}(k)$ and $\Delta_{X_2}(k)$ be the corresponding quantities for isotopic substitutions of the anion X. It follows from eqns. (15) and (16) that

$$S_{MM}(k) = \frac{\Delta_{M_1}(k)}{A_M^{(2)}} - \frac{\Delta_{M_2}(k)}{B_M^{(2)}} + 1 \qquad (27)$$

$$S_{XX}(k) = \frac{\Delta_{X_1}(k)}{A_X^{(2)}} - \frac{\Delta_{X_2}(k)}{B_X^{(2)}} + 1 \qquad (28)$$

with coefficients $A_M \cdots B_X$ given by

$$A_M^{(2)} = c^2(f_M - f_M')(f_M' - f_M'')$$
$$B_M^{(2)} = c^2(f_M - f_M'')(f_M' - f_M'')$$
$$A_X^{(2)} = n^2 c^2(f_X - f_X')(f_X' - f_X'')$$
$$B_X^{(2)} = n^2 c^2(f_X - f_X'')(f_X' - f_X'')$$

Thus a *three-pattern experiment* enables individual ion–ion correlation functions to be isolated.

In order to obtain the cross term S_{MX}, we must use four samples whose isotopic state can be represented by

$$MX_n, \; 'MX_n, \; M'X_n 'M'X_n$$

Let $\Delta_M{}^X$ represent the algebraic difference in intensity between the scattering from the first and the second samples and $\Delta_M'^X$ the difference between the third and the fourth sample. It follows that

$$S_{MX}(k) = \frac{\Delta_M{}^X - \Delta_M'^X}{2nc^2(f_M - f_M')(f_X - f_X')} + 1$$

It can be readily seen from the analysis presented in Section 3.3 that the Placzek corrections are even smaller for second-order differences than for first-order differences.

3.6. Experimental Procedures

The experimental procedures are identical with those described in Section 3.4 except that demands on the statistics and the sample definition are even more severe. Nevertheless, it is possible, with facilities like those at the ILL Grenoble, to obtain significant second-order difference scattering functions. For example, in two of the solutions studied (see Section 4) a comparison was made between the predicted values of $S_{\alpha\alpha}(0)$ [eqn. (9)] and those derived from experiment [eqns. (27) and (28)] because this represents a good test of the experimental procedures. For a 4.41 m solution of $NiCl_2$ in D_2O, $S_{NiNi}(0)$ was found to be -30 ± 5, which is in agreement with the theoretical value of -33. Similarly for a 5.32 m solution of NaCl, the experimental and theoretical values of $S_{ClCl}(0)$ were, respectively, -22 ± 10 and -16.

4. STRUCTURAL INFORMATION DERIVED FROM NEUTRON STUDIES

4.1. Introduction

We now wish to review the recent work carried out by the techniques described in Section 3 and to set the results in context both with the earlier X-ray data and with recent theory.

TABLE III. Scattering Length and Sample Parameters

Electrolyte solution	Isotopes	Abundance (percent)	Scattering lengths (10^{-12} cm)	c	Molality	$A \times 10^2$ (barns)[a]	$B \times 10^2$ (barns)	$C \times 10^3$ (barns)	$D \times 10^3$ (barns)
$NiCl_2 \cdot D_2O$	$Ni^{natural}$	—	1.03	0.0270	4.41	1.74	4.00	5.05	0.32
	^{62}Ni	94.9	−0.79	0.0192	3.05	1.26	2.90	2.52	0.15
				0.0093	1.46	0.64	1.46	0.61	0.04
				0.0056	0.85	0.385	0.885	0.22	0.013
				0.0028	0.42	0.194	0.446	0.054	0.00338
				0.00057	0.086	0.040	0.092	0.0023	0.00015
$NaCl \cdot D_2O$	^{35}Cl	99.35	1.17	0.0331	5.32	0.99	2.27	0.65	1.37
	^{37}Cl	90.4	0.35						
$CaCl_2 \cdot D_2O$	^{35}Cl	90.35	1.17	0.0275	4.49	1.60	3.68	1.16	3.77
	^{37}Cl	90.4	0.35						
$RbCl \cdot D_2O$	^{35}Cl	99.35	1.17	0.0192	2.99	0.58	1.33	0.52	0.46
	^{37}Cl	90.4	0.35						

[a] 1 barn = 10^{-24} cm².

So far four solutions have been investigated by the neutron technique. They are $NiCl_2$ (isotopically changing the Ni), NaCl and RbCl (isotopically changing the Cl), and $CaCl_2$ (isotopically changing both the Ca and the Cl). In all cases the salts were dissolved in D_2O and the concentrations and coefficients $A_1, B_1, \ldots, D_2$ are given in Table III; the statement made in Section 3 that C and D are small compared with A and B is evidently true.

4.2. Cationic Hydration (Table IV)

4.2.1. Ni^{2+}

Our evidence on Ni^{2+} hydration derives entirely from a detailed study of the $NiCl_2$ solution made over a wide range of concentrations.[649,818] Two examples of $\Delta_{Ni}(k)$ and $\bar{G}_{Ni}(r)$ relating to two values of c are shown in Figs. 3–6.

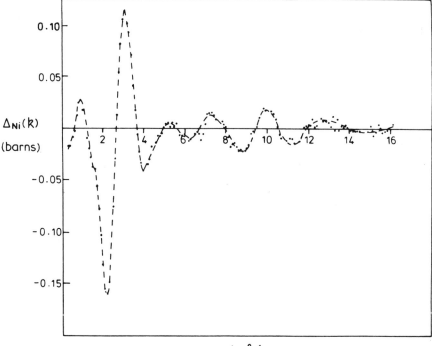

Fig. 3. $\Delta_{Ni}(k)$ for a 4.41 m solution of $NiCl_2$ in D_2O. The full circles represent experimental points and the smooth curve is the one used to calculated $\bar{G}_{Ni}(r)$. The same convention is used for Figs. 4, 9, and 16.

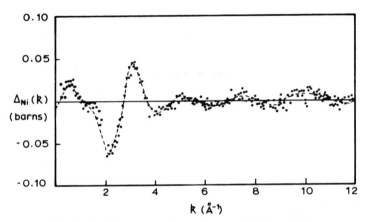

Fig. 4. $\Delta_{\mathrm{Ni}}(k)$ for a 1.46 m solution of $NiCl_2$ in D_2O.

Let us consider, for example, the form of $\bar{G}_{\mathrm{Ni}}(r)$ shown in Fig. 5. It is particularly significant that at $r \sim 3.0\,\text{Å}$, $\bar{G}_{\mathrm{Ni}}(r) \sim -E_1$; thus g_{NiO} and g_{NiD} are both small at this value of r which reflects the stability of the first hydration shell. The two peaks located at 2.07 Å and 2.67 Å we identify with Ni–O and Ni–D correlations, respectively, on the grounds that the

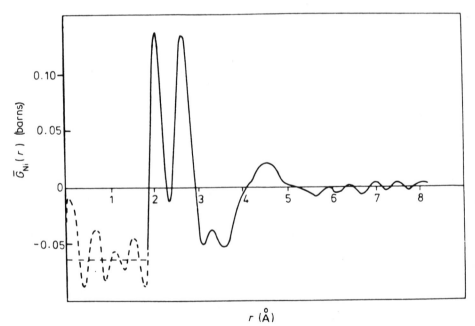

Fig. 5. $\bar{G}_{\mathrm{Ni}}(r)$ for a 4.41 m solution of $NiCl_2$ in D_2O.

ratio of the areas beneath them when weighted by r^2 is almost exactly $A_1 : 2B_1$. An integral over $4\pi r^2 \bar{G}_{Ni}(r)$ for $1.8 < r < 3.0$ Å yields a total of 5.8 ± 0.2 water molecules in the first coordination shell.

The tremendous increase in resolution gained by our technique can best be judged by comparing these data with those shown in Fig. 7, which were derived in the excellent X-ray study involving isomorphic replacement.

The first comment to make on the new data is that there is, within experimental error, no change in h (Table IV) as the concentration of $NiCl_2$ is reduced. The nickel ion, together with the six or so water molecules attached to it, can be usefully thought of as an entity on which to perform statistical mechanics.

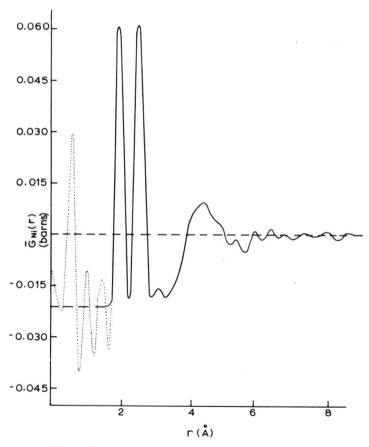

Fig. 6. $\bar{G}_{Ni}(r)$ for a $1.46\,m$ solution of $NiCl_2$ in D_2O.

TABLE IV. Hydration of Cations Obtained from Neutron Diffraction

Ion	Solute	Molality	Ion–oxygen distance (Å)	Ion–deuterium distance (Å)	θ^a ($\overrightarrow{OD}$ = 0.94 Å)	θ^a ($\overrightarrow{OD}$ = 1.00 Å)	Hydration number (h)
Ni^{2+}	$NiCl_2$	4.41	2.07 ± 0.02	2.67 ± 0.02	34° ± 8°	42° ± 8°	5.8 ± 0.2
		3.05	2.07 ± 0.02	2.67 ± 0.02	34° ± 8°	42° ± 8°	5.8 ± 0.2
		1.46	2.07 ± 0.02	2.67 ± 0.02	34° ± 8°	42° ± 8°	5.8 ± 0.3
		0.85	2.09 ± 0.02	2.76 ± 0.02	10° ± 10°	27° ± 10°	6.6 ± 0.5
		0.46	2.10 ± 0.02	2.80 ± 0.02	0° ± 10°	17° ± 10°	6.8 ± 0.8
		0.086	2.07 ± 0.03	2.80 ± 0.03	0° ± 10°	0° ± 20°	6.8 ± 0.8
Ca^{2+}	$CaCl_2$	4.49	2.40 ± 0.03	2.93 ± 0.05	45° ± 15°	61° ± 15°	5.5 ± 0.2

a θ is the angle between the plane of the water molecule and the MO axis calculated for two values of $\overrightarrow{OD}$. $\widehat{DOD}$ is assumed in both cases to be 105.5°.

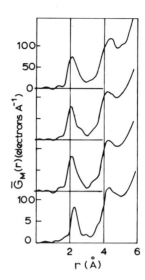

Fig. 7. First-order difference in real space derived from an X-ray experiment in which isomorphic replacement was used.[127]

The Ni–D separation is now accessible, so we are able to investigate the detailed conformation of the Ni–D_2O provided we know the geometry of the water molecule in the liquid state. Let us assume that the parameters of the water molecules are given by an OD length of 0.94 Å (a lower limit)[649] and a $\widehat{DOD}$ angle of 105.5°. In this case the plane of the water molecule must (Fig. 8) be at an angle θ of 34° $\pm$ 8° to the Ni–O axis. An upper limit for the OD bond length is assumed to be 1.00 Å, and this yields a θ value of 42° $\pm$ 8°. For a planar configuration to be consistent with the observed $\bar{G}_{Ni}(r)$, the $\widehat{DOD}$ angle would have to be 120° or $\overrightarrow{OD}$ would have to be 0.84 Å, and both possibilities seem highly unlikely. We therefore conclude that there is a significant angle of tilt which gradually disappears at low concentrations. We believe that we are seeing the first clear evidence for the distortions of the hydration spheres as the packing fraction of the hydrated ions is increased.

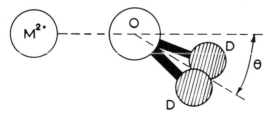

Fig. 8. The cation–water conformation consistent with the $\bar{G}_M(r)$ shown in Figs. 5, 6, and 10. θ values are given in Table IV.

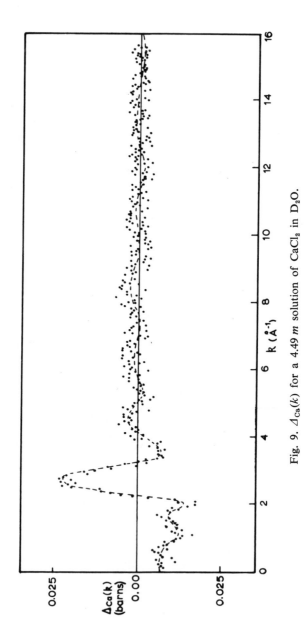

Fig. 9. $\Delta_{Ca}(k)$ for a 4.49 m solution of $CaCl_2$ in D_2O.

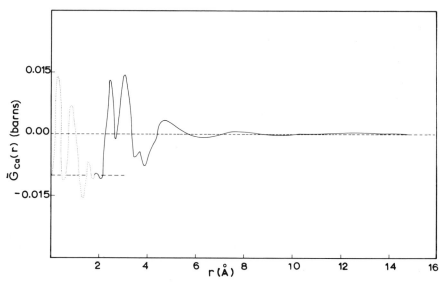

Fig. 10. $\bar{G}_{Ca}(r)$ for a 4.49 m solution of $CaCl_2$ in D_2O.

4.2.2. Ca^{2+}

So far one solution of $CaCl_2$ in D_2O has been studied.[217] The results for $\Delta_{Ca}(k)$ and $\bar{G}_{Ca}(r)$ are shown in Figs. 9 and 10. The Ca^{2+} ions are once again strongly hydrated, the two peaks being well resolved as in the Ni^{2+} case. The hydration number turns out to be 5.5 ± 0.2 and the conformation of the Ca–D_2O system is similar to that shown in Fig. 8 with a Ca–O distance of 2.40 Å and a tilt angle $45° \pm 10°$ to $61° \pm 15°$ for OD bond lengths in the range 0.94–1.00 Å.

4.3. Anionic Hydration (Table V)

4.3.1. Cl^- in NaCl Solution

The form of $\bar{G}_{Cl}(r)$ (Fig. 11) for the 5.32 m solution studied[818] indicates well-defined hydration about the Cl ions and it should be noted that $\bar{G}_{Cl}(r) \sim -E_2$ for $r \sim 2.8$ Å. One interpretation of these results is to suppose that each of the two principal peaks contain only contributions from g_{ClO} and g_{ClD} in a ratio corresponding to D_2O molecules. This yields a value for the number of water molecules in the first shell of 2.1 ± 0.2. On the other hand, it suggests a Cl–O distance of 2.7 Å, which seems, in view of the X-ray data summarized in Table I, to be highly implausible.

Soper et al.[818] therefore favored a second model (Fig. 12) which is based on the assumption that the first peak in $\bar{G}(r)$ arises solely from g_{ClD} and that there is a single hydration shell of radius 3.20 Å. There was evidence that ψ (Fig. 12) was different from zero, but the inherent low resolution makes it impossible to rule out a linear Cl–D(1) — O configuration. Integration of $\bar{G}(r)$ yields a coordination number of 5.5 + 0.2.

4.3.2. Cl⁻ in CaCl₂ Solution

One concentration of this solution[217] (4.49 m) has been studied and we first focus attention on $\Delta_{Cl}(k)$ (Fig. 13); these data, when scaled for appropriate values of $A_2 \cdots D_2$ are, within experimental error, almost identical with those for the NaCl solution. We are therefore able to assert that in spite of the different character of the cation, the nature of the hydration around the anion is similar in the two systems (see also Section 4.3.3 below). This work justifies an approach to the theory of solutions based on the principle that a reliable determination of the local coordination of a particular anion can be used in a range of solutions with different cations. We believe this to be an important and significant discovery.

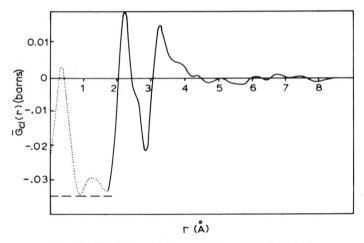

Fig. 11. $\bar{G}_{Cl}(r)$ for a 5.32 m solution of NaCl in D₂O.

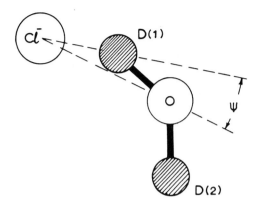

Fig. 12. The anion–water conformation consistent with the $\bar{G}_x(r)$ shown in Figs. 11 and 14. ψ values are given in Table V.

The $CaCl_2$ data are, however, intrinsically more reliable than those for NaCl because of the high values of A_2 and B_2. Furthermore, from an experimental point of view, both the statistics and the samples were of better quality and these facts allowed reliable data to be obtained at much higher values of k than before and thereby offered enhanced resolution in real space (Fig. 14). The conformation of $Cl–D_2O$ can now be given with more confidence (see Table V and Fig. 12) and accords well with the most recent quantum mechanical calculations on the isolated ions. It should be noted that the observed $Cl–D(2)$ distance is, within experimental error, equal to that calculated by assuming a bond length of 1.00 Å and a ψ value of zero. This strongly supports the view that 0.94 Å is too short a bond length for water in electrolytes. The number of water molecules coordinated to the Cl^- ion is 5.8 ± 0.2.

A final point to note is that there is no structure in $\bar{G}_{Cl}(r)$ beyond ~ 5 Å, in agreement with the earlier and less accurate studies on NaCl solutions. Thus only the first shell of water molecules can be said to be coordinated to the Cl^- ion.

4.3.3. Cl^- in RbCl Solution

The data in k space are shown in Fig. 15 for a 2.99 m solution of RbCl.* They confirm the result of the study on $CaCl_2$ solutions that the

* S. Cummings, private communication.

TABLE V. Hydration of Chloride Ions Obtained from Neutron Diffraction

Ion	Solute	Molality	$\overrightarrow{ClD}(1)$ (Å)	$\overrightarrow{ClO}$ (Å)	$\overrightarrow{ClD}(2)$ (Å)	ψ^a ($\overrightarrow{OD} = 0.94$ Å)	ψ^a ($\overrightarrow{OD} = 1.00$ Å)	Hydration number (h)
Cl^-	NaCl	5.32	2.26 ± 0.04	3.20 ± 0.04	—	0° ± 8°	10° ± 10°	5.5 ± 0.3
	$CaCl_2$	4.49	2.25 ± 0.02	3.25 ± 0.04	3.60 ± 0.04	0°	0° ± 7°	5.8 ± 0.2

[a] ψ is the angle between the Cl–D(1) axis and the Cl–O axis calculated for two values of $\overrightarrow{OD}$. $\widehat{DOD}$ is assumed in both cases to be 105.5°.

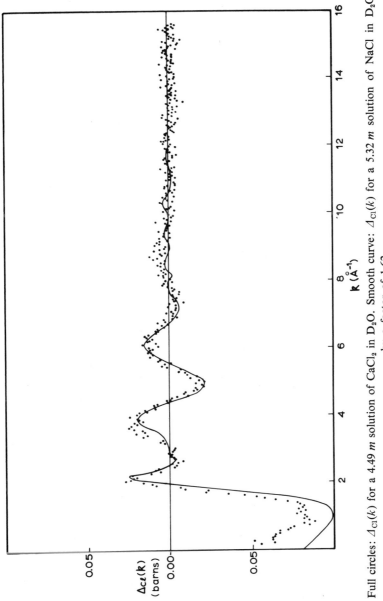

Fig. 13. Full circles: $\Delta_{Cl}(k)$ for a 4.49 m solution of CaCl$_2$ in D$_2$O. Smooth curve: $\Delta_{Cl}(k)$ for a 5.32 m solution of NaCl in D$_2$O scaled by a factor of 1.62.

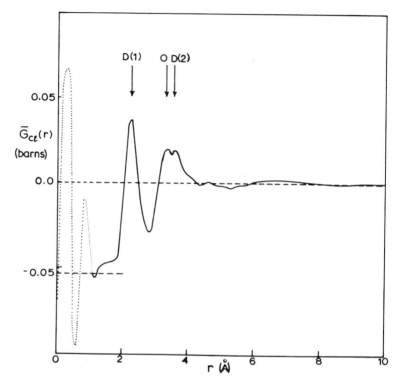

Fig. 14. $\bar{G}_{Cl}(r)$ for a 4.49 m solution of $CaCl_2$ in D_2O.

nature of the Cl^- hydration is relatively insensitive to the nature of the cation. A full account of the work, generalized to include a wide variety of cations, will be published in due course.

4.4. Cation–Cation Distribution Functions

'The technique described in Section 3.5 has been applied only to a 4.41 m solution of $NiCl_2$ in D_2O.[650] These experiments are very difficult, and the data we report must be regarded as preliminary. The sample parameters and the values of $A_M^{(2)}$ and $B_M^{(2)}$ are shown in Table VI.

The data for $S_{NiNi}(k)$ (Fig. 16) are characterized by a well-defined first peak centered at a k value of ~1 Å^{-1}. Originally, it was believed that these data supported the idea that the ions in concentrated solutions were *charge* ordered. March and Tosi,[593] on the other hand, suggested that the origin of the order was chemical and arose from the bonding of $Ni(D_2O)_4Cl_2$. More recently, Quirke and Soper[727] argued that the general form of

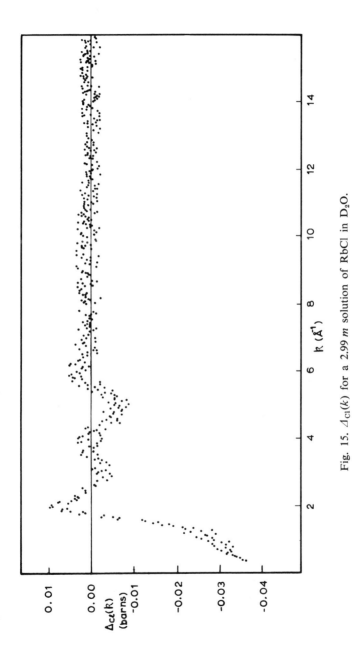

Fig. 15. $\Delta_{Cl}(k)$ for a 2.99 m solution of RbCl in D_2O.

TABLE VI. Sample Parameters for the Second-Order Differences

Electrolyte solution	Isotopes	Abundance (percent)	Scattering lengths (10^{-12} cm)	c	Molality	$A_M^{(2)} \times 10^4$ (barns)	$B_M^{(2)} \times 10^3$ (barns)	$A_X^{(2)} \times 10^4$ (barns)	$B_X^{(2)} \times 10^4$ (barns)
$NiCl_2$ D_2O	$Ni^{Natural}$	—	1.03						
	$Ni^{"zero"}$	20.41 ^{58}Ni 37.44 ^{60}Ni 41.77 ^{62}Ni	0.036	0.0275	4.41	6.21	1.14	—	—
	^{62}Ni	2.16 ^{58}Ni 2.22 ^{60}Ni 94.91 ^{62}Ni	−0.79						
$NaCl$ D_2O	$Cl^{Natural\ a}$	59.3 ^{35}Cl 40.7 ^{37}Cl	0.81						
	^{37}Cl	9.6 ^{35}Cl 90.4 ^{37}Cl	0.35	0.0331	5.32	—	—	−4.13	3.23
	^{35}Cl	99.4 ^{35}Cl 0.6 ^{37}Cl	1.17						

a Slightly modified to optimize the spread in scattering lengths.

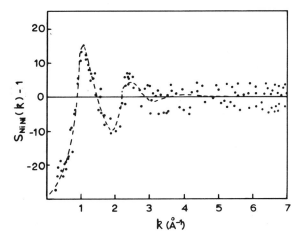

Fig. 16. The partial structure factor, $S_{NiNi}(k)$, for a 4.41 m solution of $NiCl_2$ in D_2O.

$S_{NiNi}(k)$ could be fitted by neglecting the attractive forces and considering only the random packing of spheres of radius 2.8 Å, chosen so as to simulate the hydrated Ni ion. The result of this approach is shown in Fig. 17 and it can be seen that the gross features in $S_{NiNi}(k)$ are indeed reproduced.

As we have said earlier, to resolve these various points of view, $S_{NiNi}(k)$ and $S_{NiCl}(k)$ must be measured as functions of concentration. This is a major experimental program and the only relevant results we have at present concern the position k_0 of the prepeak in $F(k)$ (see Fig. 2) as a function of molarity. Let us, following Neilson et al.,[651] identify the prepeak with the principal peak in $S_{NiNi}(k)$; it can be seen (Fig. 18) that the prediction of the hard sphere model concerning the dependence of k_0 on molarity (M) does not agree with experiment. The linear dependence of k_0 on $M^{1/3}$ implies that the closest distance of approach of the hydrated ions (which scales roughly as k_0^{-1}) and their mean separation (proportional to $M^{-1/3}$) are closely related for a wide range of concentrations. Thus, these data, though incomplete, favor the charge-ordered model. There is supporting evidence for the existence of ordering in $NiCl_2$ solutions from the work of Fontana.[314]

A direct inversion of $S_{NiNi}(k)$ yields a $g_{NiNi}(r)$ (Fig. 19) that is different from that expected on models based on the random packing of spheres. The same conclusion emerges from a recent theoretical study of the long-wavelength limit of $S_{\alpha\beta}(k)$ carried out by Cubiotti, Maisano, Migliardo, and Wanderlingh,[215] who proposed a model consistent with that suggested by March and Tosi.

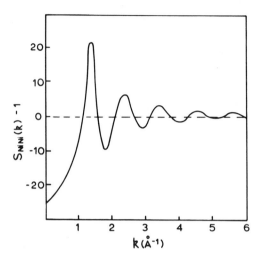

Fig. 17. $S_{\text{NiNi}}(k)$ for a 4.41 m solution of $NiCl_2$ in D_2O calculated from a hard-sphere model.

4.5. Anion–Anion Distribution Functions

The only relevant experiment so far performed to determine $S_{XX}(k)$ was carried out by Soper,[817] who considered a 5.32 m solution of NaCl in D_2O. The sample parameters and the coefficients $A_X^{(2)}$ and $B_X^{(2)}$ are given in Table VI. The results for $S_{\text{ClCl}}(k)$ are shown in Fig. 20 and are quite different from those for $S_{\text{NiNi}}(k)$. A self-consistent $g_{\text{ClCl}}(r)$ was determined from the k-space data (Fig. 21) and turns out to be structureless. The closest distance of approach of two chlorine ions is only 0.2 Å greater than that for molten sodium chloride[270]; this points to the fact that the hydrated ions are, presumably for geometrical reasons, able to penetrate each other rather readily in contrast to the situation for Ni ions. There is clearly no evidence for ordering of the anions in this particular solution even at the high con-

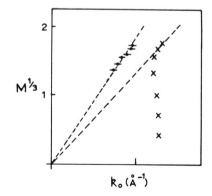

Fig. 18. The position of the prepeak (k_0) for $NiCl_2$ in D_2O as a function of molarity, M, to the power $\frac{1}{3}$. Dots: experiment; crosses: theory based on hard spheres.

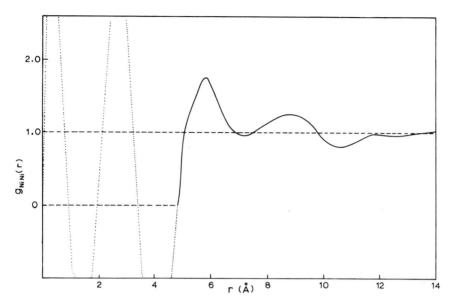

Fig. 19. The radial distribution function $g_{\mathrm{NiNi}}(r)$ for a 4.41 m solution of $NiCl_2$ in D_2O. This was derived from the smooth curve shown in Fig. 16; attempts to remove the excessive noise for $r < 5$ Å and in a self-consistent way have so far failed.

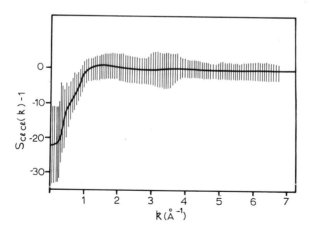

Fig. 20. $S_{\mathrm{ClCl}}(k)$ for a 5.32 m solution of NaCl in D_2O. The error band follows from a generalization of the method described by Edwards et al.[270] The smooth curve was used to derive $g_{\mathrm{ClCl}}(r)$ shown in Fig. 21.

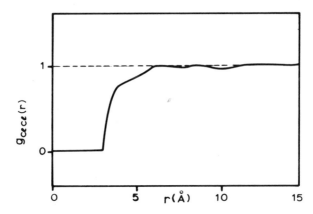

Fig. 21. $g_{ClCl}(r)$ derived from the smooth curve of Fig. 20.

centration considered. A theoretical study[737] of $g_{ClCl}(r)$ for a 1.0 m solution of LiCl and based on the HNC approximation revealed a similar
lack of structure. It will clearly be of considerable interest to measure the
anion–anion distribution function in a system in which the cation–cation
distribution function shows structure. One candidate for such a study is
$g_{ClCl}(r)$ in NiCl$_2$ solutions, and such an experiment is in progress.

5. CONCLUSIONS AND FUTURE PROSPECTS

In our survey of the structural evidence derived from X-ray studies,
two major themes emerged. The first of these concerns the nature of the
ion–water co-ordination in the vicinity of the ion. This part of the aqueous
solution problem, is, in its most general sense, an aspect of coordination
chemistry. The *first-order* difference method described in Section 4 allows
detailed ion–water conformations to be obtained directly and without
recourse to modeling techniques. We are now in a position to answer in a
definitive way specific questions relating to the number of water molecules
in the first coordination shell and the orientation of the molecule with
respect to the ion–oxygen axis.

The second theme related to ion–ion correlations over somewhat
larger distances; this part of the problem is very much the domain of the
classical theory of liquids, which is essentially concerned with long-range
short-lived structures and their dependence on the interionic potentials.
Again, we have seen how the *second-order* difference method gives direct
information on this aspect of the problem. In concentrated solutions, these

two apparently separate aspects of the structural problem become confused; as an example, we cite our discovery that the tilt between the water molecule and the Ni–O axis gradually disappears as the concentration is lowered. Clearly there is need for a fuller exploitation of the neutron method to include a variety of solutes over a range of concentrations and temperatures, and such experiments are in progress. There is also the possibility of applying the technique to look at the effect of ion hydration on water–water correlations.

There are, however, two new experimental programs which use neutrons and which will yield unique and potentially significant results. The first of these involves determining $F(k)$ as a function of pressure. There are three levels, in ascending order of difficulty, at which measurements should be made.

(a) *The Pressure Derivative of $F(k)$.* For extended covalently bonded structures in which all interparticle distances scale like $\varrho^{1/3}$, there exists an exact relationship between $\partial F/\partial P$ and $\partial F/\partial k$.[271] This relationship is, to a good approximation, satisfied by pure water. For those solutes in which water is removed from the bulk phase and attached to ions (the so-called "structure-breaking" mode) we should expect to see this relationship break down as c is increased. On the other hand, in those cases where the structure of water is changed to a new, but well-defined form (the "structure-makers"), the relationship may continue to be obeyed even at high c. Preliminary experiments carried out by Neilson[648] have indeed shown striking differences between solutions of $NiCl_2$ and CsI.

(b) *The Pressure Derivative of $\Delta_x(k)$ and $\Delta_M(k)$.* This measurement will answer the long-standing question as to whether the hydration number is pressure dependent.

(c) *The Pressure Derivative of $\Delta_{M_1} - \Delta_{M_2}$ or $\Delta_{X_1} - \Delta_{X_2}$.* It is usually accepted that the ion–ion interaction can be well represented by a pairwise potential, and thus by an extension of the method proposed by Egelstaff *et al.*,[271] a direct check can be made on the three-body distribution function for the ions. This in turn relates to the nature of the long-range correlations in concentrated solutions. For example, a quasi-lattice-arrangement of ions will yield a characteristic form for the triplet distribution function.

The second class of experiment in which neutrons are necessary is for measurements of the *time-dependent* part of the various correlation functions. Many of the long-standing problems concerned with the dynamics of ionic motion and proton exchange are susceptible to a new approach based on the techniques of inelastic neutron scattering. Until recently, the energy resolution of instruments were such that only very short time events ($\sim 10^{-11}$ sec) were observable. However, the present generation of back-scattering spectrometers yield energy resolutions of ~ 1 μeV, which corresponds to times of 10^{-9} sec. These times will be increased by a further order of magnitude with the next generation of spectrometers based on the spin-echo principle.

Coherent inelastic scattering provides information about collective motion while incoherent inelastic scattering is connected with the motion of individual nuclei. It is therefore possible to exploit the substantial incoherent cross section for protons referred to in Section 3.1 to study the motion of both free and bound water. This technique, together with the intelligent use of isotopic enrichment, will yield fundamental information about the time evolution of the basic structure which we have attempted to describe.

ACKNOWLEDGMENTS

This article summarizes the work of the Bristol/Leicester aqueous solutions group, and we wish to acknowledge the contributions made to the work by past and present members: Stewart Cummings, Alan Howe, Spencer Howells, Sue Jennison, John Newsome, and Alan Soper. The help of the staff at the ILL Grenoble, particularly, Pierre Chieux, Hans Egger, and Walter Knoll, is also gratefully acknowledged. Finally, we wish to thank Mrs. Cathy Charles for her patience in producing an excellently typed version of our handwritten manuscript.

CHAPTER 2

The Organization and Function of Water in Protein Crystals

J. L. Finney

Department of Crystallography, Birkbeck College
University of London
Malet Street, London WC1E 7HX

1. INTRODUCTION

That living tissue contains around 70% water is by now a well-worn cliché. To what extent and in what manner water is necessary to the operation of a biological system is, however, still uncertain. We *do* know that at least part of the water is *not* an *inert* solvent; it is an essential element in many biological processes.[200,307,330,527,849] The folding, structural stability, and dynamics of globular proteins are thought to be extensively controlled by solvent interactions, stress being generally placed on the poorly understood so-called "hydrophobic" or "apolar" interaction.* Similar driving forces are invoked to explain the energetics of enzyme–substrate binding, the binding of a hormone to its receptor, and protein–protein interactions in general. The structural integrity of membranes—the major component being amphiphilic lipids—depends upon solvent interactions. Within a cell, diffusion proceeds in a largely aqueous cytoplasm, the state of which *may* be perturbed by the presence of ions, large molecules, or other interfaces. The solvent close to a protein surface—the so-called "hydration shell"— has properties that differ significantly from the bulk, though how far this

---

* The role of hydrophobic effects on forces between molecules is treated in detail in Chapter 5 of this volume.

perturbation extends from the surface is still a matter of debate. In particular, a lowered water mobility would allow more rapid proton transfer, a potentially significant effect for enzyme activity. Recent studies on the photocycle of bacteriorhodopsin in purple membrane stress the importance of the immediate hydration state in the operation of the system's light-driven proton pump.[510] Similar effects may be of general importance to energy-transducing membranes.[510] The state of the solvent may be significant in electron-transfer interactions.[743] The relative insensitivity to temperature of light-induced oxidation of cytochrome in *chromatium* suggests quantum tunneling of electrons in photosynthesis[241,242,610]; the ease of any tunneling outside the protein will depend strongly on the state of the solvent region. The rate-limiting step of certain enzyme catalyses is the diffusion of a small molecule to the enzyme, e.g., in triose phosphate isomerase[502]; such diffusion rates will be dependent on the state of the solvent medium. The water molecule has been invoked as an active molecule in several enzyme catalyses.[620,726,900]

Although many of the above possibilities are recognized, few are understood in detail. At the molecular level, the details of water involvement are poorly understood; indeed the versatility of the water molecule suggests it may participate in several different guises. A major reason for our incomplete knowledge is the complexity of the systems we are examining, and the consequent difficulty—often artificiality—of separating out the solvent contribution where it appears as an integral part of a complex biomolecular system.[307] A dry enzyme is hardly an enzyme; it folds to its active tertiary structure only by interaction with its (generally aqueous) environment.

The questions we would like to ask cannot normally be answered through the simple experimental procedure of measuring the change in system response when one single variable is changed. The problems of water in biological systems thus require a multifaceted and inherently interdisciplinary approach using a variety of physical, chemical, and biochemical techniques. Adequate interpretation of the results requires a theoretical ability to handle complex interactive heterogeneous systems which, partly with the aid of computers, we are beginning to be able to do. By such a concerted attack, we hope to be able to build up a coherent picture, piece by piece, and set up a realistic model of the behavior of the biological system of interest.

Once a model has been set up, it requires extensive testing in as well-defined a state as possible before it makes any sense at all to apply it in general to less well-defined systems which cannot be fully tested. We take as our example the role of water in the folding, stability, activity, and interactions of globular proteins. A suitably well-defined system to enable us to test our ideas and their predictions in some detail is found in crystalline proteins. When we understand solvent interactions here, we can begin to examine with some confidence the more biologically relevant but less well-defined solution state.

The Use of Crystalline Proteins

Crystalline proteins contain upward of about 25% solvent, a typical value being 45%.[599,601] The structure and dynamics of this solvent exhibits at the molecular level solvent–protein (as well as solvent–solvent and protein–protein) interactions under the particular boundary conditions. Thus, provided the structure of the crystal can be solved by X-ray (or neutron) diffraction techniques to a sufficiently high resolution, we may well be able to extract significant information concerning the results of solvent interactions, especially at the protein–solvent interface. This information can then be used to test the predictions of any protein–solvent interaction model.

The likely deviation of at least some of this solvent organization from what would be observed in maximally active solution conditions should, however, be borne in mind. There will likely be severe perturbations of solvent organization under the influence of the often close interactions between neighboring protein molecules in the crystal. Moreover, the salt concentrations and pH ranges necessary for crystallizing the protein may not be those in which biological activity is maintained at its normal level. In those areas of the protein surface relatively remote from protein–protein interaction, the immediate solvent shell *may* be a good approximation to what would prevail in solution. But in general, direct extension of crystal solvent organization to solution conditions must be done with care. This is in contrast to crystal → solution extrapolations now made on good evidence for the protein itself; in the solvent region, the interactions are generally weaker, and hence more easily perturbed by environmental changes. We stress here the use of crystals in testing solvent interaction models, rather than in directly setting up hypothetical solvent models hopefully valid in solution.

In recent years, the application of refinement procedures to protein structure determination has resulted, in favorable cases, in data of sufficiently high resolution (generally better than 2.0 Å) to make investigation of solvent locations in crystals worthwhile. In what follows, we examine the use and limitations of the technique for this purpose (Section 3) and discuss, in relation to possible function, the results available so far (Section 4). Finally, current and future work is commented upon (Section 6).

We will look at crystalline proteins primarily from the point of view of the water that is the major component of the solvent. Consequently, in relation to the overall system, the discussion will be purposely unbalanced; the solvent will be overstressed at the expense of the protein, with perhaps some overimportance being attached to the water in protein operation. Such an approach is intended to point out those areas where solvent involvement seems likely to be important, and to underline the extent and nature of the questions that need to be asked and answered to justify or discount such possibilities.

First, however, we look briefly at what is known concerning relevant aspects of the water molecule and its interactions, and discuss the kinds of chemical and physical effects by which the solvent may exert influence on the macromolecule.

2. WATER–WATER AND WATER–PROTEIN INTERACTIONS

Our incomplete understanding of the structure and dynamics of liquid water in its normal, bulk phase, is a considerable stumbling block when we look at its state at both simple interfaces such as in ionic or apolar solutions, and more complex ones such as in biological macromolecules. This lack of understanding can be thought of in terms of two major problems:

(a) We have an inadequate understanding of the water–water hydrogen bond in particular, and hydrogen bonds in general.

(b) We are still unable to handle conceptually or theoretically the *dense*, molecular *disorder* which is now a generally accepted major structural characteristic of the liquid state.[97]

Thus, even if we knew *everything* about the hydrogen bond, e.g., the dependence of its energy on intermolecular separations, mutual orientations, and the extent of its non-pair-additivity we would be unable to make use of the information to construct a satisfactory analytical model of bulk water. This reflects similar, though simpler, problems in handling simpler liquids; though there is considerable satisfaction with present-day liquid theories based on integral equation approaches, the nonphysical basis of the approximations made *on the basis of mathematical convenience* to make the equations tractable are regarded by many as unsatisfactory.[898] Even these unsatisfactory approaches are extremely limited in their applicability to water. Although computer simulations can do much to clarify the nature of these disordered states[898] including water,[57,531,732,831] they are no substitute for the more satisfactory analytical theories which still require development.[97,98,148,199,405] In addition, our understanding of water–charged-group, water–polar-group, and water–apolar-group interactions is also incomplete.

These points are discussed elsewhere in previous volumes of this series and in Chapters 3 and 6 of this volume. For our present purposes in interpreting the results of protein crystal structure analyses, the following simplified picture provides an adequate starting point.

The earliest simple model of a water molecule as an approximately tetrahedral disposition of two (positive) hydrogens and two (negative) electron lone pairs has been extremely successful[99] and has been enshrined in computer simulation calculations.[57,732,831]* The molecule thus has the ability to form four approximately tetrahedral hydrogen bonds as found in the ices,[274,325] each molecule acting as proton donor in two such interactions and acceptor in the other two.

Extensive basis set *ab initio* quantum mechanical calculations give a more complex description of the hydrogen bonding interaction,[235,250,379, 712,735] of which two points are particularly significant to our discussion. First, the separation of the lone pairs is less than predicted by the classical model, suggesting that the tetrahedral organization in the ices may to some extent be a consequence of the necessity to find a molecular organization consistent with a crystalline lattice. This relatively low control by the water lone pairs of hydrogen bond directionality is borne out by crystalline hydrate data (see Section 2.1 below), and may be significant for the kinds of noncrystalline solvent organization existing in protein crystals (see Section 4.2). Secondly, the non-pair-additivity of the water hydrogen bond is significant, and has been quantified for small clusters of water molecules.[235,236,379] This is the source of hydrogen bond cooperativity that has been invoked in models of water and aqueous solutions.[322,323] This cooperativity may be responsible for the water molecule's sensitivity to environmental changes.[849] Calculations involving a free interface and based on a cooperative potential[56,58,95] have demonstrated sizable changes in "hydrogen-bonding patterns" when the cooperativity is switched on.[57,58]

---

* For a critical discussion of computer techniques applied to water, see Chapter 6 of this volume.

This strongly implies the non-pair-additivity at, e.g., a protein–water interface is a significant structural and dynamical factor.

Bearing these reservations in mind, the overall structural framework of liquid water is reasonably clear (Fig. 1). Following Bernal[97,99] and Pople,[713] the instantaneous molecular arrangement is *disordered, containing no recognizable crystalline regions.* The local organization is distorted tetrahedral, though the lower directional control by the lone pairs (see above) may allow quite significant deviations from the idealized "random tetrahedral network."[97] The characteristic lifetime of a given "hydrogen bond" is around 10^{-12} sec, although there is little evidence to suggest such an interaction "switches" on and off in anything but a continu-

Fig. 1. Reconstruction of a Monte-Carlo-generated configuration of 216 "water" molecules using a non-pair-additive potential and a periodic boundary condition. Generated by P. Barnes, Birkbeck College, London.

ous manner. This raises problems with descriptions of water properties based on the "breaking" of hydrogen bonds. This overall picture is supported by computer simulation calculations,[732,831] which are discussed in Chapter 6.

2.1. Possible Solvent Effects in Proteins

In the light of our knowledge of water and its likely interactions, *possible* solvent effects relevant to protein structure may be categorized as follows.

2.1.1. "Hydrophilic Hydration"

Direct hydrogen bonding to charged or polar groups on the protein surface is a consequence of the proton donor/acceptor capability of the water molecule. It may be relevant to the stability, dynamics, and interactions of the protein.

For example, the competing effects of ion–protein and water–protein interactions may be important in mediating polymer conformation in general.[350] Bull and Breese[154] have discussed competitive water and ion binding to proteins, and have concluded that both cation and anion effects vary in the same order as the familiar Hofmeister or lyotropic series.[413] They also conclude that protein structure denaturants tend to be those salts that bind preferentially to the protein, thus implying that direct water binding has some significance for structural stability. It must, however, be pointed out that all the binding experiments were performed at low water activities, and the interpretations must therefore be subject to some ambiguity.

As was the case for the water–water hydrogen bond, we do not have generally accepted descriptions of the details of water interactions with polar and charged groups. Some quantum mechanical calculations have been performed (Refs. 253, 254, 450, and R. B. van Duijneveldt, personal communication) for hydrogen bonding to carbonyl O and NH groups; semiempirical (Ref. 229 and our unpublished work) and *ab initio*[129,183, 186,188,796] calculations on the interactions of larger biologically important groups with water are also beginning to be made. These are discussed further in Section 6 below.

For an adequate picture for our present use, we return to reconsider the classical double-donor double-acceptor picture of the water molecule. Such a molecule is potentially capable of making hydrogen bonds to polar and charged groups as implied by the approximately tetrahedral disposition of lone pairs and hydrogens. Indications as to the validity of this picture are available from the wealth of information on hydrogen bond geometry from X-ray and neutron crystal structure analysis.[293,669] Of particular interest is the conclusion from empirical data that if a molecule contains a potential donateable hydrogen (relevant in our case to NH, NH_2, $NH_3{}^+$, OH, and H_2O groups), then normally it will participate in a hydrogen bond via that proton. Only rarely in small hydrate crystals do the hydrogen atoms of a water molecule fail to participate in hydrogen bonding. Thus, we might expect most surface NH groups of a protein to participate in hydrogen bonds with the solvent, with a strong tendency towards linearity. The structure-determining role of the lone pairs in water and NH_3 appears to be much less critical; as discussed above, molecular orbital calculations of the total electron density of the water molecule[250] fail to show two discrete regions of negative density in the classically expected positions.

In contrast, the sp^2 lone pairs on the carbonyl oxygen are seemingly well defined. That their directional influence on hydrogen bonds in which they participate is considerably greater than the sp^3 lone pairs in OH groups is supported by neutron diffraction measurements on simple hydrate crystals.[669] The average H---O=C angle is found close to the idealized lone pair-predicted angle of 120°, though significant deviations (>20°) are found in a few cases.

This simple picture of polar hydrogen bonding is consistent generally with information on solvent organization in crystalline hydrates of molecules of intermediate size, e.g., small peptides and nuclei acid components. We discuss in Section 4.2 the extent to which it is consistent with information available on larger protein molecules.

2.1.2. "Hydrophobic" or "Apolar" Effects

The so-called "hydrophobic interaction," though still only imperfectly understood, is vital to the structural stability and interactions of proteins.[307,474] Its possible origin and symptoms have been discussed in several recent reviews,[327,328,497,855] and also form the subject matter of Chapter 5 in this volume.

The following highly simplistic qualitative discussion suggests some of the aspects of the apolar interaction that may be significant, although a phenomenological description will serve in the subsequent discussion. Consider in the first instance a mixture of hard disks of two different sizes, with no *attractive* interaction between them, being highly compressed at

0 K. With respect to this driving force, assuming all volumes of phase space are accessible, the final state will be that of maximum density. It is known that the maximum density of a two-dimensional array of uniform disks is that of hexagonal close packing. The situation in the two-component system is considerably more complex, depending upon the local density at interfaces between the two components, which in general is likely to be less than that within a *single close-packed* component. However, except for particularly favorable size ratios, the highest density state is likely to be one in which the area of lower-density interface is minimized. The result implies a segregation of the two components into close-packed regions.

As soon as we allow attractive interactions between the two components, and raise the temperature, the simplicity of the above discussion is immediately lost. Our condition becomes one of minimum free energy, and the present impossibility of making reasonable estimates of entropic contributions of local regions prevents a realistic assessment of the terms involved in searching for the minimum. We now assume our two components A and B interact very differently, and are of the same size for simplicity. The A–A interaction is a relatively strong ("hydrogen-bond") angle-dependent interaction, which is likely cooperative, while the B–B interaction is a weak van der Waals' attraction. The A–B interaction is likewise weak, though slightly stronger than the B–B; quantum mechanical calculations have shown a very shallow well for this mixed interaction[571] and "classical" dipole-induced dipole arguments would suggest this also.

Consider first a single "apolar" B molecule in a bath of "water" A. The assembly requires to minimize its free energy. Simple arguments imply that this might be done by maximizing the number of A–A interactions, and the preferred geometry of water molecule interactions suggests that this can be done particularly easily by forming some kind of "clathrate-type" cage. In this, each water molecule can potentially make four relatively strong hydrogen bonds with its neighbors, the relatively weak A–B interaction doing essentially nothing to reduce the hydrogen-bonding complement of the surrounding waters. Such geometrical groupings of water molecules are also favorable to an enhancement of their interactions through cooperative effects. The extent to which this kind of water "structuring"—of which the above is an extreme example and may well overstress what actually happens—depends upon how much the possibly increased enthalpy contribution can compensate the unfavorable—and presently uncalculateable—entropy decrease which would consequently occur. In the real example of dissolving an inert gas molecule in water, there is a positive excess free energy change, the source of which is thought to be the solvent "structuring." This argument might be applied to any extended, weakly interacting interface in contact with a bath of more strongly interacting molecules. To some degree, the volume in phase space made inaccessible to the solvent bath by simply the presence of the interface (so-called "excluded volume") will itself lead to some localization in phase space of the water molecules. The directionality of the water–water interaction may well result in the average water–water interaction in the slightly confined phase space volume being marginally stronger than in the bulk.

In looking for solvent organization around apolar groups, we should not be surprised to find little evidence of very stable cages confined sufficiently in phase space to make them visible to X rays or neutrons. In small molecule clathrate hydrates, there are strong crystal lattice forces acting to localize the water cage arrangement, while the guest apolar molecule is in relatively free rotation within the host (water) cage. The registration between host and cage would be small. This is hardly surprising considering the weakness of the host–guest interaction, but in addition, the geometry of at least the smallest clathrate cage results in an extremely low and symmetrical central field.[226] In a protein molecule, although the water around an exposed apolar group may be significantly localized in phase space, the very limited *specific* interaction between group and solvent may well be too low to localize particular molecules sufficiently in *real* space to make them visible to our time-averaged probings. If such clathrate-type groupings are seen, it is likely they are further stabilized by additional specific interactions within the neighborhood.

Returning to our simple model, when we consider more than one B (apolar) component, the situation becomes even more complex, with several possible models of the structural and dynamical outcome.[327] Again, the assembly minimizes its free energy. The simple packing model above would suggest that some kind of segregation may be thermodynamically favorable, though this would depend upon the balance of the entropic contributions. The standard concept—which is far from being accepted[327]—imagines essentially "contact aggregation" of the B component, with a consequent entropic driving force as water molecules, previously phase-space localized in the apolar hydration region (be it clathratelike or not) are released into the bulk solvent. The sign of the consequent enthalpy change is unclear. The details of the interaction at the molecular level are clouded by our lack of knowledge of the intermolecular interactions involved and our inability to deal with the statistical mechanics.* In the following

* Attempts to cope with this problem form the subject matter of Chapter 5 of this volume.

discussions we will retain little more than the simple and vague phenomenology of some kind of longish-range tendency to aggregation of apolar groups.

2.1.3. Hydrophilic Surface Ordering

An extended hydrophilic surface may affect the dynamics of nearby water molecules.[192,238,315] The water close to clay surfaces may be affected to a depth of 30 Å,[360] and it is possible that the depression of the freezing point of water close to minerals and biomolecules may occur by a similar mechanism.[49,604,605] For proteins, suitable arrangement of hydrogen donor/acceptor sites on the molecular surface may "structure" the interfacial solvent sufficiently to affect both the dynamics of the macromolecule and diffusion processes across or close to the interface. Such effects have been invoked to explain the action of so-called "antifreeze" glycoproteins in arctic and antarctic fishes; a suitable geometrical arrangement of polar sites on the (extended) protein may produce a strong affinity for the ice structure, which may thus be prevented from further growth by a protein coating on the ice front.[246,390,738] However, a somewhat different interpretation, based on the relative orientations of sugar -OH groups and H_2O molecules in water and ice, is that the glycoproteins inhibit ice nucleation, i.e., they favor supercooling.[331]

It is also of interest in this context that some steroids (and amino acids to a lesser extent) are excellent ice nucleators.[344] Again the relative location of hydrogen bonding groups is invoked: e.g., in cholesterol, the OH groups form a 7 Å × 9 Å lattice, which fits fairly well with the 4.52 Å × 7.37 Å hydrophilic lattice found on the prism plane of hexagonal ice. Moreover, most steroids crystallize with tightly bound water; there are also suggestions that water may be involved in the binding of steroid hormones to receptors (W. Duax, personal communication).

2.1.4. Charge-Transfer Effects

The relatively high proton mobility in ice was explained by Bjerrum,[114] who invoked an H_3O^+ defect mechanism, coupled with transfer of protons along hydrogen bonds. Thus the high proton conductivity of ice depends upon the existence of a relatively stable hydrogen-bonded network. Liquid water has a less perfect, less rigid, and more mobile network structure, and hence a lower proton conductivity. Any "ordering" effects that result in a "firming up" of the network (a longer characteristic relaxation time) would be expected to result in an increase in proton conductivity, with possible consequences for relevant enzyme mechanisms and turnover rates.

Parallel mechanisms have been proposed for electron transfer, which may be of relevance to oxidation/reduction proteins. There are several reports of probable tunneling of electrons in both ice[520] and other media such as glassy methanol,[621] with suggestions that such processes may be important in, e.g., photosynthesis.[610] Tunneling would also be strongly dependent upon the mobility of the hydrogen-bonding network.

2.1.5. Dielectric Effects

Liquid water has a high dielectric constant ε resulting from its high molecular polarizability. Changing the medium and temperature but *keeping ε constant* does not appear to disrupt the activity of an enzyme and would therefore appear not to cause any significant denaturation of the active tertiary structure.[108,472,812] The serine proteases trypsin and α-chymotrypsin retain the same catalytic pathway in both 65% aqueous DMSO and 70% aqueous ethanol at low temperature where the dielectric constant is about the same as in its room-temperature aqueous phase.[304,305] Thus, whatever is maintaining the protein's active structure and substrate affinity, only some—if any—of the water is specifically necessary. However, the molecular ratios in the solvent should be borne in mind: in a *crystal*, 70% aqueous ethanol translates to about a 1:1 ethanol/water ratio. This would still be enough to give a first layer of water close to the protein. Although proteins can thus be shown to remain stable and active in a less aqueous medium than normal once they have folded and reached their equilibrium active conformation, there is no information as to their ability to find their normal folding path through phase space in such altered solvents.

Dielectric effects (or molecular polarizability at the molecular level) may be useful in aiding structural stability by smearing the undesirable effects of local charges within a protein.

2.1.6. Water as an Active Molecule in Catalysis

Water is found to act as a general acid (proton donor) or general base (proton acceptor) in many model reactions, and thus may be so active in actual catalytic processes. In such cases, the presence of relatively strongly "bound" water molecules near active sites, and the possibility of rapid proton transfer (Section 2.1.4 above) may become important. Its possible active role in photosynthesis has been alluded to above.[610,885]

2.2. Conventional Wisdom of Protein Hydration

No single technique is capable of adequately characterizing the structural and/or dynamical nature of the solvent region associated with a protein. From the results from a variety of techniques (mainly spectroscopic and calorimetric) the following three-zone picture has been built up.[200,527]

(a) A small number of water molecules may be found internally bound to protein molecules. Some of these may be strongly, irrotationally bound with very long lifetimes, and could be thought of as an integral part of the protein itself. Others may bind less strongly with shorter lifetimes. These solvent molecules may be important in maintaining local structural stability, and some may be directly involved in protein function.

(b) A larger number of water molecules appear to be bound at specific charged and polar surface sites. They appear to have a wide spread in relaxation times, averaging around say 10^{-6} sec.

(c) There is also a surrounding "perturbed solvent shell" region of unknown and probably variable extent (depending on the protein and conditions); this region comprizes molecules less mobile than in the bulk, with relaxation times perhaps 10–100 times longer, and merges into bulk solvent. The existence of a significantly perturbed solvent shell that can be realistically separated from both the directly bound surface region and the bulk remains to be fully demonstrated (see below).

The extent of the various regions is difficult to characterize, depending critically on the experimental method used. So-called "bound water"—presumably meant to include regions (a), (b), and possibly some of (c)—is a somewhat indefinite term; its extent is defined operationally by the measuring technique used.[356,527] For example, the nonfreezing water fraction as measured by infrared and differential scanning calorimetry for the same protein (lysozyme) prepared under as nearly as possible identical conditions both yield values of about 0.3 g/g.[356] This figure interestingly corresponds to the amount that would be expected if all accessible surface charged and polar hydrogen-bonding sites were filled.[308]

That this strongly bound water region is very limited in extent has been shown for several protein crystals, where several techniques have suggested that channels of essentially "bulk" water remain.[112,113,152,627,774]

It is not clear how far this should be generalized to all protein crystals, nor how far the measuring techniques themselves (e.g., small molecule diffusion) perturb the solvent region.

Fractions of "bound water" attached to proteins in crystals have been obtained from measurements of crystal density and unit cell parameters.[8,196,696,786] In Scanlon and Eisenberg's very neat approach, where "free solvent" is defined operationally as having the density of the free bulk solvent, values of 0.13–0.27 grams of bound solvent per gram of unsolvated protein were obtained. Determination of the partial specific volumes of the bound fractions were much less accurate, giving values indistinguishable from the bulk. The physical picture consistent with these observations is again that of a solvated protein interpenetrated by channels of essentially free solvent.

Any "ordering" of the liquid phase of a strongly directional molecule like water would be expected to lead to an increase in partial specific volume, although complicating effects at the interface may compensate for this to some degree. Work by Low and Richards[573] using protein crystals to examine the upper end of the composition range concluded that, within the accuracy of the data, the partial specific volume of water was the same as in the bulk. Recent very accurate work by Bernhardt and Pauly[101] on solutions of BSA and bovine hemoglobin suggest an increase in partial specific volume of water above 0.1 and 0.2 protein mass fraction, respectively. This increase is small, amounting to no more than 0.3%, outside the detectability limits of Low and Richards' method. Similar increases in water partial specific volume were observed by Bøje and Hvidt[135] with polymer solutions.

The very accurate data of Bernhardt and Pauly[101] are not straightforward to interpret. For bovine hemoglobin, the partial specific volume of water in a solution of 0.42 weight fraction of protein was found to be 1.006 cm^3, in contrast to 1.003 cm^3 for bulk solvent. Taking a value of 0.25 g bound solvent to 1 g unsolvated protein (about the mean of the values given for Scanlon and Eisenberg's four proteins[786]), and assuming the "nonbound" fraction to be identical to the free bulk solvent, we obtain a partial specific volume of the bound fraction of about 1.020 cm^3, 1.7% greater than the bulk value of 1.003 cm^3.

If this is a reasonable interpretation, how these two regions merge together is of some interest. Were we to think of the bound fraction as a water phase, the minimum in the volume–temperature curve of pure water suggests that the interface between the two regions—"bound" and "free"—would be somewhat disturbed, as suggested around ions by the work of

H. S. Frank. How far this is reasonable depends very much on the detailed molecular nature of the surface of the "bound" region. Assigning a partial specific volume to a very thin layer about one molecule deep is probably pushing thermodynamics too far. The value of such a quantity will be determined in large part by the geometrical nature of the protein surface itself. We are unlikely to pick up a volume change of less than 2% from X-ray coordinate measurements.

Thus the nature of the solvent region away from the immediate protein–solvent interface remains uncertain. Nuclear magnetic resonance measurements suggest it may be significantly perturbed from the bulk, while density and X-ray measurements are in most cases inadequate tools. If reduced mobility and increased relaxation times do extend significantly from the interface, the implied "ordering" need be considered only in dynamical terms. There is *no reason* to contemplate local crystalline icelike structures. Rather, the liquid network—perhaps perturbed by the proximate interface— might be thought of as being partially stabilized, akin perhaps to a supercooled water. There is some restriction of the phase space accessible to the solvent in this region, perhaps a real restriction in that certain points in phase space become inaccessible, or merely the rate of exploration of phase space proceeds at a slower pace. "Ordering" in this situation means "softening" modes of vibration, with consequent increased lifetimes of particular (noncrystalline) molecular configurations, and perhaps an increase in the probability of occurrence of certain of these configurations at the expense of others.

3. CRYSTAL STRUCTURE ANALYSIS OF PROTEINS AND ITS SUBSEQUENT REFINEMENT

Of the kinds of water expected in association with a protein (Section 2), crystal structure analysis should be capable in most cases of giving direct information concerning the internally bound fraction (if any), and in favorable, reasonably well-ordered crystals, those molecules strongly bound at the molecular surface. For less well-ordered solvent regions, the best that we can really hope for is a "probability density distribution." This is only possible in certain cases where very high resolution refinement procedures can be applied (Section 3.2). Information on the relative locations of particular groups within the protein is also obtainable, and can give indirect, though often very useful, information on solvent effects (Section 4.1).

X rays are the radiation probe usually used. This essentially limits us to the location of only the oxygen atom in the water molecule, the scattering cross section of hydrogen for X rays being relatively low. In contrast the neutron scattering cross section of hydrogen is much larger in relation to other nuclei, and thus, provided other effects such as temperature factors do not reduce the visibility of hydrogens, neutron diffraction is potentially capable of greater detail concerning solvent. There are, however, major disadvantages. Much larger crystals are required than for X rays, even using the intensities available in the very few high flux neutron reactors. As growing even small good crystals is often the limiting step in protein structure analysis, this can be a serious problem. The cost and time required to collect the data is much higher than for X rays. Finally, the incoherent neutron scattering cross section of hydrogen is high, giving rise to extensive background scatter. Substitution of exchangeable hydrogens by deuterium can help reduce this problem; to remove it altogether requires the synthesis of the fully deuterated protein from scratch.

A structure analysis can be considered in two parts:

(1)   The collection of X-ray or neutron data, and their interpretation in terms of a plausible atomic model, followed by

(2)   refinement of the model to improve the atomic coordinates and their reliability.

3.1. Theory, Data Collection, and Initial Solution of the Structure

A crystal is an *ordered* array of molecules, the smallest convenient unit of which is described by the *unit cell*. The whole crystal can be generated conceptually by translating the unit cell contents through the *unit cell translations a, b,* and *c* along a set of general *coordinate axes, x, y,* and *z.*

Figure 2 illustrates the essential geometry of scattering, and defines the basic quantities used. The direction of the incoming X ray is defined by a unit vectors s_0, modulus $1/\lambda$, where λ is the wavelength of the radiation used. The direction of radiation scattered through an angle 2θ is defined

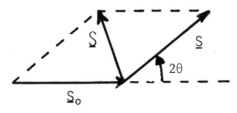

Fig. 2. General scattering diagram. The direction and wavelength λ of the incoming radiation is denoted by vector s_0, while that of the scattered radiation is denoted by s. $|s| = |s_0| = 1/\lambda$. The *scattering vector* $S = s - s_0$.

by **s** (again with modulus $1/\lambda$). As only one vector is required to describe the change in direction between incident and scattered radiation, the scattering vector **S** is defined as

$$\mathbf{S} = \mathbf{s} - \mathbf{s}_0$$

Clearly $|\mathbf{S}| = (2 \sin \theta)/\lambda$. Several alternative notations for the scattering vector exist, depending on the field of study. All—e.g., **Q**, **K**—are essentially the same as **S**, with perhaps scale differences of 2π.

When considering scattering from a crystal, we observe a diffracted intensity only when the three Laue conditions are fulfilled, viz.,

$$\mathbf{a} \cdot \mathbf{S} = h, \qquad \mathbf{b} \cdot \mathbf{S} = k, \qquad \mathbf{c} \cdot \mathbf{S} = l$$

where h, k, and l are indices. Each of these three equations defines a plane in so-called reciprocal or Fourier space. Only where these planes intersect are all three Laue conditions fulfilled and a diffraction "spot" observed. Each such spot is labeled by the applicable set of indices (hkl). These indices also define sets of crystal planes for which each spot is sometimes considered a "reflection." These planes are separated by a distance d_{hkl} given by $\lambda/(2 \sin \theta)$ (Bragg's law). This nomenclature is, however, misleading, and obscures the basic nature of the scattering process.

Thus for integral values of (hkl) there exists a diffracted beam at an angle $2\theta_{hkl}$ to the incident beam, of amplitude $|\mathbf{F}_{hkl}|$ and phase α_{hkl}. In essence, the diffracted intensity from any object is given by the Fourier transform of the function describing the density distribution of the scattering centers—electrons for X rays, nuclei for neutrons. The diffracted amplitudes, or *structure factors* $|\mathbf{F}_{hkl}|$ can be regarded as a *sampling* of the Fourier transform of the unit cell contents—i.e., the electron density distribution in the unit cell.

Making use of the reciprocity property of the Fourier transform we can, in principle, recover the electron density distribution $\varrho(x, y, z)$ by performing the triple summation

$$\varrho(x, y, z) = \sum \sum \sum |\mathbf{F}_{hkl}| \, e^{-i\alpha_{hkl}} \tag{1}$$

Our structure—the $\varrho(x, y, z)$—can thus be reconstructed by simple Fourier synthesis of the observed (and suitably corrected) diffracted amplitudes weighted by their phases. This equation raises two important points.

(*a*) *Resolution.* The structure factors $\mathbf{F}_{hkl}$ are only samples of the underlying transform of the unit cell contents, taken at those points in reciprocal space described by the intersection of the (reciprocal space) planes defined by the indices (hkl). Equation (1) requires a summation of each integer index to infinity; this would correspond to the sampling of the transform at every point, and our information would be complete. Such is not possible, and we are limited by several factors to only finite values of (hkl). Thus, we can obtain structural information—$\varrho(x, y, z)$—to only a limited *resolution*, which is measured by the maximum (hkl) values accessible and/or measurable. Considering our diffraction experiment in terms of "reflections" from crystal planes separated by a distance d_{hkl} (see above), the maximum resolution achievable is given theoretically as $0.72 d_{min}$. In reality the resolution of an experiment is taken as d_{min} itself—the smallest separation of "scattering planes" for which observable intensities are measured. We note here in passing that the unavailability of amplitude and phase data beyond a given resolution has unfortunate effects on the Fourier transform, in that spurious detail may be introduced merely by the necessity of truncating the Fourier series at finite (hkl). Such "termination of series" errors have been examined in detail in liquid diffraction experiments.[891] Handling their effects in crystal diffraction, particularly for high-resolution refinements, is more problematical.[140,566]

(b) *The Phase Problem.* The Fourier summation [eqn. (1)] requires the knowledge of both the amplitude and phase of each diffraction spot. Unfortunately, unlike in a lens imaging process, we have no direct access to the phases in a diffraction experiment. We can measure *only* the *intensities*, which are proportional to $|F_{hkl}|^2$. The phase information is lost. It is the recovery of this phase information that is the major problem in most structure analyses. Several methods are available for small molecules. For larger molecules such as proteins, use is generally made of the method of *multiple isomorphous replacement*. This procedure depends upon the preparation of, and data collection for, several crystals in which different heavy atoms have been bound to the molecule. To be successful, the binding of the heavy atom must cause only limited local changes to the structure of the molecule. Ideally, the native and heavy atom structures must be as nearly as possible *isomorphous*. The degree to which this condition is *not* fulfilled is a limitation on the accuracy of the resultant phases, and hence of the effective resolution of the structure analysis. In favorable cases, such phase information may be usable to a resolution of about 2 Å.

Once reasonable phases have been assigned to all observed $|F_{hkl}|$, an electron density map of the unit cell (or part of it) is constructed. At this stage, we try to trace the topology of the polypeptide chain(s) and then proceed to fit what we think the chemical structure is (if we know the sequence of amino acids along the chain) to the various regions of electron density, by either manual or computational means. The coordinates of the model can then be read off, and finally adjusted if necessary to remove any gross chemical inconsistencies (such as overlapping atoms). At this stage we make use of what we think we know of the stereochemical constraints; if we did not think we knew what a protein should look like, we would have difficulty in finding out solely from a crystal structure analysis.

3.2. Refinement of Structure

Although the above procedures may produce a reasonable atomic model, with the topology of the polypeptide chain(s) reasonably well established, the list of atomic coordinates obtained by fitting to the electron density map of ~ 2.5 Å nominal resolution tends to instill a false sense of confidence in their reality. At this resolution, we might expect most atomic coordinates to be within about 0.5 Å root mean square deviation from their actual positions, and perhaps up to several angstroms in the least well-defined regions.[599] In addition, no atom in the crystal is stationary; our atomic coordinates are of the estimated mean positions of each atom. To account for thermal vibrations, each atom is assigned a "temperature factor," which assesses the resultant "diffuseness" of position, being proportional to the mean square displacement from the expected position. Within this "B value" is hidden also the effects of structural disorder; our diffraction experiment is an average over many unit cells, in which a given group may occupy one or more different positions. Such B values are generally much higher for terminal groups and highly exposed side chains. Coordinate values should be read in conjunction with estimated temperature factors.

Hence, the reliability of the coordinates may well be inadequate to allow us to draw confidently conclusions about the detailed structure in, say, the active site region, or to assign with confidence hydrogen bonds in the structure. Solvent information—except perhaps very strongly bound molecules—is unlikely to be at all reliable. For more accurate information, we need to institute some kind of *refinement procedure* to take the structure to a higher resolution.

This is not possible for a large number of proteins. The disorder in the molecule may well be too high for the higher-order reflections which contain the necessary intensity information to be observed. For certain particularly favorable cases, intensity data can be collected out to higher resolution, and the structure further refined.

Assuming these weak high-angle data can be collected with sufficient accuracy, there are still problems concerning the phases of both these and the lower-angle reflections phased with the help of the heavy atoms. The phases obtained from multiple isomorphous replacement techniques are usually very inaccurate.[231] If we wish to refine the structure, we use these only as a starting point and replace them by phases calculated during the refinement procedure. The ultimate detail of protein maps is very dependent upon the accuracy of these improved phases. The problem is essentially one of producing a better atomic model from experimentally phased electron density maps of resolution not really sufficient for the purpose.[784]

The problem of protein refinement is presently a very active field, and reference should be made elsewhere for details concerning both the general problem and applications to specific cases.* Refinements have been performed both in real and reciprocal space; both have advantages and disadvantages, the balance of which is presently not clear. The problems of termination errors are in principle less relevant to the reciprocal space methods, though it can be argued that difference Fourier methods in real space minimize these also. In most cases, we may have only about 2–3 observations per parameter to be refined. For refinement of, say, 6000 independent parameters, this is a procedure not entirely devoid of danger.

Whatever the method used, an accurate description of the solvent region becomes important for a reliable refinement of the protein itself. The low-order intensities $[(\sin \theta)/\lambda \lesssim 0.05]$ are seriously affected by neglect of the solvent[212]; thus these are often omitted in phasing and refinement work. More seriously, changes made in a proposed solvent structure are found to significantly affect the electron density (difference) maps within the *protein region itself*.[12] The converse is also the case: errors in placing the atoms within the protein (perhaps by forcing the structure to fit a necessarily imperfect density map, or dealing inadequately with disorder in the protein, is liable to affect electron density in the solvent region. Such artifacts may easily lead to erroneous solvent interpretation.

The goodness of fit of a structure to the measured data is estimated by means of a residual R defined as

$$R = \frac{\sum \left| |\mathbf{F}|_{obs} - |\mathbf{F}|_{calc} \right|}{\sum |\mathbf{F}|_{obs}}$$

which can be thought to represent an "average error." For small molecules, a well-defined structure would be expected to give an R value of less than

* See Refs. 197, 231, 247, 248, 336, 448, 449, 627, 634, 783–785, 823, 873, and 892.

0.1, and often very much less. It is tempting in the protein case to regard reduction in R as leading to a more accurate structure, but this may not be the case, depending upon whether we are fitting to real or artifactual detail in the electron density maps. For example, for carp muscle calcium-binding parvalbumin,[627] a significant reduction in R could be obtained by an almost random placing of hypothetical solvent molecules. For well-ordered solvent molecules the problems are less. However, when considering the relatively disordered solvent region away from direct contact with the macromolecule surface, the peaks representing solvent become really a probability density function, with heights that may well represent only partial occupancy over time or different unit cells. Such information rapidly drops to the level of noise, or below. If a perfect diffraction experiment yields a disordered region with solvent positions not fully occupied, then our necessarily imperfect experiment and refinement techniques will be in danger of overinterpreting noise as if it were real low-level information. It would be quite feasible to locate all density peaks in the solvent region, estimate their intensity, refine occupancies and temperature factors, and thus list a set of solvent sites of variable occupancy. Such a procedure would inevitably produce better agreement with experiment, but we will be deluding ourselves about the reality of the solvent assignments made. In this connection the comments of Moews and Kretsinger in their 1.85-Å refinement of parvalbumin are highly relevant[627]; their caption to a table of "*possible* solvent molecules" states that "these (coordinates and temperature factors) *may* represent solvent molecules." Elsewhere in their paper, they estimate only about half of their assigned 138 solvent positions are likely to be real.

In the protein itself, one can make use of known stereochemical constraints during the refinement; certain chemical parameters (e.g., bond lengths) are known with a precision better than can be deduced from the available data. These are normally constrained during refinement. In the solvent region—in particular where extensive disorder is suspected—this cannot be done. We must rely much more on the quality of the electron density map itself, and look for ways of extracting signal from noise; independent checking of sites is possible on stereochemical grounds only when the solvent region is highly ordered. Several ingenious ways have been developed to tackle the signal/noise problem, such as removing one-eighth of the structure in slabs in turn, and refining on the remaining seven-eighths (G. Dodson and D. C. Hodgkin, personal communication) A further obvious test of these methods is to check the results of *different* refinement procedures against each other. In those cases where such comparisons have been attempted, significant disagreements in the solvent region were found.

The accuracy of the *protein* coordinates from several refinements can be estimated fairly well. In the 1.5-Å refinement of rubredoxin,[892] a final R value of 0.126 was obtained; the estimated standard deviation of the C_α–C_β bond lengths was 0.19 Å; a further refinement to 1.2 Å[894] including anisotropic temperature factors gave an R value of 0.110. The 1.85-Å refinement of carp muscle calcium-binding parvalbumin yielded a final overall residual of $R = 0.25$. Estimated upper limits of errors in atomic position varied from 0.15 Å (Ca^{2+} ions) to 0.35 Å for surface side-chain atoms; surface lysine side-chain atoms were less reliable (0.50 Å). Solvent positions would be no better than the surface side chain figures, and likely significantly worse. Selected intramolecular hydrogen bonds were within the limits of length and off-linearity found in small molecules, though several inter-molecular bonds showed an unlikely combination of short length (2.6–2.65 Å) and off-linear NH– – –O angle (38°–47°).[627]

3.3. Summary: Reliability of Atomic Coordinate Data

The above discussion illustrates the problems in obtaining accurate data for both the atomic coordinates of the protein and solvent positions, even in favorable cases where a high (good) resolution is in principle achievable. Data of resolution lower than 2 Å should be used with great care for solvent information, restricting interpretation to only unambiguously placed strongly bound molecules. Subsequent refinement of even higher resolution structures has resulted in significant "changes" in solvent structure.

In higher-resolution refined structures, *for the atoms in the* protein molecule, (root mean square) errors in positioning are unlikely to be less than 0.20 Å for well-defined atoms. For surface molecules—especially highly exposed side chains and terminal ends of polypeptide chains—the uncertainties are likely to be greater (say 0.5 Å or more).

Strongly bound solvent molecules (those internal, or strongly bound at the surface) may be placed in favorable cases as reliably as surface side-chain atoms. As we proceed further into the less well-ordered solvent region, the problems increase dramatically as errors (in phasing, in dealing with the protein itself, and from termination effects) become of the order of the effects we are trying to observe. There are dangers in forcing the map to agree with a large number of fully ordered solvent molecules; though this may in reality be the case, it is not easy to assess how much is reality, how much artifact. Conversely, there are dangers in assigning fractional occupancies to apparently small solvent-region peaks to reduce R. Such partially occupied, high-temperature factor sites *may* be realistic, representing the physical realization of phase space restriction discussed in Section 2, but it is difficult to know whether excessive reality is being assigned to errors or noise. Unlike the protein molecule itself, we can only with difficulty—if at all—use independent chemical information to help sift the real from the artifactual. In at least one protein structure a well-ordered chain of solvent molecules was later reassigned as protein.

4. SOLVENT INFORMATION FROM CRYSTAL STRUCTURE ANALYSIS

We consider information obtained from two sources.

(*a*) *Indirectly from Protein Coordinates.* In essence, by knowing where the protein *is*, we know where solvent is *not*. Although apparently a trivial statement, excluded volume arguments based on this information have led to many proposals concerning the significance of solvent in protein folding, stability, and interactions. The protein coordinates themselves, in particular the relative location of particular kinds of groups—e.g., hydrophobic side chains—illustrate the *results* of solvent effects at the molecular level, and may suggest possible mechanisms of operation of particular systems. They also give us hints of the likely molecular structure of the first hydration shell, which more direct information (Section 4.2) appears to support.

(*b*) *Directly from Water (or Other Solvent) Molecule Coordinates.* Assuming the reliability (see Section 3) of the information gleaned from structural studies—in particular from high-resolution refinements—we discuss the organization and likely functions of identified discrete solvent molecules.

4.1. Information from Protein Coordinates

4.1.1. Folding and Stability

In general, the solvent is thought to exert extensive thermodynamic control on protein folding and stability. The importance of so-called "hydrophobic" solvent effects was discussed by Kauzmann in 1959.[474]

Assuming groups within a protein molecule can be assigned a degree of hydrophobic or hydrophilic character, the loss of entropy on folding was argued to be compensated by a gain in "hydrophobic energy" resulting from a reduction in the number of contacts between water and nonpolar groups; these groups are consequently buried inside the folded molecule, away from solvent contact. An alternative to this "grease in" principle was proposed by Klotz,[498] who argued for a favorable enthalpy change accompanying the formation of water ("clathratelike"?) cages around exposed apolar groups. In contrast, the polar groups are considered to contribute little to the molecule's stability. Hydrogen bonds made within the protein are thought unlikely to be significantly stronger than those made with solvent in the extended chain conformation;[499] thus the only significant contribution to the folding process expected from this hydrophilic source is a favorable entropy contribution from the release to bulk solvent of extended-chain bound water.[179]

The first few protein structures solved appeared consistent with the "hydrophobic in, hydrophilic out" proposal. Charged side groups tended to stick out into the solvent; many of these polar groups which were not at the surface made extensive internal hydrogen bonds, in particular within the α-helix and β-sheet regions of secondary structure.

However, as more protein structures were solved, and were looked at in more detail, the situation appeared to be considerably more complex. Solvent accessibility of groups in proteins were quantified by Lee and Richards[540] and others,[179,308,808] and attempts were made to correlate such quantities with hydrophobic energy contributions.[178] In addition, the constraints imposed on tertiary structures by hydrogen bonding and internal packing efficiencies have been discussed.[179,306,745,746] The following generalizations can consequently be made.

(a) Formal charges on protein side groups are almost invariably exposed to solvent. Where they are not, salt bridges are usually formed.

(b) There is much less bias between internal and external positions for polar groups. The efficiency of internal hydrogen bonding of buried polar groups seems very high, generally over 80%. It should be noted that hydrogen bond assignments are made purely on geometrical criteria, and may well reflect the known significant errors in atom coordinates.

(c) Of the solvent accessible area of known native structures, 40–50% is composed of nominally apolar groups. Although assignment of polar/apolar character is hardly a clear-cut process, these figures seem approximately invariable to reasonable changes in apolar assignment criteria.[308] Moreover, during folding, the changes of accessible area of apolar and polar groups are similar. Thus the simple picture of "grease in" appears an oversimplification. Quantitative estimates of the hydrophobic energy contribution to tertiary structure stability have been made.[179]

(d) Internal packing in globular proteins is essentially as good as is found in crystals of smaller molecules.[179,306,745,746]

The impression of a globular protein as an *extremely well-fitting* three-dimensional jigsaw puzzle is very strong. Although the hydrophobic contribution to the stability is clearly large, the efficiency of internal hydrogen bonding, salt bridges, and simple packing implies that both internal charge–charge, hydrogen-bonding, and van der Waals' interactions must also have a significant structural influence. This point is discussed in more detail below, and with respect to water molecule involvement.

Recent hydrogen-bonding and molecular-area studies on ribonuclease-S underline the likely efficiency of the internal hydrogen bonding, and suggest relationships to solvent accessibility that may be significant in determining the first hydration shell of the protein.[308] Within the limited accuracy of the data, and using geometrical hydrogen-bonding criteria similar to those known empirically from crystalline hydrates and computer-simulated water, the degree of internal hydrogen bonding was remarkably high with respect to predictions based upon the "classical" number of hydrogen donors and acceptors expected for particular groups (see Section 2.1.1). Thus, NH groups were found almost always *either* to participate as a hydrogen donor in a good internal hydrogen bond, *or* to contribute to the surface of the protein an area which, from calculations on liquid water, *would be sufficient to accommodate a solvent hydrogen bond*. This strongly suggests that hydrogen donors—which almost always find acceptors in small crystal structures (Section 2.1.1 and Ref. 669)—almost always form a good hydrogen bond within the much more complex protein molecule or are sufficiently exposed to be capable of forming one with a solvent molecule. The early X-ray work on α-chymotrypsin[120] suggested that the hydrogen-bonding capabilities of internal NH groups were almost always satisfied, and calorimetric measurements[500] also support the satisfaction of the demands of proton donors. In carp muscle calcium-binding parvalbumin[627] all side-chain hydrogen donors were found to be hydrogen-bonded, or in solvent contact.

Studies on the coordinates and accessibilities of several proteins show that turns in the polypeptide chain are frequently associated with a sequence of 3–8 polar residues. Moreover, at such turns, the internal hydrogen bonding is poor, resulting in a particularly high exposure of potential hydrogen-bonding groups.[525] These groups were assumed solvated, implying a solvent influence on turn generation or stabilization. It is perhaps coincidence that to denature many proteins, an average of 1–2 urea molecules per turn is required.[525] Early studies on carboxypeptidase-A suggested that more than 85% of those unoccupied backbone hydrogen-bonding sites (38% of the total) were at the surface, presumed hydrogen-bonded to solvent.[526] The implications of these observations for the first hydration shell are perhaps significant.

Although the control asserted by lone pairs on the geometry and number of apparent hydrogen bonds is less than that of hydrogen donors (see Section 2.1.1), it is known that in small crystals $C={=}O$ groups tend to participate as hydrogen acceptors in two hydrogen bonds with geometries in reasonable agreement with electrostatic predictions from lone-pair orientations (Section 2.1.1 above, and Ref. 669). Within the protein also this tendency is reflected: in RNase-S, either two such bonds are made internally, or the oxygen is exposed to solvent sufficiently to allow making up the deficit with solvent. It is known that most proteins have more potential hydrogen acceptors than donors, the ratio in carp muscle calcium-binding parvalbumin being 1.88.[518] That solvent interaction would appear a likely mechanism for evening out this imbalance is supported by the significantly higher exposure of carbonyl oxygens compared with main chain NH groups.[308,540]

As Fig. 3 shows, this relationship between "unmade potential hydrogen bonds" and exposed area is fairly general for all groups. If n_H is the sum of the number of donatable hydrogens and lone pairs on a group, n_I the number of hydrogen bonds made internally, and a_0 the exposed area assigned to a potential solvent hydrogen bond, then to a good approximation the molecular area contribution A of the group is given by

$$A = (n_H - n_I)a_0$$

As expected, the agreement is least good for OH groups, where the relatively poor directional control of the lone pairs is known from small crystal work and theoretical calculations. Expected exceptions also occur for the carbonyl oxygens within regions of helix and sheet (Fig. 3). For the antiparallel sheet in particular, extra stability may come from the antiparallel lineup of dipoles, the twist in the sheet possibly reflecting quadrupole interactions.

This extensive hydrogen bonding, either internal or to solvent, cannot be energetically totally irrelevant to the structure and stability of the

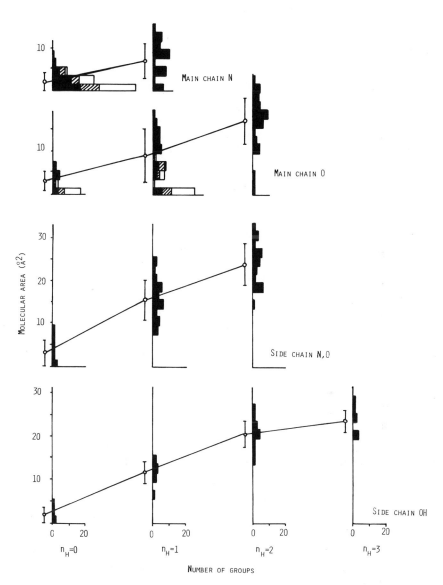

Fig. 3. Exposed molecular area of groups in RNase-S as a function of the number of internally unsatisfied donatable protons and lone pairs (n_H). Unshaded areas denote groups within β-sheet regions, hatched areas groups within α-helices, filled areas all other main-chain and side-chain groups.

molecule. Recalling the high packing efficiency within the protein,[179,306, 745,746] it would appear that van der Waals contributions are also significant. Similar conclusions can be drawn from the protein folding studies of Levitt and Warshel.[549,551,552] Using several sets of empirical potential functions, the relative contributions to the total energy of the folded structure of hydrogen-bonding, van der Waals, and solvent interactions (a "hydrophobic" term) fell within the ranges of 28–38%, 18–24%, and 41–54%, respectively. In their studies of the folding of pancreatic trypsin inhibitor, simulations performed in the absence of hydrogen-bonding contributions were still able to approach the correct volume of phase space. However, although the topologies of the simulated and native structures were similar, the computer-folded structure was excessively "loose" around the hydrogen-bonding regions. This suggests we need to then "switch on" the more specific, shorter-range polar interactions to shuffle the model into its final energy minimum.

The relative contributions of each of these types of interaction can be simply discussed as follows. Assume initially that an intramolecular protein hydrogen bond (PP) is equally as strong as one between a polar group and a water molecule (PW), and between two water molecules (WW). In fact for good internal bonds, we would expect* PP > PW > WW, an order of relative strength which would further increase the contribution of hydrogen bonding to the total stability of the folded conformation with respect to extended chain. Were all *charged* groups fully solvent-accessible in both extended and folded structures, their contribution to the stability of the folded structure would be zero.

In the extended chain, *all types* of group are considered fully exposed. In particular, *all* polar groups will be potentially capable of forming some kind of solvent hydrogen bond, though its strength is open to argument.† Exposed *apolar* groups interact weakly with solvent, and may induce some solvent ordering in their vicinity.

On folding, some, *but not all* the apolar groups are removed from

---

* For example, within the protein, a given polar interaction is likely to be in a neighborhood of considerably lower dielectric constant than a water–water interaction or a polar-group–water interaction on the surface.[308] This of itself is likely to result in internal hydrogen bonds being generally significantly stronger.

† There are suggestions from calorimetric measurements[109] that not all exposed polar groups are fully hydrogen bonded. The crystal data suggest there is adequate space for such hydrogen bonds to occur. Any reduction in the extent of polar-group–solvent hydrogen bonding from, e.g., dynamic effects will be expected to be similar for the extended chain; the essential argument is not affected.

solvent contact, resulting in a large hydrophobic contribution to the stability of the folded structure *with respect to extended chain*; this folding will result also in large conformational entropy losses, which will considerably offset—or even overcome—this hydrophobic gain. That some of the apolar groups—in fact up to about half—remain exposed indicates that other considerations prevent a complete "burial of the grease," which would maximize the hydrophobic contribution. Now focusing attention on the polar groups in the folded structure, we find nearly all either are participating in hydrogen bonds internally, or appear to be geometrically capable of interacting with the solvent. Moreover, what evidence we have (see above) suggests this total hydrogen bonding is likely about the same in extent as in the extended chain, so what energy difference there may be from this source between unfolded and folded conformations will depend upon the relative strengths and populations of PP, PW, and WW hydrogen bond interactions. In addition we expect an entropic contribution from the release of water molecules that were bound to the protein in the extended chain conformation. For every hydrogen bond that *cannot* be made in the folded conformation, however, there will be a penalty to offset the hydrophobic gain; this would be the case both for protein–protein and protein–water interactions. The unsatisfied polar groups thus removed from solvent contact will contribute as hydrogen-bond satisfied polar groups to the free energy change on folding both in terms of the entropy gain from release of water bound to the groups in the extended chain, and the configurational entropy loss consequent upon the folding itself.

Thus, although the hydrogen-bonding contribution to the energy may to a first approximation change little on folding (though in actual fact the difference—albeit possibly small—between PP and PW interactions may give some positive contribution to native conformation stability), this does not imply the irrelevance of the specific interactions. Any attempt to increase the supposedly dominant hydrophobic free energy term by "burying more grease" which would result in extensive disruption of hydrogen-bonding possibilities would lead to a negative contribution to stability from the broken, unsatisfied polar groups. The folded structure appears to be a very sensitive balance between these several driving forces (with the addition of the apparently significant van der Waals contribution), with the solvent acting as some kind of "hydrogen bond sink," keeping up the hydrogen bond contribution by potential interactions with otherwise unsatisfied polar (or charged) groups.

In the absence of realistic estimates of configurational entropic contributions to the free energy change on folding, it is difficult to put these

arguments into quantitative terms.* We can, however, indicate the orders
of magnitude involved, and underline the fineness of the balance between
the different driving forces. For ribonuclease-S, using data on the change
in accessible surface area,[179] we can estimate the hydrophobic contribution
to the change in free energy on folding to be about 320 kcal mol^{-1}. Offsetting
this will be configurational entropic losses of uncertain magnitude; estimates
of 2–5 cal (residue mol)$^{-1}$ K^{-1} [142] suggest a $T\Delta S$ for this 124-residue protein
of about 70–180 kcal mol^{-1}. Turning to hydrogen bonding, we assume a
value of about 4 kcal mol^{-1} for the energy of a polar-group–water hydrogen
bond, this being an approximate average of several *ab initio* calculations
for idealized geometries,[253,254,450] which is consistent with several estimates
from spectroscopy.[142,269] Ignoring all charged groups, which are assumed
fully hydrated in both extended and folded conformations, we obtain (from
the data in Fig. 3) a polar-group–solvent contribution to the energy of the
extended chain of about 2000 kcal mol^{-1}. Compared to the very low precision of these estimates of the contributions to free energy and also their
expected orders of magnitude, the additional $P\Delta V$ term is negligible.

In the folded conformation, were all polar groups still optimally
hydrogen bonded, with each one of identical strength as those made both
with solvent in the extended chain and between solvent in the liquid, the
total hydrogen-bond energy would be unchanged. Two factors now need
to be considered. First, the strengths of the internal hydrogen bonds may
be significantly greater, giving an increased hydrophilic contribution.
Assuming a figure of, say, 0.5 kcal mol^{-1} additional energy for internal
interactions (the average of values suggested in the literature), would give
on this hypothesis about an extra 150 kcal mol^{-1}. This is not insignificant
compared to the estimated 320 kcal mol^{-1} from hydrophobic sources,
even though only a small energy difference between internal and solvent
hydrogen-bond interactions has been postulated.

Secondly, for every internal polar group *unable to make a good hydrogen
bond either internally or to solvent*, a loss of stabilization free energy with
respect to extended chain of about 4 kcal mol^{-1} could be expected. Thus

---

* Taking an average value of 4 cal K^{-1} (mole of residue)$^{-1}$ for the configurational
entropy of an extended polypeptide chain and approximating the entropy of the folded
state as zero suggests that the configurational entropic loss on folding will be about
150 kcal mol^{-1} at 300 K. This and other estimates in the text can be little more than
order-of-magnitude calculations, but they do underline the large values of the individual terms compared to the final relatively small ΔG of stabilization of the folded
conformation of 10–20 kcal mol^{-1}.

about 80 unsatisfied polar groups would be required to completely wipe out the hydrophobic contribution. In ribonuclease-S, about 20 such unsatisfied interactions are found. Even in such an apparently well-fitting jigsaw puzzle, this carries a fairly dangerous penalty of about 25% of the hydrophobic free energy gain, especially when the uncalculated unfavorable entropic term from loss in translational and rotational motions is remembered. It is easy to imagine this destabilization contribution could become much larger if we tried to bury more of the exposed hydrophobic groups. The criticality of the need to maximize the hydrogen-bonding efficiency with respect to the hydrophobic contribution is underlined by the measured free energy of denaturation of RNase. Thermal denaturation at physiological temperatures results in a free energy change of about 10 kcal mol^{-1},[721] which is a *very* small difference between several contributions of more than an order of magnitude greater. Unfolding in strong denaturants such as guanidinium chloride and urea, which might be expected to yield a denatured state closer to the fully extended chain conformation, does not increase this ΔG by more than 50%–100%. Thus, in the light of this *very fine balance* between opposing forces, a high hydrogen-bonding (and packing or van der Waals) efficiency both internally and to solvent seems not unimportant. The maximum hydrophobic contribution from burying *all* the "grease" would be of little significance if more than about a third of polar groups could not make hydrogen bonds either internally or to solvent.

The contribution of polar interactions to the exact shape and depth of the potential well in which the native protein is sitting is also likely significant. The folding process might be regarded as being directed by the hydrophobic, longer-range interactions, the final step consisting of local adjustments to optimize the internal hydrogen bonding. These latter specific interactions will be important in controlling the rigidity or flexibility of local regions within the protein, which are likely important for the exact positioning of groups during catalysis. The relatively long-range nature of the hydrophobic contribution implies that the *details* of the native structure's potential well will be largely controlled by the specific hydrophilic interactions. Thus the details of the dynamics around the native conformation will depend mostly upon these interactions. If hydrophobic contributions are indeed proportional to accessible areas, fluctuations about the global minimum will perturb accessible areas much less than internal lengths; the "force constant" for hydrophilic interactions would be expected to be greater than the "apparent force constant" with respect to the hydrophobic contribution. Taking this argument further suggests one might consider the hydrophobic effects to control the path taken through phase space during folding, but not the precise nature of the final conformation. This is supported by recent experiments on partially dehydrating proteins or changing the solvent *after* folding in which no significant denaturation was found[17,108,260,304,305,472,702,812] (see Section 5 below). Once the protein is folded, the resultant global potential well appears deep enough to withstand significant change in solvent surroundings, and it is likely that the specific internal interactions are important in this, rather than the overall hydrophobic interactions which may have directed the major part of the initial folding. The extent and degree of internal hydrogen bonding thus does appear important; it may well be not without evolutionary significance. These points are further discussed below with respect to the likely specific role of water interactions.

4.1.2. Amino Acid Substitutions

Small sequence differences between proteins from different sources need to be accommodated without significant disruption of stability or active site conformation. Considering the extremely good hydrogen-bonding and packing fitting observed, such changes are likely to be found on the molecular surface, where the solvent is the major element that needs to accommodate the change. This is indeed the case for some of the enzymes of halophilic and thermophilic microorganisms. At other times such changes may be apparently nonfunctional but nevertheless crucially important as in the hypervariable regions of immunoglobulins.[711] The hydrogen-bond versatility of water is likely important in accommodating these changes, together with its trivial property of fluidity.

Any changes within the molecule would be expected to be limited, except where either the change is consistent with hydrogen-bonding *and* packing constraints, or there are complementary changes that maintain the efficiency of these internal constraints. A particularly beautiful example of the latter situation is provided by the comparison of the insulin and postulated relaxin structures,[65,437] where the internal packing in the two molecules is maintained with very little (21%) absolute sequence homology. The possibility of internal water molecules assisting in the maintenance of complementary surfaces is discussed below.

4.1.3. Surface Polar Group Distribution

Work on the hydration of polysaccharides,[335] DNA,[93] and of collagen and other fibrous proteins[93] has led to models in which the separation

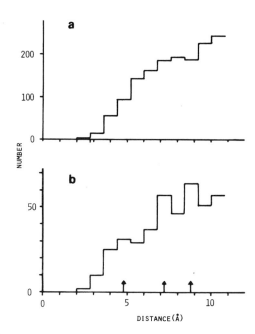

Fig. 4. Distribution of distances between accessible (a) apolar and (b) polar groups on the surface of the insulin molecule (P. A. Timmins, unpublished data).

of hydrophilic sites is thought to be consistent with standard hydrogen-bonding distances in water.[890] Such interactions are thought to be relatively long-lived. The statistics of separation distances of polar groups on the surface of insulin were examined by P. A. Timmins.* Figure 4 shows the distribution of both interpolar-group distances and distances between surface apolar groups. Although the statistics are poor, comparison of the two figures does suggest some peaking around 4.8 Å, and also about 7.2 Å and perhaps 8.8 Å for the polar group plot, though better statistics might well merge these two latter features. A similar plot restricted to peptide groups also shows a peak around 4.5–5.0 Å. Bearing in mind the expected errors in the X-ray coordinates, the similarity of these figures to the second and third peaks in the liquid water pair distribution function (4.6 Å and 6.9 Å, respectively[643,645]) suggests that single and double water bridges between particular polar surface groups might be expected. Examination of models of reasonable water networks between surface polar groups gives similar figures with peaks around 4.6 Å, 6.6 Å, and 7.5 Å; in addition, variability in third and higher neighbor distances could smear out the peaks in distributions such as observed in Fig. 4. The above data are suggestive, though inconclusive. Direct evidence for such surface bridges is presented

* Unpublished data.

in Section 4.2 below. The significance, if any, of the weak maximum at about 8.0 Å for apolar group separations (Fig. 4) is not clear.

4.1.4. Protein–Protein and Protein–Substrate Interactions

Examination of the nature of contact surfaces of proteins suggests that aggregation—e.g., between molecules in the crystal, or subunits in multimeric proteins—is also driven largely by hydrophobic effects. For example, the insulin dimer appears to aggregate by two apolar surface patches coming together, excluding solvent from close proximity to the apolar regions.[122,691] Prealbumin dimers aggregate to tetramers across a hydrophobic interface.[117] Many hormone–receptor interactions are thought to be similarly controlled.[781]

A more detailed examination of the nature of these contact surfaces shows that the situation is not quite so simple.[180] As with the protein folding process, protein–protein interactions involve in general some specific charge–charge or polar interactions across the interface; moreover, the packing across the interface is as good as that observed within proteins themselves.[180] Thus although the apolar interface makes up 55%–75% of the total area, compared with 15%–40% for polar and 5%–20% for charged surface,[180] the complementarity of the nonapolar surfaces appears crucial. The same argument would seem to apply as in the protein stability case (Section 1.1 above); although making only a minor contribution to the total stabilization energy, the absence of complementary polar and/or charged interactions and poor packing of the interface would give rise to a significant energy deficit. Such short-range interactions are likely also important for specificity of an interaction and, as discussed above, the details of the potential well close to the minimum.

Similar considerations are thought to apply to protein–substrate and other protein–ligand interactions. Structure analyses of carboxypeptidase-A[726] and -B,[791] ribonuclease-A,[172] adenylate kinase,[795] and liver alcohol dehydrogenase[277] are but a few examples suggesting essentially apolar binding clefts. In ferredoxin,[11] the iron–sulfur cluster sits in a highly apolar environment, while the hem groups in, e.g., cytochromes-b_5, -c, myoglobin,[603] and lamprey hemoglobin[399] are found in hydrophobic crevasses.

More complex binding sites occur for example in carbonic anhydrase-B, where its geometry is described as a cone, one half of which is lined with apolar, the other half with polar groups.[466] In concanavalin-A, the saccharide binding site is thought to consist of two distinct subsites. The deeper one is predominantly apolar, the one nearer the surface more polar.[115] The thyroxene binding sites in prealbumin are found in a central channel which has polar, apolar, and charged regions strung linearly along it.[117] The solvent-excluding cavity hosting the reduced iron–sulfur cluster in reduced *chromatium* high potential iron protein is of mixed character, making direct hydrogen bonds to the cluster.[173]

In summary, *from considering the protein coordinates alone*, we can usefully discuss the possible solvent-involved mechanisms relevant to folding, stability, dynamics, and interactions of proteins. Although many of these processes appear to be driven by so-called hydrophobic forces, the relevance of efficient direct hydrogen bonding is established; otherwise stability would be lost in the unmade hydrogen bonds previously made to water. Complementarity of charged and polar surfaces is important for specificity between molecules, and local structural rigidity within a molecule. The internal hydrogen bonding may be the major determinant of vibrational motions around the free energy minimum of the native protein, and hence control the molecule's dynamics. The surface accessibility of polar groups not internally hydrogen bonded seems a significant constraint, and solvent exposure at hairpin turns may be necessary for their stabilization. The relative distribution of polar groups on the molecular surface suggests that simple solvent bridges may be made across them. Direct evidence concerning and extending some of these points is presented in the following section, with particular emphasis on water participation in hydrogen-bonding interactions.

4.2. Direct Location of Water Molecules

The data available from crystal structure studies are generally consistent with the conventional wisdom (Section 2.2). In some protein molecules, especially larger ones, water molecules are found to be bound internally, and can be considered an integral part of the protein. In all high-resolution refinements, a first hydration shell is seen, making hydrogen bonds as expected to surface polar and charged groups. The extent of solvent hydrogen bonding of accessible surface groups appears to be high, consistent with the above discussion on the basis of the protein coordinates alone. In only a few cases does this ordered water region extend significantly into the solvent channels between the molecules. Whether or not this is a real effect, or a consequence of the problems of extracting signal from noise in this region, is presently not clear.

We consider each of these three regions in turn, paying particular attention to the detailed geometries and likely function—if any—of the located water molecules. This distinction by "region" is not made rigidly; e.g., cases of water attached to metal ions are considered together, whether or not they are *strictly* internal. Table I lists those structure analyses that have produced solvent data. Note that the extent, quality, and reliability of the solvent information is extremely variable. The data from some structures, particularly those of limited resolution (say >2 Å generally) should be treated very carefully, concentrating on very strongly bound, fully occupied solvent sites.

4.2.1. Internal Water

(a) *Water Bound to Metal Ions.* Strongly bound water molecules are found frequently as one or more ligands of the coordination shell of a metal ion. The geometries of such coordination shells are generally normal, containing other polar groups, usually side chains. These water molecules may often be displaced during binding of substrate. Since X rays in these large structures will see only the oxygen atom of a water molecule, at suitable pH some of these sites will be occupied by hydroxyl ions. Complementary information on ion-bound solvent has been obtained from nmr measurements.

Approximately tetrahedral coordination shells containing one water ligand are found for the catalytic zinc ion in horse liver alcohol dehydrogenase,[278] carbonic anhydrase,[465] and carboxypeptidase-A[564] and -B[791] (Fig. 5). The other three ligands are generally polar side chains, often histidine. In carboxypeptidase-A,[564] the zinc-bound water, displaced on substrate binding, links to several ordered waters in a hydrophobic substrate-binding pocket. In carbonic anhydrase[465] and liver alcohol dehydrogenase,[278] the water links to other polar side chains, forming hydrogen-bond networks of possible significance to the enzyme's catalytic activity (see below).

In subtilisin novo,[262] there appear to be three water molecules close to the alkali-metal binding site, one being an ion ligand. One of the calcium binding sites in carp muscle calcium-binding parvalbumin has a single water ligand,[627] while the coordination sphere of bovine β-trypsin[124] is an almost regular octahedron, with two water molecules bridging to a side chain (Glu 77 O_ε) and further waters (Fig. 6). In staphylococcal nuclease,[202] the calcium binding site of the nuclease–inhibitor complex is formed by a square array of carboxylate ions, with "a sort of secondary coordination sphere, complex, including threonines and water molecules."[202] One pair of

TABLE I. Proteins for which Solvent Data are Available[a]

Molecule	Maximum resolution (Å)	Reference
Actinidin	<2.0[b]	—
Bence–Jones REI	2.0	283
Carbonic anhydrase	2.2[b]	466, 561[b]
Carboxypeptidase-A	2.0	726
Carboxypeptidase-B	2.8	791
Chromatium high-potential iron protein	2.0	173
α-Chymotrypsin	2.0	110, 874
Chymotrypsinogen	2.5	921
Concanavalin-A	2.0	64
Cytochrome-c	1.5	448
Deoxyhemoglobin (human)	2.5	300
Erythrocruorin	1.4	824
Ferredoxin	2.0	11–13
Flavodoxin	2.0	893
Insulin	1.9	122, 691
	1.5–1.3[b]	—
Liver alcohol dehydrogenase	2.4	141, 277, 278
Lysozyme, tetragonal	2.0[b]	116, 435, 704, 705[b]
Lysozyme, triclinic	2.5	635
	2.0[b]	448, 449[b]
Methemoglobin (horse)	2.0	532
Myoglobin (X-ray and neutron)	2.0	793, 895
Metmyoglobin (X-ray and neutron)	2.0	792, 794, 851, 852
Papain	2.8[b]	94, 263, 264[b]
Pancreatic trypsin inhibitor	1.5[b]	231[b]
Penicillopepsin	2.8	422, 442
Prealbumin	1.8[b]	117[b]
Carp muscle calcium-binding parvalbumin	1.85	627
Ribonuclease-A	2.5[b]	171, 172[b]
Ribonuclease-S	2.0	747, 923
Rubredoxin *clostridium pasteurianum*	1.2	634, 892
Rubredoxin *desulforibrio vulgaris*	2.0	14
Staphylococcal nuclease	2.0	43, 202, 203
	1.5	198
Subtilisin novo	2.8	262
Subtilisin BPN	2.5	21
Thermolysin	2.3	478, 601, 602, 900
Triose phosphate isomerase	2.5	50, 51[a]
Trypsin and complexes	1.4	124–126, 423, 519, 836
Trypsinogen	1.8	296

[a] Not all of these proteins have been subjected to refinement procedures. The extent of data is highly variable between different molecules.
[b] Presently known to be under further refinement.

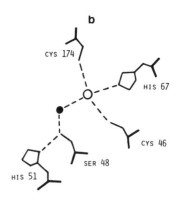

Fig. 5. Diagrammatic sketch of the zinc coordination in (a) carboxy-peptidase-A, (b) horse liver alcohol dehydrogenase, and (c) carbonic anhydrase-B and -C. Open circles denote metal ions, filled circles bound waters (or OH⁻).

features in thermolysin, thought to be two of the four calcium-binding sites, are surrounded, probably octahedrally, by a cluster of ordered water molecules, backbone carbonyl oxygens, and acid groups which participate in a network of interactions.[600] The calcium and manganese binding site in concanavalin-A can be thought of as two edge-sharing octahedra (Fig. 6).[64] Two water molecules act as bridges from the Ca^{2+} to side-chain oxygens. Of the two waters in the Mn^{2+} coordination sphere, one acts as a bridge to both a side-chain oxygen and hydroxyl, while the second links to the external solvent, forming a solvent channel to the surface.

In methemoglobin, a water molecule is bound to each Fe(III) ions.[300,697,700,895] A 1.8-Å-neutron-diffraction study of metmyoglobin[794] locates a water as the sixth iron ligand, hydrogen-bonded to the N_ε of His 7E; this feature is absent in CO-myoglobin. A 2.0-Å refinement of myoglobin[851] finds the same water, although it appears to move away from the iron in the deoxy form.[300] The equivalent water ligand in erythro-cruorin, an insect hemoglobin, is only loosely bound at a distance of ~2.8 Å; in the reduced deoxy form, the site is marginally further away (~2.9 Å) and only partially occupied.[824] Similarly in human deoxyhemoglobin, the water is much further from the Fe^{3+}.[300] The significance, if any, of these changes is unclear.

NMR techniques have suggested metal–water ligands in a variety of other enzymes, including pyruvate kinase, pyruvate carboxylase, fumarase, and acotinase.[620]

These ion-bound water molecules have been discussed in terms of their likely (a) stabilizing (perhaps only of local conformation) and (b) enzymatically active roles. Concerning the first, we might say at the simplest

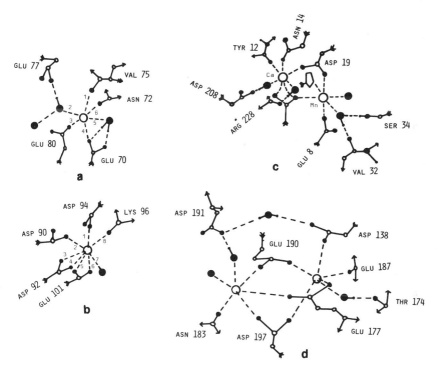

Fig. 6. Metal binding sites involving one or more water molecules. (a) The single calcium-binding site in β-trypsin.[124] Labeled distances are 2.5, 2.6, 2.4, 2.4, 2.6, and 2.4 Å, respectively. Corresponding distances in trypsinogen are generally significantly smaller.[296] (b) One of the calcium sites in carp muscle calcium-binding parvalbumin.[627] Labeled distances are 2.6, 2.9, 2.5, 2.1, 2.4, 2.4, 2.4, and 2.2 Å, respectively. Note both oxygens of the carboxylate groups of Asp 92 and Glu 101 are coordinated to the metal. (c) The double site in concanavalin-A.[64] One of the waters in the manganese shell leads to the surface. (d) The double calcium site in thermolysin.[900] Water molecules provide two and one of the octahedral ligands. In this and subsequent topological sketches, the following notation is used: ●—●—●—water oxygen with two hydrogens. Hydrogens are shown explicitly only where they can be unambiguously placed. ○—metal ion. ○—carbon. ●—oxygen. ⊖—nitrogen. —●—hydrogen where it can be placed unambiguously.

level that the water is merely completing the ionic coordination shell. In all these examples, however, there is some degree of apparently unbalanced internal charge. In such cases as carboxypeptidase-A, where the charge on the zinc ion is thought to be necessary for attacking the carbonyl oxygen of the substrate (a polypeptide whose carboxy-terminal peptide bond is to be hydrolyzed), the water ligand is displaced on substrate binding, together with those associated solvent molecules occupying the empty hydrophobic binding cleft.[564] The local dielectric constant (molecular polarizability) is thus considerably reduced, and the polarizing effect of the charge of the zinc on the C=O of the substrate thereby enhanced. In the absence of substrate, the high dielectric constant of the water ligand and the associated solvents in the cleft help dissipate the additional charge on the Zn^{2+}, thus helping to stabilize the structure. Similar stabilizing roles have been suggested in other cases. In staphylococcal nuclease, the second (water-containing) Ca^{2+} coordination sphere has been invoked as aiding structural stability, "perhaps providing a transitional medium between an extremely ionic and a nonpolar environment."[202] In bovine β-trypsin one of the waters bridges between the Ca^{2+} and the ionic side chain of Glu 77 (Fig. 6), resulting in a "water-mediated salt bridge."[124] Substitution of Eu^{3+} for Ca^{2+} in one of the calcium-binding sites of thermolysin results in an *additional* metal–water ligand.[601] As the calcium ion is important to the stability of the protein, additional local stabilizing mechanisms such as water bridges may well be significant also. The stability of local arrangements may be enhanced by strong hydrogen bonding of the metal to the molecule through a water bridge as in β-trypsin.[124] Without the water to fill a gap the protein cannot occupy, the local structure is likely to be somewhat floppier.

Direct catalytic roles have been argued for the metal-bound water (or OH^-), in particular for processes involving hydrolysis. The hydrogen-bonding network at the active site of liver alcohol dehydrogenase is shown in Fig. 5. The water molecule (or hydroxyl ion depending on pH) which completes the zinc's tetrahedral coordination may be involved both in the proton release on binding of NAD^+,[807] and directly in the catalytic mechanism by assisting substrate polarization.[276]

The possible water involvement in the mechanism of carbonic anhydrase is of particular interest. This enzyme catalyzes the hydration of carbon dioxide and the dehydration of bicarbonate (or carbonic acid), and has one of the highest measured turnover rates among all enzymes, especially those involving acid–base catalysis. The interpretation of the kinetics has thus been problematical, and has led to mechanisms being postulated that involve rates greater than the diffusion limit.

It is known that an ionizing group on the enzyme with pK_a about 7 is involved in the carbonic anhydrase catalysis mechanism; titration of this group results in changes in the immediate environment of the Zn^{2+}.[562] The simplest proposal assigns this function to a zinc-bound hydroxyl as the basic form active in hydration of CO_2, with the acid form $ZnOH_2^+$ active in dehydration of bicarbonate. Alternative suggestions that an ionizable side chain such as an imidazole group might be the active group have been made, though what evidence there is strongly supports the hydroxyl mechanism.[562] Sulfonamide inhibition removes the OH, and leads to a *direct* linkage between the ion and Thr 199.[661]

A major problem remains concerning the necessary rapid exchange of protons between the catalytic group at the base of the 12-Å cavity and the external solvent. A possible mechanism which could satisfy this constraint[272] is a system of strongly bonded water molecules (see Section 2.1.4) within this cavity. Such a proposal is involved in the push–pull mechanism proposed by Khalifah.[479] This mechanism [Fig. 7(a)] minimizes the development of uncompensated charge during catalysis, and, by virtue

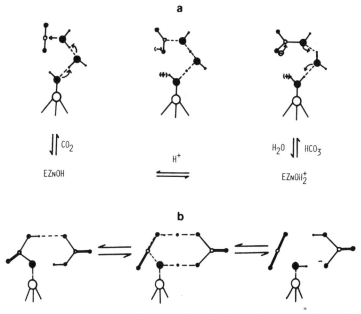

Fig. 7. Two proposed mechanisms for carbonic anhydrase. (a) Cooperative push–pull mechanism with the zinc-bound water serving both donor and acceptor functions, using water bridges to facilitate rapid proton transfer between the solvent and active site.[479] (b) Mechanism proposed by Yeagle *et al.* involving attack by the zinc-bound OH^-.[928]

of the proposed linked water molecule chain, would allow the enzyme to undergo reverse protonation to return to its active form sufficiently rapidly to be consistent with the very high turnover rates observed. At least one other mechanism has been proposed[459] which also involves rapid proton transfer along such linked water networks. More recently, both Khalifah[480] and Linskog and Coleman[562] have stressed the likely contribution of the buffer to ionization processes in the active site, perhaps mediated through water molecules. In the crystal there is no evidence at the present resolution (2.2 Å) for such stable groups of water; the carbonic anhydrase structure is currently being refined (K. K. Kannan, personal communication).

An alternative scheme also involving the zinc-bound OH has been proposed on the basis of ^{13}C NMR.[928] Evidence is given for two binding sites. A weak binding site is proposed as the site of the substrate which is bound directly to the metal ion. The proposed scheme [Fig. 7(b)] involves OH^- attack on HCO_3^-, with the second, tightly bound noninhibitory HCO_3^- facilitating catalytically important proton transfers. Again, rapid exchange with buffer components is suggested.

A variety of other instances of direct water activity has been proposed, often involving ion activation of the bound water. In the nucleotidyl transfer enzyme staphylococcal nuclease, the calcium ion may enhance the nucleophilicity of the bound water, which may then attack the 5'-phosphate group of the substrate. Moreover, from the structure of the enzyme–substrate complex, the bound water molecule would seem to be in an ideal location to displace the observed leaving group.[202,203,266,620] Similarly, the divalent ion in yeast enolase may activate the metal-bound water for nucleophilic attack, with perhaps the substrate phosphate acting as a general base, assisting the metal in deprotonating the water.[662] In pyruvate carboxylase and transcarboxylase, the metal ion has been proposed as promoting the acidity of its bound water, which then protonates the carbonyl oxygen of pyruvate.[620] The driving force for the conformational change required by acotinase seems to be provided by a water ligand substitution on the enzyme-bound iron.[351]

Most of these examples are to varying degrees speculative, though strongly suggestive of metal ion promotion of (bound) water activity in certain enzyme mechanisms. The possibility of a linked water chain along the active groove to facilitate rapid proton transfer is particularly provoking.

(b) *Close to Internal Charged Groups.* In general, X-ray diffraction studies have shown that the large majority of charged groups are highly exposed to the solvent. For example, in carp muscle calcium-binding parvalbumin, 83% of charged groups were found at the molecular surface.[659] Those charged side chains that do occur internally generally form salt bridges, thus avoiding the problem of internal uncompensated charge. Such relatively strong internal linkages have been invoked to explain the exceptional temperature stability of certain proteins[699] and in discussions of function, e.g., as possible sources of the Bohr effect in hemoglobin.[483,698]

However, cases have been found of buried charged groups, inaccessible to solvent, which do *not* participate in salt linkages. Electrically, we have a similar situation to internal metal ions, and as there, a water molecule (or molecules) tends to be found close by. In carboxypeptidase-A,[389] the carboxylate ion of Asp 104 is too far away from the side group of Arg 59 to make a direct salt bridge; a water molecule is conveniently situated in line to mediate one. Similar water-mediated salt bridges have been found in other molecules. In papain, a rather complex water-mediated double salt bridge is found, where a water molecule is linked to three charged groups, in addition to a further water which leads to an interesting communication network to the protein surface [Fig. 8(b)][94] The necessarily strained geometry resulting from a fourfold ring of interactions is noteable here. In carboxypeptidase-A,[389] the charged Glu 108 is isolated within the molecule, with no possibility of even a mediated salt bridge. In this case, a nearby water molecule (postulated to be H_3O^+) provides a hinge point in a hydrogen-bonding network to two backbone carbonyls [Fig. 8(a)]. In penicillopepsin, the carboxylate of Asp 32 is linked to a bound water.[442] A water-mediated salt bridge proposed to link the amino- and carboxy-terminal groups of opposite α-chains in deoxyhemoglobin was thought to be broken on oxygenation.[32,64] Direct evidence for this has, however, not been found in high-resolution studies.[300]

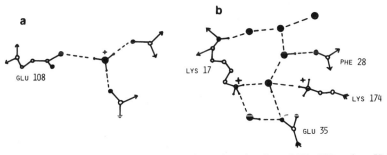

Fig. 8. (a) An internal salt bridge between the carboxylate ion of Glu 108 and an H_3O^+ ion in carboxypeptidase-A.[389] The H_3O^+ connects also to two main-chain carbonyls. (b) Water-mediated salt bridges in papain, with a water network leading to the molecular surface.[94]

A particularly interesting buried charged group (at active pH) is the Asp 102 of the serine proteases such as α-chymotrypsin[111] and trypsin.[125] This group supplies an electron to the charge-relay system, a strongly hydrogen-bonded linking of Asp 102, His 57, and Ser 195, which is an essential component of the likely mechanism of these enzymes.[427,764] Although the carboxylate ion is *not* linked directly to a water molecule, it does appear to participate in an extensive hydrogen-bonding system through Ser 214 which includes two strongly bound water molecules (Fig. 9). These molecules have been located directly both in tosyl-α-chymotrypsin and bovine β-trypsin. It has been suggested[110] that these internal waters may be important, through a raising of the local dielectric constant (molecular polarizability), in "spreading" the buried charge over a larger volume, and thus aiding stability, at least locally. They may also help to rigidly position the carboxylate group so as to optimize the effectiveness of the charge-relay system. These two water molecules appear to be absent in the pro-enzymes, and become trapped during the movement that occurs on activation.[296,921]

This likely charge-spreading, local stabilization function of buried water molecules [cf. close to ions, Section 4.2.1(a)] is well illustrated in the aggregation of the three dimers in insulin.[122] The consequences of this aggregation are to bring the six charged B13 glutamates into an approximately circular cluster about the trigonal axis. Crystal structure analysis showed the charged groups were all in approximately the same plane within 5 Å of the threefold axis, with consequently a very high concentration of negative charge. Close examination of the nearby electron density indicated the presence of loosely bound water molecules, which were thought possibly important in shielding the unfavorable charge distribution. Three molecules were found with trigonal symmetry above the plane, while below there appeared to be two sets of molecules in symmetrical sites.

(c) *Internal Polar Groups.* Polar groups tend to be found either on the molecule surface, presumably hydrogen-bonded to solvent (see below)

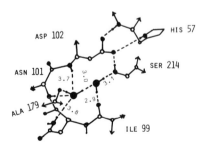

Fig. 9. Two water molecules close to the buried charged group of Asp 102 in α-chymotrypsin.[110] One water molecule appears to accept three hydrogens, two of these being at relatively long distances. Polar-group–water distances not labeled in the figures are from Ala 179 O (3.5 Å) and Asn 101 NH (3.1 Å).

or internally, making hydrogen bonds within the molecule—e.g., within regions of secondary (α-helix and β-sheet) structure. As mentioned above for mediated salt bridges between internal charged groups, water molecules are often found mediating hydrogen bonds between polar groups which otherwise would be too distant to interact significantly. The larger the molecule, the greater the extent of fully internal waters, though even for the smaller proteins such as pancreatic trypsin inhibitor, such molecules are sometimes found.[231]

The kinds of internal water organization found are many and varied. Molecules may occur singly, others in pairs, and so on up to complex internal networks. Within the limits of the data, by which hydrogen bonds are generally inferred by the existence of "reasonable" bond lengths and angles, most internal water molecules make two to three hydrogen bonds, with sometimes four or apparently more. We may recall here the earlier suggestions (Section 2.1.1) concerning the likely deviation from full fourfold coordination consequent upon the relatively low control of the lone pairs in water hydrogen bonding. There is much linking of main-chain NH and CO groups, both close to and distant from one another along the chain; without the intervening water these groups would be unable to form reasonable hydrogen bonds. Side-chain polar groups of all possible types may also form water-mediated hydrogen-bond bridges.

Figure 10 shows schematically the kinds of topology observed, generally consistent with what would be expected from our knowledge of water in small molecule hydrates (see Section 2 above). At least one water is internal in insulin (G. Dodson, personal communication), while three internal molecules are located in the same positions, according to independent studies[116,635] of triclinic and tetragonal hen egg-white lysozyme (129 residues). Two of these are close together, making hydrogen bonds to glutamine and serine side chains, together with main-chain NH and CO groups. One internal water per monomeric variable domain is found in a Bence–Jones protein REI.[283] In the 58-residue bovine pancreatic trypsin inhibitor,[231] there are four internal water molecules. One is in a distorted tetrahedral environment, making four hydrogen bonds of reasonable lengths to main-chain groups [Fig. 10(a)], while the other three form a network between two remote regions of the chain [Fig. 10(b)]. All four internal water molecules appear to be an integral part of the structure, and would be expected to be in similar positions in solution.[231] In *chromatium* high potential iron protein (85 residues),[173] one internal water makes three hydrogen bonds to main-chain NH and CO groups, and a glutamine side-chain carbonyl. In carboxypeptidase-A, a much larger

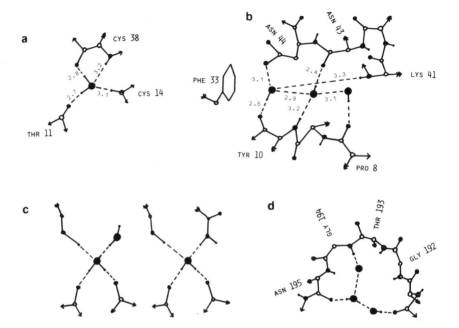

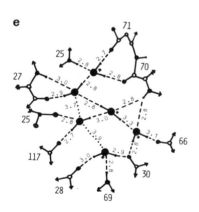

Fig. 10. Examples of internal waters hydrogen-bonded to polar groups. (a) A single four-coordinated molecule in pancreatic trypsin inhibitor, connecting to peptide groups only.[231] (b) A group of three water molecules internal to pancreatic trypsin inhibitor, connecting to main-chain and side-chain polar groups.[231] (c) Two four-coordinated internal waters in carboxypeptidase-A.[389] Two main-chain carbonyls and a serine OH are involved in each case, the fourth position being taken by a further water molecule or an asparagine terminal NH₂. (d) Three water molecules linking nearby main chains in papain.[94] (e) An extensive internal network in β-trypsin. The four water molecules marked "+" occur in similar positions in α-chymotrypsin.[125]

protein with over 300 residues, ten internal molecules have been located. Two of these make four approximately tetrahedral links with main-chain and side-chain polar groups and another trapped water [Fig. 10(c)]. Seven other water molecules appear to make three hydrogen bonds, while only one makes as few as two. In papain, a significant amount of water bridging occurs between close and distant parts of the backbone [Fig. 10(d)], and a considerable amount of water–water bonding is observed.[94]

In addition to several isolated water molecules and smaller groups, an extensive internal water network is found in bovine β-trypsin[125] [see Fig. 10(e)]; this network is closely connected to other internal solvent regions. The assigned hydrogen-bond lengths are generally acceptable, and most of the solvent molecules appear to make four hydrogen bonds. Figure 11 shows the distribution of hydrogen-bond lengths in the trypsin–PTI complex.[126] The data refer to water–polar-group interactions, though they are indistinguishable from the limited data on water–water distances. Within the accuracy of the data, the distribution is consistent with expected behavior. Little published information is available on angular distributions. A similar network is found in α-chymotrypsin.[110]

Recent work on the trypsin proenzyme trypsinogen shows that of the 25 internal water molecules found in benzamidine-inhibited β-trypsin, 18 are found in similar locations in trypsinogen.[296] Those missing are in the flexible regions of the proenzyme. This relates to the possible relevance of solvent in the activation of this enzyme [Section 4.2.2(c)].

The role of these internal molecules is discussed below, in context of the relevance of hydrogen bonding in general to global and local structure stability. We can note here that where two or more polar groups would otherwise be placed unfavorably for making internal hydrogen bonds, water molecules may bridge the gap. This can occur even when two polar groups may both be hydrogen donors or both hydrogen acceptors; the double-donor–double-acceptor potentiality of the water molecule allows it to assist in making such a bridge. Thus, with the presence of water, otherwise unmakeable hydrogen bonds can be made, increasing the *overall* depth of the potential well in which the protein sits, as well as giving enhanced local stability. In this context, we may note that in α-chymotrypsin,

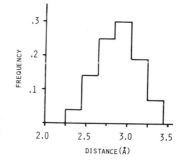

Fig. 11. Distribution of first-neighbor polar-group–water distances ≤ 3.4 Å in the trypsin–pancreatic-trypsin inhibitor complex refined at 1.9 Å.[126]

many of the observed internal waters bridge across those polar groups which are external to the largely hydrophobic-surface cylinders into which the molecule may be regarded as folded.[110] Bode and Schwager have argued[125] that in larger molecules such as trypsin, the inclusion of rigidly bound water is necessary in order to maintain a stable structure, especially in cases (like trypsin) where the extent of secondary helix and sheet structure is limited. Reference to Fig. 10(e) suggest that, were the internal water removed to the solvent, *at least* 14 potential hydrogen donors or acceptors would be left without hydrogen bonds. The consequences for the local structure would likely be quite significant. Assuming each $Å^2$ of accessible area buried on folding contributes 24 cal mol^{-1} of hydrophobic free energy,[179] the total hydrophobic contribution to the folding of trypsin is about 600 kcal mol^{-1}. Using the figures discussed in Section 4.1.1, 14 unmade internal hydrogen bonds would carry a deficit with respect to the (assumed fully hydrated) extended chain of about 64 kcal mol^{-1}. This is not an insignificant fraction of the hydrophobic contribution; with respect to a total free energy of denaturation of about 10–20 kcal mol^{-1} it is catastrophic.

4.2.2. Surface Water

In most protein crystals that have been studied to a reasonably high resolution (say around 2.5 Å), regions of electron density are often evident close to the molecular surface. Closer investigation reveals that these peaks tend to lie at reasonable hydrogen-bonding distances from surface polar groups. This is the "first hydration shell" of relatively strongly bound water molecules, and a relatively large amount of information is available concerning it. Close to protein–protein contacts, where they exist, the solvent will presumably be perturbed considerably from what would be the case in solution.

(a) *Attachments to One Molecule.* A general impression is that most exposed polar groups are hydrogen bonded to reasonably well-localized water molecules. In triclinic lysozyme, "practically all the accessible polar and charged groups interact with solvent."[635] Almost all these solvent molecules have at least three identifiable ligands, almost invariably at least two from the lysozyme molecule itself. Thus, this molecule presents a picture of a *very* strongly[635] bound first hydration shell, based on "practically all" accessible hydrogen-bonding groups, with single and multiple bridges across polar surface groups. This is consistent with the earlier predictions from surface polar-group distributions[307] (P. A. Timmins,

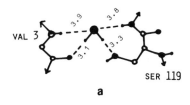

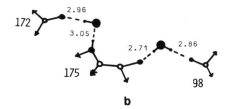

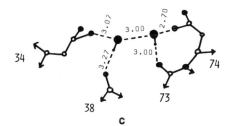

Fig. 12. Typical surface bridges. (a) Apparently four-coordinated single bridge in α-chymotrypsin[110]; (b) two simple single bridges; and (c) a double bridge in β-trypsin.[125] More complex bridges are found, e.g., see Fig. 14.

unpublished data). It is not possible to assert from published data the extent to which the number of water–protein interactions relates to accessible area and the number of hydrogen bonds made internally[308] (Section 4.1.1 and Fig. 3).

Other molecules show similar patterns, although except in a few cases (e.g., horse methemoglobin,[532] sperm whale myoglobin[851]) the data have not been presented in such a way as to see whether or not this picture of single water molecules with at least two attachments to the same molecule is general or not. However, of the 36 water molecules thought to hydrogen-bond to the surface of the Bence–Jones protein REI, only 16 show more than one direct protein link.[283] In rubredoxin from *clostridium pasteurianum*, recently refined to 1.2 Å resolution, only 26 water molecules made more than one hydrogen bond to protein, while 42 made only a single contact.[891a] A significant number of polar groups with potential multiple hydrogen-bonding capabilities do show multiple attachments as postulated in Section 4.1.1 (see Fig. 3). A full analysis of water-linked polar group accessibilities is needed.

In many molecules, single and multiple bridges of water across the surface have been identified; these are presumably of some importance to the local structural stability and dynamics of the protein molecule in a way similar to that argued above for internal water molecules, though probably less so. Water molecules are found in surface crevices in α-chymotrypsin and trypsin; in at least one of these crevices in trypsin, the water is expelled on inhibitor binding.[519] Figure 12 illustrates some of these simple and more complex bridges across the surface of one molecule. Where the structure of the same molecule has been determined by two or more groups independently, strong similarities have generally been found in the location of this strongly bound solvent fraction; for example 60 of the 110 water molecules postulated in triclinic lysozyme occur in similar surface locations to those found in the tetragonal form, particularly those in the active site region.[635] This gives us considerable confidence in the ability of present refinement procedures to place most of the strongly bound molecules reasonably accurately. Similar agreement in bound solvent locations is found between different hemoglobins,[300,340,532] and independent X-ray and neutron studies of myoglobin.[793,851] On the other hand, there are significant differences between the surface hydration found in bovine β-trypsin and α-chymotrypsin.[125] This may be regarded as illustrating the ability of the water molecules to adjust themselves to accommodate changes in polar site distribution caused by amino acid substitution (Section 4.1.2).

Table II shows the distribution of protein–water hydrogen bonds between different types of polar group. Though there is considerable variation between proteins, a general pattern emerges of a preference for main-chain CO groups over main-chain NH by a factor of an average 2.8. These figures are broadly comparable with the molecular area and accessibility

TABLE II. Statistics of Polar-Group–Water Hydrogen Bonds for Selected Proteins, Compared with Accessibilities and Molecular Areas

Protein	Main chain NH	Main chain CO	Side chain N, O
Concanavalin-A[64]	0.20	0.51	0.29
β-trypsin[125]	0.25	0.39	0.36
Parvalbumin[627]	0.11	0.29	0.60
REI[283]	0.10	0.46	0.44
Rubredoxin[891a]	0.18	0.37	0.45
Myoglobin (neutrons)[793]	0.10	0.36	0.55
Lysozyme accessibility[540]	0.04	0.28	0.69
Myoglobin accessibility[540]	0.04	0.22	0.75
RNase-S accessibility[540]	0.05	0.25	0.71
RNase-S molecular area[308]	0.12	0.28	0.60

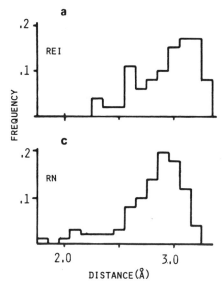

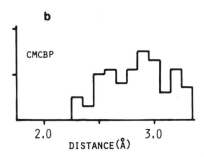

Fig. 13. Distribution of first-neighbor polar-group–water distances on the surfaces of (a) Bence–Jones protein REI (2.0 Å)[283]; (b) carp muscle calcium-binding parvalbumin (1.85 Å)[627]; and (c) rubredoxin C. *pasteurianum* (1.2 Å).[891a]

ratios, though with perhaps slightly more NH bonding than is reflected in the accessibility data. This is consistent with the known strong tendency of proton donors to participate almost invariably in hydrogen bonds; in contrast, the hydrogen-bonding control exerted by proton acceptors is less strong. These figures are consistent with the suggested relation between accessible area and the number of solvent hydrogen bonds.[308]

Figure 13 shows distributions of polar-group–water hydrogen-bonding distances for three proteins: Bence–Jones REI,[283] carp muscle calcium binding parvalbumin,[627] and rubredoxin C. *pasteurianum*.[891a] The smooth distribution of the rubredoxin data is particularly noticeable, together with its peak at around 2.9 Å; such a distribution would be expected from our knowledge of the nature of the hydrogen-bond interactions and the expected errors in placing solvent. The very close approaches below, say about 2.4 Å, are surely unrealistic and give some assessment as to the reliability of solvent positioning even in a very highly ordered crystal at very high resolution. As four of these distances relate to poorly defined protein atoms (Section 4.3 and Table III), their expected accuracy will be significantly reduced. No real space constraints were placed on the refinement; the hydrogen-bond length distribution is thus essentially unforced, resulting largely from the data themselves.

The other two protein histograms give some idea of the likely quality of data at significantly lower resolutions. The peak of the REI distances is shifted to higher values, while that for parvalbumin is very broad. The close approaches are probably not significant considering the estimated error in distance of $\sim$0.35 Å.

An analysis of the location of water molecules near hairpin turns suggests the water may be a significant factor in their stabilization. In *chromatium* high-potential iron protein,[173] 31 of the 85 residues are near these turns; the backbone atoms of these residues (36% of the total) bind 56% (28 out of 50) of the water molecules bound by the main chain. In ribonuclease-A[172] there are at least two examples of water molecules bound on such turns. This direct evidence is consistent with Kuntz' suggestions based on examination of the *protein* coordinates of carboxypeptidase-A[526] (see Section 4.1 above).

Two further points of interest can be made. First, there is little evidence for any highly ordered "clathrate-type" cages around exposed apolar groups, implying—consistent with the discussion in Section 2.1.2—that such structures are apparently not highly localized in *space*. However, ordered water has been found in hydrophobic clefts, such as around the haem in myoglobin,[851] and in many substrate binding sites (see below). Secondly, for two closely related proteins, the sequence changes may lead to *no reduction* in total hydrogen bonding *if the water is taken into account*. For example, in insulin, Arg B22 appears to be disordered over two alternative sites; a water molecule occupies one of the sites the side group vacates and appears to maintain the local hydrogen bonding. In 4Zn-insulin, part of the sheet structure at the end of one of the chains in one of the monomers in the 2Zn dimer transforms to a helix, extending the helix already there. Interestingly, those hydrogen bonds freed from the sheet and not taken up in the helix hydrogen bonding form an alternative network with a chain of solvent molecules. Moreover, a water mediating a polar group interaction in the 2Zn form is not present in 4Zn; instead, the two polar groups form a direct hydrogen bond linkage.* This underlines a tendency noted earlier for water to accommodate structural changes by altering the hydrogen-bonding scheme to minimize the loss in hydrogen bonds.

(b) *Bridges between Molecules or Subunits.* Many simple and complex water bridges have been observed between *different* protein molecules in the crystal, and between subunits in the protein. Some examples are shown in Fig. 14. The same characteristics observed in simpler surface attachments and internal networks are evident. In addition, for example in Fig. 14(b), the solvent network may be consistent with the symmetry of the arrangement

---

* G. Dodson, personal communication.

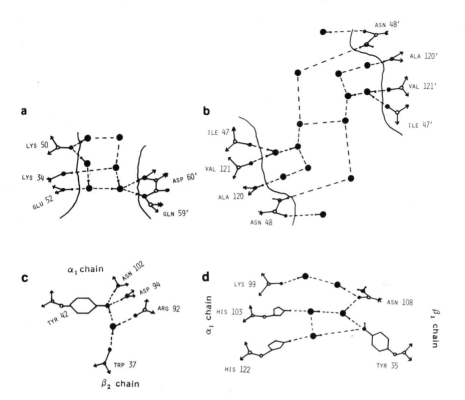

Fig. 14. Examples of intermolecular and intersubunit bridges. (a) Relatively extensive network linking two myoglobin molecules.[793] (b) Extensive bridging network between two α-chymotrypsin molecules.[110] The dyad symmetry relating the two protein molecules is retained by the solvent network. (c) Simple single molecule bridge linking $\alpha_1\beta_2$ chains in horse methemoglobin.[532] (d) An $\alpha_1\beta_1$ contact in horse methemoglobin.[532]

of the protein molecules. Similarly, single water bridges between arginine and glutamine side chains are found to link each of the dyad-related subunits in triose phosphate isomerase.[50] Such well-ordered networks tend to be found understandably close to the contact region. Triose phosphate isomerase is an illustration of a subunit interaction across a largely hydrophobic region, where the specific charge and/or polar interactions thought to form a complementary interface[180] require water mediation to avoid unmade hydrogen bonds. Similarly, three water molecules are found in sites between aggregated Bence–Jones protein (REI) molecules.[283]

Particularly interesting situations have arisen with recent refinements of human deoxyhemoglobin,[300] human fetal deoxyhemoglobin,[340] horse methemoglobin,[532] and sperm whale myoglobin[851] and deoxymyoglobin.[852] Bound solvent molecules were found at the same positions between neighboring subunits in both human adult and fetal hemoglobin. Similarly, despite differences in amino acid sequence, the $\alpha_1\beta_1$ subunit contact in horse methemoglobin included solvent bridges in the same positions as found in adult deoxyhemoglobin. The $\alpha_1\beta_2$ contact was, however, quite different.

Refinement of horse methemoglobin led in fact to a wholesale change in the picture of the hydrogen-bonding patterns at the $\alpha_1\beta_1$ and $\alpha_1\beta_2$ contacts. The number of hydrogen bonds at the $\alpha_1\beta_1$ contact rose immediately from 5 to 17 or 19 once solvent positions were included, a rise of over a factor of 3; similarly the number of hydrogen bonds in the $\alpha_1\beta_1$ contact rose from 2 to 6 or 7. Examples of some of these solvent–bridge networks are shown in Figs. 14(c) and 14(d). These additional hydrogen bonds through water molecules are now considered to help stabilize the tertiary structure.[532] Moreover, several abnormal hemoglobins show shifted oxygen equilibria and an increased dissociation into subunits. Some of these altered properties are thought to be due to the disruption of hydrogen-bonded networks of water molecules at the subunit interfaces.[532]

(c) *Active Sites, Substrate Binding and Mechanisms.* As noted in Section 4.1, when considering the coordinates, active site substrate binding clefts tend to be hydrophobic in character, being lined mainly by apolar side groups. Thus most of the driving force for substrate binding is generally believed to be hydrophobic in nature. Specificity likely depends upon direct charged or hydrogen-bond interactions with hydrophilic groups in or close to the cleft.

In the uncomplexed enzyme, relatively well-ordered solvent molecules are often directly observed in the binding cleft. These may be wholly or partly expelled on substrate binding, the resultant entropy gain of the expelled water being thought a significant contribution to the free energy change. There are examples where a bound water isolated in the binding clefts after substrate binding may be involved in the catalytic mechanism (e.g., thermolysin, see below). Water bridges between enzyme and substrate or inhibitor are sometimes found, both within and around the surface of the cleft. Such bridges can be considered in similar terms to inter subunit and surface water bridges. These may perform a similar structural role by permitting specific hydrogen-bond linkages to be made over otherwise impossible distances, and perhaps adding to the stability of a local configuration—here the interface between the enzyme–substrate complex at the protein molecular surface.

The following examples illustrate many of these points, and the possible direct involvement of water in particular proposed catalytic mechanisms.

Carboxypeptidase-A.[726] The largely hydrophobic binding pocket exhibits electron density that has been assigned to ordered water molecules. As discussed above [Section 4.2.1(a)], these have been ascribed a charge-spreading function with respect to the buried Zn^{2+} ion. On binding of glycyl-L-tyrosine, a poor substrate, some of these molecules are displaced; in particular the zinc-bound water is displaced by the carbonyl oxygen of the susceptible peptide bond, and the end of the pocket converted into a largely hydrophobic region. Figure 15(a) illustrates the most significant interactions, which include a water molecule bridging the substrate and Glu 270. This latter water bridge is possible only with dipeptide substrates, and has been invoked as an explanation for the relatively low rate of dipeptide cleavage.

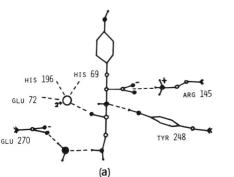

(a)

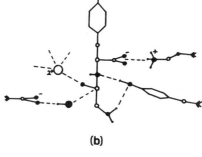

(b)

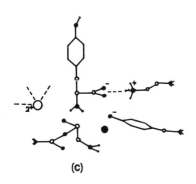

(c)

Fig. 15. Schematic sketches of (a) the binding of Tyr-Gly to carboxypeptidase-A and two proposed catalytic mechanisms; (b) proposed general base mechanism: the lone pair of the bound water, promoted by the carboxylate of Glu 270, attacks the susceptible carbonyl carbon, the OH of Tyr 248 providing a proton; (c) proposed nucleophilic pathway in which the Glu 270 attacks directly. The resulting acyl-enzyme intermediate is then broken down by either the water molecule promoted by the tyrosine, or directly by the OH⁻ formed by regeneration of the tyrosine OH.[726]

Two mechanisms have been proposed for the enzyme[726] which catalyzes the hydrolysis of carboxy-terminal peptide bonds. Both involve water molecules somewhere in the process, and similar mechanisms have been proposed for related enzymes. Thus, we will consider these possibilities in some detail.

The hydrophobic environment resulting from water expulsion from the binding pocket enhances the polarizing ability of the now-buried Zn^{2+} ion which attacks the substrate carbonyl, thus rendering the group's carbon atom more prone to nucleophilic attack [see Fig. 15(a)]. This substrate binding results in large conformational changes of the enzyme. In the possible general base path, illustrated in Fig. 15(b) the Glu 270 side-group, probably ionized at the optimum pH of 7.5, promotes the attack of a water molecule lone pair on the susceptible carbonyl carbon. The undissociated side group of Tyr 248 is well placed to supply a proton to the NH group adjacent to the cleaved peptide linkage. The proton donor may be another water molecule, but in view of the direct linkage between substrate and Tyr 248, and the probable inaccessibility of this region to solvent, it is more likely to be Tyr 248.

The second possible mechanism is illustrated in Fig. 15(c) and invokes direct nucleophilic attack of the carbonyl carbon by Glu 270. The resulting acyl–enzyme intermediate is then attacked by either a water molecule, promoted by the O^- of Tyr 248 resulting from transfer of its proton to the inner NH of the substrate, or directly by the OH^- formed by regeneration of the Tyr OH.

Which of these mechanisms is more likely is uncertain, though negative evidence favors the nucleophilic pathway; it appears to be the simplest mechanism that accounts for all the data, and is probably preferable on stereochemical grounds.[458] It may also be possible that different mechanisms could operate for different substrates: peptide and ester hydrolysis seem to occur by different paths.[620]

Crystallographic studies of the binding of the inhibitor phosphoramidon to thermolysin suggest this enzyme may operate via a path similar to the general base mechanism proposed for carboxypeptidase-A[478,900] [Fig. 15(b)]. As in that enzyme, the substrate binds in a hydrophobic pocket,[195] at the bottom of which is a zinc ion bound to a water molecule. Again as in carboxypeptidase, this zinc-bound water is displaced on inhibitor binding by the carbonyl oxygen of the scissile peptide; a water bridge is also observed between the phosphite hydroxyl and peptide NH of Trp 115. Inhibitor binding buries the zinc ion, Glu 143, and an associated water molecule. The zinc ion helps to align the substrate, and to polarize the carbonyl

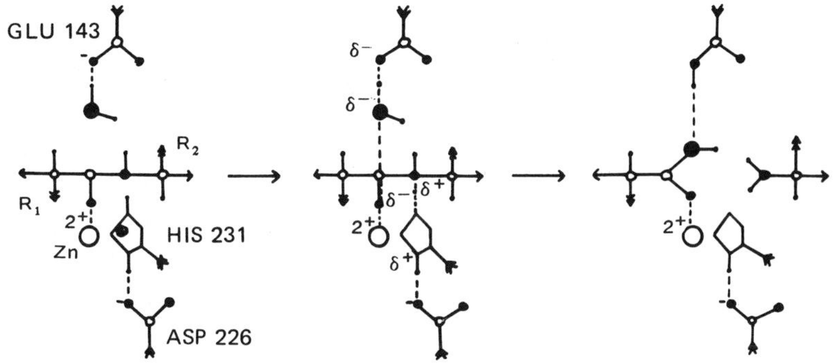

Fig. 16. Proposed thermolysin general base mechanism.[478,900] The susceptible carbonyl carbon undergoes nucleophilic attack by the buried water, promoted by Glu 143.

group. Figure 16 illustrates the proposed mechanism, in which the carbonyl carbon undergoes nucleophilic attack by the buried water; the water attack is promoted by Glu 143, which occupies a position analogous to that of Glu 270 in carboxypeptidase-A. The position of Tyr 248 as the proton donor in carboxypeptidase-A is probably taken by His 231 in thermolysin. In contrast to carboxypeptidase-A, this general base path appears from the data to be much preferred over the alternative mechanism involving direct attack of the carboxyl carbon by Glu 143, resulting in the formation of an anhydride intermediate.

A similar mechanism has been proposed for acid protease catalysis,[422,442] based upon the crystal structure of penicillopepsin. Again the hydrophobic substrate binding site contains ordered solvent; substrate binding displaces two water molecules, while one remains between the substrate and the carboxyl of Asp 32. The proposed mechanism is illustrated in Fig. 17. The carbonyl bond of the substrate is polarized by the proton shared between the carboxyl groups of Asp 215 and Asp 32, a function undertaken by the Zn^{2+} in carboxypeptidase-A. The electrophilicity of this proton is argued to be a function of the nature of the substrate. It is also of interest that the two water molecules displaced on substrate binding were within hydrogen-bonding distances of Asp 215; the activity of the buried proton is likely enhanced by their removal. The remaining bound water is proposed to perform activated (by Asp 32) nucleophilic attack of the polarized carbonyl carbon. Tyr 75 is in a position to protonate the NH group of the peptide bond, in a similar manner to the Tyr 248 in carboxypeptidase-A.[595] As in the case of thermolysin, this general base mechanism is preferred to a direct nucleophilic attack by the COO^- of Asp 32.

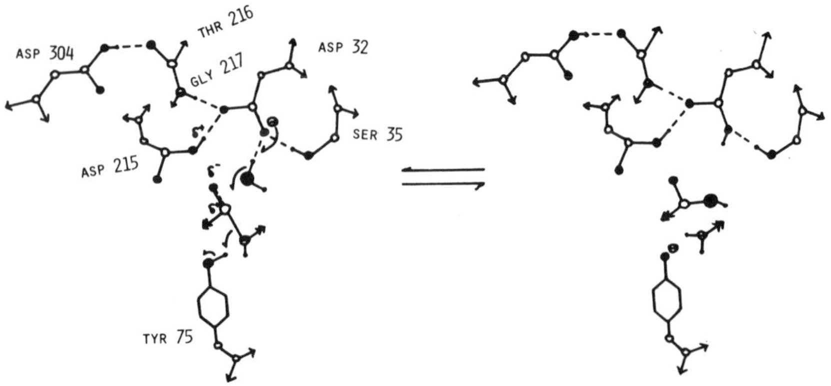

Fig. 17. Proposed general base catalytic mechanism for the acid protease penicillopepsin, in which the carbon atom of the substrate is attacked by Asp 32 through a bound water molecule.[442] The proton shared between Asp 215 and Asp 32 is proposed to polarize the carbonyl bond of the substrate (cf. the role of Zn^{2+} in carboxypeptidase-A).

Serine Proteases. These hydrolytic enzymes, such as trypsin, chymotrypsin, and elastase, have been extensively studied at high resolution with much data on native, complexed, and proenzyme forms available. Again, the specificity pockets are generally hydrophobic, containing ordered and partly ordered water molecules which are wholly or partially removed on substrate or inhibitor binding.[125,396,825,836] Thus again it seems likely that desolvation of the specificity pocket provides a significant entropic contribution to the stabilization of the complex.

The water pattern in the trypsin–pancreatic-trypsin inhibitor (PTI) complex has been particularly well defined.[125] Eight water molecules link the enzyme and inhibitor. Six of these are surface groups, making in general three hydrogen bonds each to polar surface groups on both molecules and other surface waters. Again, these (here simple) intermolecular bridges can be discussed in similar terms to intermolecular or intersubunit bridges in general [see Section 4.2.2(b)]; they are thought to contribute to the stability of the enzyme–inhibitor complex.[125] Of additional interest are the two bound water molecules which remain in the inhibitor pocket, as shown in Fig. 18. Although the carboxylate of Asp 189, found deep in the specificity pocket, is considered essential for the specificity of trypsin for charged side chains such as lysine, the geometry of the trypsin–PTI complex does not permit an adequately close approach of the NH_3^+ of Lys 15I (of the inhibitor) to Asp 189. The electron density map, however, revealed a feature between the two charged groups, indicating a salt bridge *mediated by a water molecule.* Thus, the enzyme's specificity for this inhibitor is not via a simple *direct* salt bridge. The second water molecule provides a hydrogen-bond bridge between the inhibitor NH_3^+ and two main-chain carbonyls. Similar *water-mediated* salt bridges have not been found in other trypsin complexes so far studied.

Extensive studies of various inhibitor complexes of α-chymotrypsin have suggested the mechanism of peptide substrate hydrolysis involves an acyl-enzyme intermediate (Fig. 19). Such intermediates have been observed in experiments with trypsin–inhibitor complexes.[303,773,844] Deacylation is viewed as a reverse of the acylation process, with the leaving group replaced by a water molecule. This molecule is one of two which appear to be bonded between substrate and enzyme, and is linked to the $N_{\varepsilon 2}$ of His 57, one of the side chains participating in the charge-relay system. One of the water protons is partly removed by the charge-relay system, initiating nucleophilic attack of the carbonyl of the acyl-enzyme. The second water molecule, located between the carbonyl oxygen and the peptide carbonyl of Phe 41 of the enzyme, may assist the stabilization of the substrate carbonyl in this attack-prone position, allowing deacylation to proceed by general base-catalyzed removal of the proton of the first bound water molecule, with the formation and subsequent breakdown of a tetrahedral intermediate.[396,397] Further support for the active role of the first bound water molecule comes from high-resolution structure analyses of benzamidine, PTI, and diisopropylfluorophosphate complexes with trypsin.[125,231,396,836] Studies on the α-chymotrypsin-PTI complex suggest its stability is partly due to the shielding of the acyl group from water.[121]

A recent structure analysis of the trypsin proenzyme trypsinogen has been refined to 1.8 Å.[296] Of the 25 internal water molecules found in the benzamidine–trypsin complex,[125] 18 are found with similar occupancies in trypsinogen. Those "missing" are in flexible regions of the proenzyme, and several hydrogen bonds in the activation domain are mediated by bound water molecules. The surrounding solvent networks were concluded to

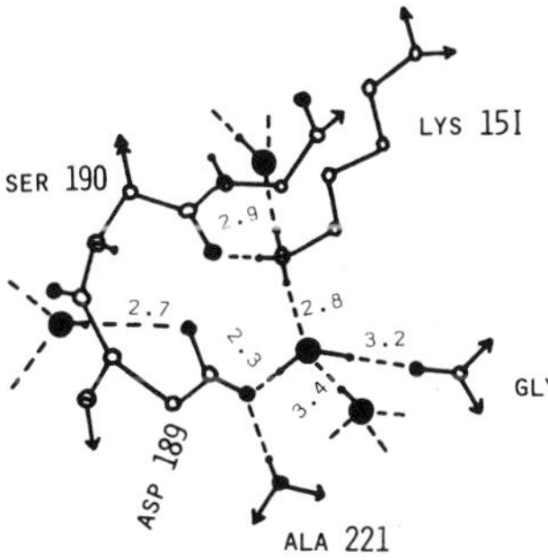

Fig. 18. Water molecules in the binding of pancreatic trypsin inhibitor to β-trypsin.[126] Of particular note is the water molecule between the NH_3^+ of Lys 15 of the inhibitor and the carboxylate of Asp 189, to which latter group the specificity of the binding site is assigned.

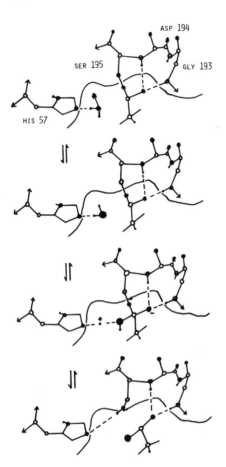

Fig. 19. Proposed deacylation step of α-chymotrypsin.[397] The leaving group is displaced by a water molecule, and the reaction proceeds by general base-catalyzed removal of the water proton, with the formation and subsequent breakdown of a tetrahedral intermediate.

influence the conformation around the active site in both free trypsin and free trypsinogen; the Ser $195O_\gamma$–His $57N_{\varepsilon2}$ hydrogen bond became sterically optimal and functional for proton transfer only during complexing. Moreover, the trypsinogen → trypsin conversion was seen as one between a nonfunctional partially flexible structure, and a highly ordered rigid and functional one, rather than being a transition between two different rigid structures. The highly cooperative nature of the activation domain was also stressed. Remembering the change in internal solvent structure between the proenzyme and enzyme, the relation of the water networks to the flexibility of regions of the structure, and their importance to the enzyme structure[125] [see Section 4.2.1(c)], the solvent network appears as a very important structural and dynamical feature in the achieving of a fast, effective switch from one state to a functionally different one, and in maintaining the structural integrity and functional capabilities of the latter state—the active molecule.

Other Proteins. In other hydrolytic enzymes, we might also expect water to be involved in the overall reaction scheme. Those involving metal ions have been discussed above. In the proteolytic enzyme papain, which requires a sulfydryl group for its activity, water, as well as alcohols and amines, may be active in attacking the postulated acyl-enzyme intermediate.[263] In lysozyme, which catalyzes the hydrolysis of glycosides, the glycosyl enzyme undergoes nucleophilic attack by water.[703,881]

Water bridges between enzyme and substrate are found in several further cases, such as between ribonuclease and 3'CMP.[747] Similar situations may occur in hormone–receptor interactions, especially bearing in mind the high degree of affinity between steroids and water reflected in the observed strong water binding in the crystal forms, and their apparently high ice-nucleating efficiency.[344] In prealbumin, a human plasma protein which binds the water-insoluble thyroid hormone thyroxine, the binding site involves a well-defined water molecule linking a hydroxyl of the hormone with the side-chain hydroxyls of serines and threonines in the hydrophilic region of the site.[117] The likely significance of water bridges in steroid hormone binding has been pointed out by W. Duax.*

In electron-transfer proteins, water has been invoked in postulated mechanisms. In flavodoxin, the binding of riboflavin-5'-phosphate occurs through mediating water molecules at both the flavin and ribityl portions of the molecule (see Fig. 20).[893] In addition, the OH3' hydroxyl of the ribityl group hydrogen-bonds to two surface water molecules. Water is directly implicated in a mechanism proposed for flavodoxin,[578] and also in the oxidation–reduction *chromatium* high potential iron protein.[173] In the latter case, the acceptance of a proton by a surface-bound water from a nearby tyrosine side chain may stabilize the resulting phenoxy radical, necessary if the tyrosine ring is mediating electron transfer to and from the Fe_4S_4 cluster within the molecule. A similar mechanism involving the tyrosine side chain is thought unlikely for flavodoxin, where studies of the oxidized form of the protein from *clostridium* reveal that this group is absent from the immediate environment of the flavin ring.[158] A proposed mechanism for apoferritin[581] in oxidizing Fe(II) involves the accumulation of hydrous ferric oxide. A cooperative two-electron transfer from two close Fe(II) to oxygen is thought to occur, followed by the formation of an oxo-bridge by elimination of bound water protons.[52]

---

* Personal communication.

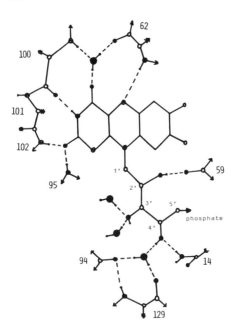

Fig. 20. Water molecules in the binding of ribityl and flavin portions of riboflavin-5′-phosphate to flavodoxin *desulfovibrio vulgaris.*[893] The two molecules linked to hydroxyl OH3′ are on the surface of the protein.

4.3. Water in the Outer Solvent Region

The precise nature of the solvent region beyond the surface "first hydration shell" is unknown. Experimental evidence on the extent to which the ordered bound region observable at the surface affects the structure and dynamics at greater distances from the interface is not clear. Considering that any deviation from the properties of the bulk solvent may be small, resulting perhaps in slightly enhanced relaxation times (see Section 2.2), and bearing in mind the problems of separating signal from noise in crystal structure refinements (Section 3.2), X-ray and neutron diffraction methods may be of very limited use in this region. Although placing specifically bound, highly occupied water sites closely related to the protein may be, with care, a relatively unambiguous procedure (Section 4.2), any assignment of coordinates, temperature factors, and occupancies outside this region must be treated with reserve. An assigned solvent network may give a reduction in the crystallographic residual, but the reality of the actual coordinates and parameters may be no more than hypothetical (Section 3.2). Skepticism is therefore necessary, and lack of evidence for structuring further away from the molecule demonstrates no more than that our technique is unable to pick up a restriction to phase space occupation that is less than or comparable with the noise level.

What evidence we do have from protein crystals is variable and in-conclusive. It does suggest, however, that in a few particularly favorable cases, generally of lowish solvent content, ordering beyond the first shell may occur. In carboxypeptidase-A the electron density maps (2.0 Å) show distinct features through about 50% of the solvent region.[565] Ribonuclease-A—a particularly favorable case, with a potential resolution of close to 1 Å—appears to have some distinct solvent peaks away from the immediate molecular surface region.[172] Although the data are still at an inadequately low resolution (2.8 Å) to be confident in making statements about the outer solvent region, the maps of carboxypeptidase-B also show features through-out the intermolecular channels.[791] Peaks were identified within 8–10 Å of the surface—well beyond the first hydration layer—although they faded out towards the middles of the channels. In triclinic lysozyme, no second shell of solvent was found, except in a few small regions, e.g., around two of the lysine side chains.[635] In carp muscle calcium-binding parvalbumin, the question is surprisingly a nonstarter: presumably because of the shape of the molecule, an estimated 97% of solvent is in first-shell contact with the protein surface.[627]

No extensive networks were found in the 2.0-Å refinement of horse methemoglobin.[532] In contrast, in myoglobin and metmyoglobin, where the protein molecules are more highly packed, extensive solvent networks have been found in independent X-ray[851,852] and neutron[793] refinements (Fig. 21). An extensive network links two α-chymotrypsin molecules across a dyad axis [Fig. 14(b)]. In rubredoxin (from *Clostridium pasteurianum*), a particularly well-ordered protein crystal recently refined to 1.2 Å,[894] only about one-third of the 137 water molecules included in the refinement were outside the first directly bonded shell. Positions were thought generally not detectable to more than the second shell, although there were extensive networks between protein molecules where they approached each other closely. Rubredoxin from *desulfovibrio vulgaris* has been taken to 2.0 Å.[14] Some of the highly occupied *C. pasteurianum* water sites were similarly occupied, though there were differences due to differences in molecular packing and changed side groups. This crystal is apparently capable of further refinement down to about 1 Å, and in general it appears the solvent is more highly ordered than in typical protein crystals.

Insulin is another case of an exceptionally well-ordered crystal, and evidence of relatively well-ordered solvent is coming out of a refinement of the 2-zinc form to about 1.5 Å.* The crystal contains about 270 water

---

* G. Dodson, D. C. Hodgkin, and M. Vijayan, personal communication.

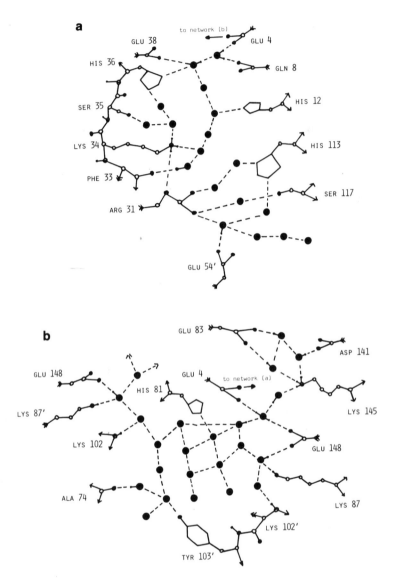

Fig. 21. Two extensive solvent networks in sperm whale metmyoglobin.[851] Despite the relatively large numbers of water molecules shown, network (a) is really a surface network, a water molecule rarely being more than one molecule distant from a protein polar group. Network (b) shows several water molecules considerably further from the surface. Note also the significant number of fourfold rings, implying considerable deviation from ideal tetrahedrality in the water–water bonding. Both networks connect through the carboxylate of Glu 4.

molecules, about 70% of which are regarded as securely placed. This includes considerable numbers away from the first shell. The exceptional care being taken in this refinement should result in a very reliable set of solvent data. An independent refinement to 1.3 Å using low-temperature data is also in progress.* Comparison of the two sets of data could be an important test of the reliability of the refinement procedures being used.

As discussed above in Section 3.2, the resulting solvent locations will depend critically on the methods used, and the selection and rejection criteria of possible positions. In some cases, occupancies have been refined; in others, only coordinates and temperature factors were adjusted during the refinement procedure, nonunit occupancies being allowed only exceptionally. Figure 22(a) is a plot of occupancy against temperature factor for solvent molecules located in rubredoxin *C. pasteurianum* at 1.5 Å.[892] B values for the protein vary between 4 and 20, though the termini show much higher values of 40–50. The 20 or so molecules with high occupancy and low B values indicate the most localized—presumably most strongly bound— water molecules. The bunching of points with low occupancy (0.3–0.5) and lowish temperature factor (20–40) is a perhaps surprising feature. It implies well-localized sites with only relatively low thermal motions (and/or static disorder across several unit cells), *yet with only low occupancies.* Although such a situation is physically possible, its occurrence seems somewhat strange. It underlines the problems of refinement procedures, what constraints should be put in, and what parameters can be realistically refined.

Figure 22(b) shows the same plot for the recent 1.2-Å refinement of the same molecule.[891a] The differences are significant, in that there is a noticeable shift away from the low-occupancy–low-temperature factor sites implied by the 1.5-Å data. It does suggest that at these very high resolutions, occupancy refinement of solvent may be more realistic; at lower resolutions —even as high as 1.5 Å—the occupancy data probably have reduced physical significance.

Table III lists B values and occupancies of those assigned water positions likely too close (<2.5 Å) to the protein. It is interesting and perhaps worrying that some of these doubtful positions [also marked on Fig. 22(b)] are of high occupancy, others showing reasonably low temperature factors. Three of these positions (marked R in Table III) were subsequently rejected on stereochemical grounds, though on occupancy and temperature factor

* N. Sakabe and K. Sakabe, personal communication.

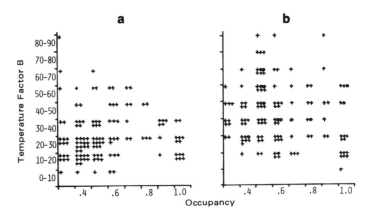

Fig. 22. Plots of refined occupancies and temperature factors for rubredoxin *C. pasteurianum* at (a) 1.5 Å[892] and (b) 1.2 Å.[891a] The open circles in (b) refer to close approach distances listed in Table III.

TABLE III. Distances <2.5 Å between Surface Polar Groups and Assigned Water Molecules in Rubredoxin *C. pasteurianum* Refined at 1.2 Å[a]

Distance (Å)	Temperature factor *B*	Occupancy	Water molecule sequence number	Comments[b]
1.2	41	0.5	86	R
1.8	44	1.0	38	PD
1.9	75	0.5	121	R
2.0	65	0.5	111	
2.1	33	0.8	28	
2.1	37	0.6	56	PD
2.1	56	0.5	102	
2.1	51	0.5	95	R
2.2	36	0.7	43	PD
2.2	25	0.4	55	
2.3	58	1.0	53	
2.3	18	0.6	20	PD
2.4	31	0.6	42	

[a] From Ref. 891a.
[b] R indicates positions rejected on stereochemical grounds; PD indicates poorly defined protein groups.

grounds only number 121 might be regarded with extra suspicion. The protein–water distances of four molecules (marked PD) relate to poorly defined protein groups, and hence larger errors would be expected here. Of the remaining molecules, the temperature factors and occupancies of at least half would seem reasonable; rejecting all other molecules with similar or worse values would result in removal of over half the assigned solvent positions. Even at this exceptionally high resolution, there is no clear-cut temperature factor or occupancy criterion to assist selection of real from artifactual solvent positions.

Few coordinate data are yet available to allow the extraction of bond length and angle statistics within this outer solvent shell in cases where it appears well ordered. The meaning of such data would be related to the constraints applied during refinement. If "unreasonable" approach distances and angles were not allowed, then the resultant distribution functions, assuming only full occupancy sites, would merely reflect those constraints. Moreover, if variable occupancies are allowed—as is very likely the real situation in a system highly disordered in crystallographic terms, though highly ordered to the liquid state physicist—the alternative coordinate nets would first have to be disentangled before the statistical data would look anything but ridiculous. This latter would be a very severe disorder problem in a system whose stereochemical constraints we cannot fully specify; the possible disentanglement is discussed in Section 6 below.

Figure 23 shows the distribution of water–water interactions in carp muscle calcium-binding parvalbumin.[627] As little beyond the first shell is seen in this protein, this hydrogen bond length plot shows only water–water interactions within this shell. The refinement (at 1.85 Å) did not assign occupancies. Within the estimated precision of about 0.35 Å for

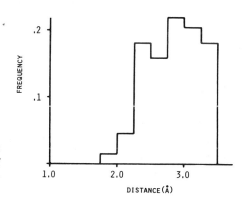

Fig. 23. Close water–water distances ≤3.5 Å from the 1.85-Å refinement of carp muscle calcium-binding parvalbumin.[627]

interatomic distances, this distribution peaks at about the distance expected for water networks (2.75–3.0 Å). The histogram is rather broad, and shows —even within the estimated errors—significant very close approaches ($\lesssim 2.5$ Å). These must either be artifacts, or relate to partially occupied sites. Again, the problems of solvent network interpretation are underlined, even for the first hydration shell in a well-ordered protein at a resolution as high as 1.85 Å.

In summary, the crystallographic evidence on the nature of the solvent region beyond the strongly bound component of the first shell is inconclusive. Several possibilities can be discussed.

(a) The region is generally disordered, being essentially indistinguishable from bulk solvent. A "perfect" diffraction experiment would see a flat region of electron density in this area. A good experiment and refinement would see some "structure" at a level of noise which might be reasonably estimated.

(b) The region is "significantly ordered." Whether or not a diffraction experiment would see the resultant phase space localization would depend upon the size of the resultant peaks in relation to the unavoidable noise level. Where the real system can be described realistically only in terms of partial occupancies, the problem of extracting signal from noise is even more difficult and subject to error.

What the diffraction experiment sees depends upon the extent of space localization compared to the general noise level. High-resolution refinements can, with great care, reduce and allow a realistic estimate of the level of noise, but by only pursuing those systems where this is possible, we are effectively selecting against the more disordered solvent system. We are in danger of expecting to see more than may be there, and of putting too much credence to our solvent structure.

Looking at the data available and becoming available, it seems that the extent of solvent ordering may vary with the protein. This would hardly be unexpected; we know some protein molecules themselves are more ordered than others in the crystal, and solvent is likely also to be so. We might reasonably expect very strong solvent localization for very well-ordered protein crystals with relatively low solvent content—e.g., rubredoxin *D. vulgaris*, insulin, and ribonuclease-A. Such ordered solvent would presumably in the extreme case approach the nature of a "glass" or supercooled liquid, with deviations of structure from the ideal which are related to specific geometrical effects at the interface. Such extreme systems would be of use in assisting our understanding of the nature of solvent interactions at complex, but biologically significant, interfaces.

5. SUMMARY: WHAT WE THINK WE NOW KNOW

It is clear that great care must be taken in interpreting solvent data from crystal structure analyses. As protein structure refinements become more common, we will need to have a good model of the solvent to be able to fully refine the protein. The fact that we need to know the protein structure very accurately *before* we can be reasonably certain of the solvent assignments underlines the problems involved. It is not easy to separate effects of protein from those of the solvent. Nevertheless, a flood of solvent data will no doubt be forthcoming in the near future. Interpreting its reality will require great care, especially in the region not directly adjacent to the protein. Except in the unachievable perfect diffraction experiment, it would seem unrealistic to refine solvent positions, occupancies, and thermal parameters to give a lower residual. We know an almost random solvent distribution can do that. Away from the protein surface we have a signal:noise ratio problem of great complexity, with little in the way of stereochemical information to help us.

The information we have so far is, however, fairly extensive within and close to the protein surface. Within certain proteins, we find isolated groups of one, two, or more molecules, forming "hydrogen-bond bridges" between polar groups too far apart to otherwise link. Without the intervening water molecules, there would be energetically unfavorable "dangling polar groups" with a corresponding enthalpy penalty *with respect to the solvent interactions in which these groups would presumably participate in the extended chain conformation.* In some larger proteins such as trypsin and chymotrypsin, fairly extensive water networks are found. Apart from the relevance to local stability, such internal molecules may be important in maintaining the critical alignment of groups that may be necessary for optimal activity. They are presumably also important to the molecule's dynamics. If, as is argued, specific hydrogen bonding *is* important to stability and dynamics, especially close to the energy minimum occupied by the native conformation, then water bridges will fulfill a similar role. The geometry of these internal groups and networks are as expected, having reasonable bond lengths and angles, with each water molecule making generally three or four hydrogen-bond contacts. Similarly, water mediated salt bridges are found. Sometimes small groups of water molecules are found near internal charged groups, either ions or charged side chains. Such groups may perform a stabilizing role by spreading the buried charge through a region of higher dielectric constant.

Water is generally found as one or more ligands to ions. Where the ion is involved in substrate binding, a "protective" water molecule is easily displaced, perhaps with other ordered or partly ordered water molecules in the binding cleft, thus enhancing the polarizing effect the ion may need to exert on the substrate.

Surface-exposed polar groups appear generally solvated, as predicted from hydrogen-bonding and accessibility studies. As such, this polar group solvation is likely structurally significant. Again, the geometry of these hydrogen bonds contains no surprises. Often short surface bridges are found between polar groups in the same molecule, as suggested from the distribution of surface polar groups. Such molecules often make two or more hydrogen bonds to the same molecule and are thus likely of significance in the mechanical system that is the protein. In some particularly well-ordered crystals, some solvent structure is seen beyond this first shell (see below). There is no direct crystallographic evidence for the existence of localized "clathrate cage" structures about exposed apolar groups. Unless the localization is specific—which would seem unlikely for apolar group interactions—the time scale of X-ray and neutron experiments may be so long that the time-averaged results of such cages may be no more than a uniform smear with respect to the noise level.

On the other hand, well-ordered groups of water molecules have been found in hydrophobic clefts, e.g., in trypsin and myoglobin. Expulsion of such molecules on enzyme–substrate binding may contribute significantly to the binding energy. Protein–protein and protein–ligand (including substrate) interactions often involve water bridges across otherwise undesirably separated polar or charged groups. Although the major driving force for such interactions may be hydrophobic, the specific interactions—sometimes water-bridged—may be necessary for good specificity of the resulting interface, in addition to preventing a destabilizing enthalpy deficit from unpaired polar groups.

There are examples of likely water involvement in enzyme mechanisms. In the serine proteases, a bound water molecule, probably activated by the charge-relay system, initiates nucleophilic attack of the carbonyl of the acyl-enzyme intermediate, with a second bound water possibly assisting in stabilizing the substrate carbonyl in the attack-prone position. Similar mechanisms proposed for thermolysin, the acid protease penicillopepsin, and carboxypeptidase-A involve nucleophilic attack of the susceptible carbonyl carbon by an activated buried water. In carbonic anhydrase, the zinc-bound water (or OH^-) may supply both the acidic and basic functions, with perhaps a linked water network providing a rapid proton transfer path to the solvent. Other water-involved mechanisms have been proposed for several electron transfer proteins, while the increased dissociation into subunits of some abnormal hemoglobins may be due to the disruptions of intersubunit water networks by certain unfavorable amino acid substitutions. Electron tunneling through ordered water networks—perhaps within the hydration shell—has been suggested in the context of photosynthetic mechanisms. The relevance of internal water molecule networks to the molecule's flexibility and dynamics has been invoked as an important aspect of the activation of trypsinogen.

Only limited light is shed on the solvent region outside the first hydration shell. In several proteins, elements of a second shell have been reported, but only in a very few cases, such as myoglobin and insulin, is there reasonable evidence of extensive networks more remote from the protein surface. The difficulties of interpreting this region are further underlined by the case of triclinic lysozyme. One study fails to find anything significant outside the first shell,[635] while another suggests the solvent region is fully ordered.[449]

A full analysis of this region in the few apparently well-ordered crystals remains to be performed. Such is made more difficult if partial occupancies are assigned—as would be expected to be physically realistic. We may have to somehow distribute n molecules among N sites, each of which may have only partial occupancy, such that each "instantaneous configuration" is internally consistent with a not very well defined set of stereochemical constraints.[307,798] Sorting out such problems in small crystals, *with n and N both small, and strong stereochemical constraints*, is difficult enough.

For reasons similar to those mentioned in discussing the search for clathrate like cages, crystallographic methods may be powerless in tackling this problem except perhaps in one or two very favorable cases where the solvent is highly ordered. Such systems would thus appear untypical of most proteins, and therefore may seem of limited interest. They would, however, be of immense value in testing our models of solvent interactions before extending them to more interesting solution studies (see Section 6).

The nature of this region may or may not be particularly interesting. Any partial ordering in the sense discussed in Section 2.2, with consequently reduced mobility and higher partial specific volume (but *not* isostructural with a crystalline ice, as is implied by the often-used term "icelike"), would in principle affect substrate diffusion, electron and proton transfer. With respect to stability and dynamics of the protein molecule, this region may, once the protein has folded, be of limited further significance. Recalling the discussion in Section 4.1, if the effectively long-range hydrophobic interactions are mainly involved in directing the molecule's path through phase space to a region close to its global energy well, after which the specific charge–charge and hydrogen-bonding interactions (which may involve solvent mediation) take over to determine the details of the depth and shape of the well, then we may have further need only of a relatively thin hydration layer to maintain those significant surface interactions and perhaps what organization there is around exposed apolar groups. It is possible that these functions could be discharged by another solvent away from the immediate surface. In fact, we know that much of the original solvent can be replaced after folding, and even after crystallization, by suitable alternative solvents[17,260,303–305,472,702,812] without loss of crystalline order or enzymatic activity. This suggests that *all* the original solvent may not be vital to maintaining stability of the active conformation. The necessity to maintain constant dielectric constant and ionic strength of the solvent may be indicative of the kinds of forces necessary to maintain stability. We should remember, however, that we are discussing the highly simplified system of a single protein in a given medium; replacement of 50% of the water by methanol *in a living system* is likely to have serious consequences!

6. CURRENT WORK AND FUTURE DEVELOPMENTS

6.1. Experimental Work

Although interest in biomolecule activity ideally would concentrate on operation in solution, the existence of a well-defined solvent–macromolecule interface in the crystal is of immense attraction as a test bed of theories and models of such an interface, and the effect of solvent on structure, dynamics, and interactions. In the crystal, the constraints at the protein–solvent interface are clearly different from those in solution, especially where close approaches between protein molecules occur. Nevertheless, studies of the solvent in such situations do give valuable pointers to the nature of the molecular interactions and their structural and dynamical consequences.

Unfortunately, there is likely to be an inverse relationship between the amount of solvent in the crystal and its degree of order. Thus, only in cases where protein–protein interactions are strong are we likely to obtain reasonably reliable, high-resolution solvent information away from the immediate protein–solvent interface. In such cases, the value of crystal structure analyses by themselves will be relatively low; other methods are being applied to investigate solvent in crystals, e.g., IR, calorimetry, diffusion, dielectric, and density measurements.[356,909] The possibilities of quasielastic neutron scattering to probe solvent dynamics in crystals will also likely be developed.

More solvent data will come naturally from high-resolution refinements of proteins. Several are currently under way, and more will no doubt be undertaken. It is stressed again that an accurate refinement *of necessity* requires a model of the solvent; it is less clear how realistic this solvent model needs to be to give apparently satisfactory "improvements" in the protein area of the map and the crystallographic R value. The reservations made earlier concerning refinement of occupancies and temperature factors without considerable human intervention should be emphasized further, as should the problems of such intervention when the stereochemical constraints are weak and poorly understood.

As more refinements are performed, we may compare independent results on the same crystal, using both the same and different refinement procedures on different or the same data sets, respectively. A case in point is insulin, where the same data have been used in direct phase refinements, difference Fourier, and other approaches.[219,237] Furthermore, a low-temperature refinement of insulin is also under way (N. Sakabe and K. Sakabe, personal communication). Results from related proteins will also be of interest, for instance the 2Zn and 4Zn forms of insulin. It is of interest that the 2.0-Å refinement of horse methemoglobin[532] showed significant peaks at locations which in the 2.5-Å refinement of human deoxyhemoglobin[300] were present but too close to background to be reliably accepted.[532] Triclinic and tetragonal lysozyme showed many solvent peaks in similar locations; moreover, both sets of data were concatenated to give a plausible model of the first hydration shell in solution.[635] There is need for more such overlap or near-overlap to help assess more closely the problems attached to incompletely ordered solvent location by present refinement techniques.

Changing the solvent composition allows many protein crystals to be studied at subzero temperatures. Although the temperature reductions possible are considerable—up to 100 K in favorable examples—in only a couple of cases of originally exceptionally well-ordered crystals is there significant improvement in achievable resolution.[702] This suggests that most of the "temperature factor" is structural rather than dynamic disorder. Nevertheless, such improvements should be exploited. Disorder in particularly "floppy" surface groups might be frozen out, thus decreasing the uncertainty concerning nearby solvent interaction. Although the solvent in such cases would contain a large proportion of nonaqueous component, the partition of the solvent species within the crystal would give interesting information concerning their likely functions. Knowing the sites the DMSO, glycol, ethanol, or methanol prefer, could throw light on where and why water is *not* essential. In addition, a significantly more accurate protein model would assist in solvent refinement in the room-temperature crystal with more normal solvent.

Extension of neutron diffraction would also be useful. Current work is proceeding on triclinic lysozyme to 1.7 Å resolution,[92] though the crystals diffract to considerably higher resolution. Further work is needed, though the cost and time involved in such extensive neutron data sets are considerable. Flux limitations also restrict the technique to those proteins that can be grown to large crystals (several mm³). Such do exist, and it is to be hoped that such neutron diffraction measurements will eventually be undertaken.

6.2. Theoretical Work

We still lack the tools to handle the dynamics of complex heterogeneous systems, though techniques based on new approaches to simple liquids may, in conjunction with the electronic engineer's tools, eventually be of help. In the meantime, extensions of computer simulation techniques are being developed to begin to tackle such systems.*

Unfortunately, given a large computer, it is too easy to plug in a set of "reasonable" potential functions and watch the evolution of our complex system—say our "protein" in a bath of "solvent." We require (a) advances in technique, to allow us to tackle longer time scales than are currently possible,[633] and (b) reliable potential functions both for the macromolecule component, the water or other solvent component, and the interaction between them.

The problem of realistic potentials is crucial if our simulations are to have any value beyond imagination. It is here that crystals can be used to check the predictions of hypothetical potentials. This must be considered at several levels. For example, as we have seen in Section 4.1, with solely our classical knowledge of polar interactions and the relative exposure of polar groups on the surface, we can likely predict the structure of the first solvent shell. Similarly, with relatively crude potential functions and a large computer, we would not expect to do any worse. It is beyond this region that the quality of the potentials we use may be much more critical.

---

* For an up-to-date discussion of such developments see Chapter 6 of this volume.

For example, low-temperature stability of crystals requires the changed solvent to match the dielectric constant of the solvent at room temperature; this strongly suggests that our water–water potential function must be capable of reproducting dielectric effects in the simulated pure solvent before it can expect to be successful in the protein situation. Similarly, the large area of protein–solvent interface in our system of interest suggests again that our water potential should be able to handle effects that become important at interfaces.

There is evidence from quantum mechanical studies that cooperative effects are significant in the water–water interaction.[235,379] Attempts have been made by Barnes[56] and Berendsen[95] to develop simple potential functions that are capable of reproducing this non-pair-additivity, using relatively simple polarizable electrostatic models of the water molecule. These models have been remarkably successful in reproducing the energy, equilibrium geometry, and force constants of the water dimer, together with the non-pair-additivity in small clusters of molecules.[57,58] Simulations of *isolated* clusters of tens of molecules made to explain results of molecular beam experiments have demonstrated the extensive effects of non-pair-additivity at extended interfaces.[58] That such models are also able to reproduce the expected average (enhanced) dipole moment in hexagonal ice are further encouraging signs that such a potential may be of particular value in complex interfacial systems such as our protein.*

The problems of the interaction between solvent molecules and groups on the protein is no less severe. Progress will no doubt be made on the basis of simulations of polar and apolar solutions. In addition, quantum mechanical calculations on chemically significant fragments of macromolecules of interest are being made, at both the semiempirical (our unpublished data and Ref. 229) and *ab initio*[129,183,186,188,796] levels. Dealing with the *intra*molecular potentials for the macromolecule requires a somewhat different approach; extensive semiempirical data are available.[550]

Actual attempts to simulate the solvent organization in macromolecule crystals are presently being made. Using the simple four-point-charge Rowlinson[770] model for water in a Monte Carlo calculation, Hagler and Moult[374] were able to reproduce the positions of only 49 of the 80 water molecules thought to be within 4.2 Å of the surface of lysozyme. This underlines the problems of potentials, discussed above.

A molecular dynamics study has also been made of the water in pancreatic trypsin inhibitor[400] using the ST2 four-point charge model.[831]

---

* See Note Added in Proof on page 122.

Monte Carlo calculations using the non-pair-additive water potential have been made on the solvent in crystals of coenzyme B_{12},[655] a non-protein macromolecule (molecular weight $\sim$1749) with a very well-defined solvent structure. Although these latter calculations—and others in progress on solvent in small polypeptides and oligonucleotides with well-defined solvent—are not completed, the predicted solvent structure is encouragingly consistent with crystal structure data. Further stringent tests against 0.9-Å refined neutron data are being made,[309] and the calculations extended to those proteins where the solvent data are thought most reliable. It is in this way that any crystallographic disorder of the solvent in the crystal is most likely to be rationalized and stringent tests of potentials and methods made before we can reliably extend them to static and dynamic solution studies. The crystal solvent structure is providing an indispensable test of predictions in the complex heterogeneous systems in which proteins exist *in vivo*.

ACKNOWLEDGMENTS

I thank particularly Professor D. C. Hodgkin, Dr. G. Dodson, and Dr. M. Vijayan for discussions both general and specifically concerning the insulin refinement in progress, and Professor L. H. Jensen, Dr. J. Deisenhofer, and Dr. R. Huber for prepublication data on rubredoxin and trypsin. Discussions with Dr. E. N. Baker, Dr. P. Barnes, Professor T. L. Blundell, Dr. W. Duax, Professor J. T. Edsall, Professor F. Franks, Dr. N. Isaacs, Dr. B. Jacrot, Dr. D. S. Moss, Mr. J. D. Nicholas, Dr. D. Sayre, Professor F. M. Richards, Dr. P. A. Timmins, Dr. A. V. Westerman, and Dr. G. Zaccai, are especially appreciated.

NOTE ADDED IN PROOF

Further recent work on this potential has satisfactorily explained a variety of other gas- and condensed-phase properties, including the second virial coefficient of steam, and (in contrast to ST2 and central-force models) the structural features of the pair-correlation function as measured by neutrons [J. L. Finney, *Faraday Discuss. Chem. Soc.* **66** (1979)]. It has also been able to account for the angular relations of cation and nearest-neighbor water molecules discussed in Chapter 1 of this volume (J. Quinn, P. Barnes, and J. L. Finney, unpublished data).

CHAPTER 3

Ab Initio Methods and the Study of Molecular Hydration

W. Graham Richards

Physical Chemistry Laboratory
South Parks Road, Oxford

1. INTRODUCTION

In the heady days when "polywater" was believed by many to exist and even to represent a threat to life, theoretical chemists were quick to risk their reputations and lend credence to the myth purely on the grounds of calculations. Since the bursting of the polywater bubble the work of quantum chemists on water and on solvation has been more realistic. Nonetheless it is still true to say that the achievements of *ab initio* quantum mechanical calculations in this area have been only moderately successful.

The phrase "in principle" punctuates the conversations of theoreticians. The art of *ab initio* molecular orbital computation has reached the point that in some areas the calculations are more accurate than experiment. In principle, treating a polymeric water complex or a biological molecule including water of solvation is only an extension to a larger molecular system. In fact, even a system such as $(H_2O)_3$ has so many possible variables in terms of the positions of the constituent nuclei that it is not possible to do sufficient calculations to satisfy a statistical mechanician without using grotesque amounts of computer time.

Calculations based on a brute force approach have not contributed anything of outstanding significance, while even more thoughtful attempts to use *ab initio* calculations as the lead to more approximate treatments still seem promising rather than definitive.

In this chapter we will review briefly the work on monomeric water and its polymers and then take a closer look at how *ab initio* calculations have been used to study the solvation of ions and molecules.

2. MONOMERIC WATER

The structure of the water molecule was extensively reviewed in 1972 by Kern and Karplus.[477] The two volumes of the bibliography of *ab initio* molecular orbital calculations[749] list over 280 references to such calculations prior to 1979 and a good coverage of the literature of the same period has been supplied by Goddard and Hunt.[352]

For the water molecule in its experimental geometry it is now possible to calculate extremely accurate wave functions incorporating the effects of many thousands of excited configurations. With such a wave function virtually any question about the isolated molecule can be answered to an arbitrary level of accuracy. Properties such as dipole moment, charge distribution, and spectroscopic constants have been calculated many times and in all the different variants of the molecular orbital method, but since there are not any outstanding structural questions about isolated H_2O, the methods have not been pushed to their ultimate limit is this case. Far more interest has centered on polymeric forms of water and solvated species where the theoretical problems multiply and the results are much less convincing than those for the monomer.

3. THE WATER DIMER

Recently the attention of quantum chemists has been directed more towards the dimer $(H_2O)_2$ rather than the isolated molecule. The dimer is of special interest both as a constituent of liquid water and also as the source of a potential function for intermolecular interactions which can be used in statistical calculations.

The dimer has twelve internal coordinates, which results in what is superficially a simple problem becoming very complex. A molecule with two main atoms and four hydrogens is on the small side in terms of molecules currently being treated by accurate *ab initio* methods well beyond the Hartree–Fock limit. In such a case, however, the geometry is frequently known in advance from spectroscopic studies and accurate calculations are only made on structures that resemble the experimental arrangement

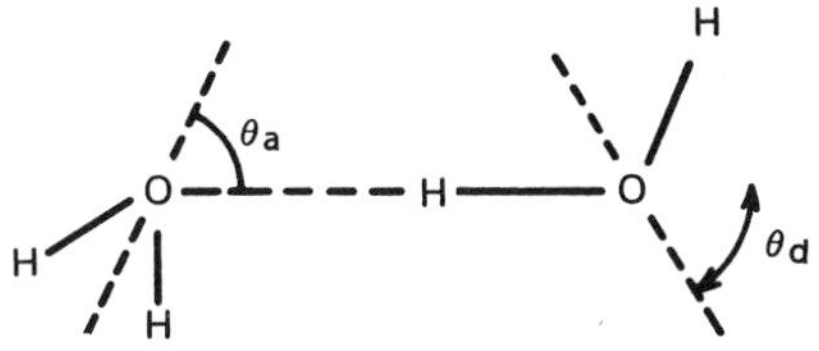

Fig. 1. The structure of the water dimer.
$R_{OO} = 2.98$ Å; $\theta_a = 58°$; $\theta_d = 50°$.

rather closely. From the variation of calculated energy with bond lengths and angles accurate spectroscopic constants can be obtained. The case of the water dimer was quite different. Until the last few months the structure of the isolated dimeric species was unknown spectroscopically and the theoretician had to assume a structure and then search for a minimum on the energy hypersurface. Most calculations* were done assuming that the monomer experimental geometry is appropriate and that the linking hydrogen bond is linear.

The water dimer has now been studied spectroscopically by Dyke *et al.*[267] Molecular beams of the hydrogen-bonded dimers were generated by expanding water vapor through a pinhole nozzle and radiofrequency and microwave transitions were observed arising from changes in the rotational energy of the rotating dimer. These spectra and the dipole moment of the dimer are interpreted to give the structure and relative orientations of the two water molecules as illustrated in Fig. 1.

The structure is the so-called "*trans*-linear" complex and is in most respects in quite good agreement with the most recent *ab initio* calculations.[560] In terms of the lengths and angles shown in Fig. 1 the calculations are in excellent agreement with experiment for the favored O–O distance R_{OO}; good agreement for θ_d but only poor accord for θ_a.

The importance to theoreticians of this experimental study is that it permits an objective assessment of calculations done in the absence of prior structural knowledge. The comparison is mildly encouraging; the latest results do seem to be reliable, but obviously great care is necessary. It is insufficient merely to run standard programs without taking precautions about basis sets, including configuration interaction and incorporating a wide-ranging search of the energy hypersurface. The calculations could, in principle, have predicted the correct structure, but a 20-electron problem is far more complex if it is a dimer of two stable components rather than a simple molecule. Despite the fact that molecular orbital methods do not

---

* See Refs. 29, 130, 134, 144, 218, 233, 234, 236, 249, 250, 252, 294, 378–380, 443, 450, 490, 495, 507, 547, 560, 598, 631, 632, 654, 677, 712, 805.

take into account where the bonds would be drawn, it does make a difference to the ease of finding the most stable atomic arrangement.

Given this qualified encouragement it is attractive to use calculations on $(H_2O)_2$ to provide an intermolecular potential for subsequent use in statistical thermodynamic computations. This problem has been considered in some detail by Clementi in a recent review.[184] Clementi took the calculations[547,631] which report almost 200 different geometrical arrangements of the two constituent water molecules and added an additional 26 points on the energy surface in the neighborhood of the minimum. The resulting numerical potential surface was then fitted to a variety of model potentials. A critical discussion of potential function is given in Chapter 6.

The best fit was found to be given by a simple point charge model similar to that proposed by Bernal and Fowler.[99] The analytical potential obtained is given in atomic units by the expression

$$\begin{aligned}
E = {}& q^2(1/r_{13} + 1/r_{14} + 1/r_{23} + 1/r_{24}) + 4q^2/r_{78} \\
& - 2q^2(1/r_{18} + 1/r_{28} + 1/r_{37} + 1/r_{47}) + a_1 \exp(-b_1 r_{56}) \\
& + a_2[\exp(-b_2 r_{13}) + \exp(-b_2 r_{14}) + \exp(-b_2 r_{23}) + \exp(-b_2 r_{24})] \\
& + a_3[\exp(-b_3 r_{16}) + \exp(-b_3 r_{26}) + \exp(-b_3 r_{35}) + \exp(-b_3 r_{45})]
\end{aligned}$$

where $a_1 = 582.277054$; $a_2 = 0.143789$; $a_3 = 5.470184$; $b_1 = 2.520593$; $b_2 = 1.221765$; $b_3 = 1.936626$; $q^2 = 0.449387$ and the O_5–M_7 distance is 0.436 a.u., the labeling being as shown in Fig. 2. The fit is very good (standard deviation 0.0002 a.u.) for attractive regions of potential but less accurate for the less important repulsive regions.

This potential can be compared with the empirical potentials of Rowlinson[770] and Ben-Naim and Stillinger.[87] All three potentials show that the hydrogen bond is linear for the most stable conformations, but the Hartree–Fock potential predicts a longer O–O distance than the empirical forms. This is in line with the recent molecular studies that show that the

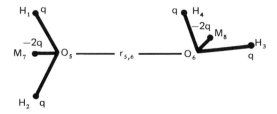

Fig. 2. The point charge model used for the analytical potential for water–water interaction.

O–O distance in the dimer is longer than that in liquid water, as is also true for $(HF)_2$. The empirical potentials have, of course, been derived as effective pair potentials for the condensed phase.

This type of extension of *ab initio* molecular orbital calculations to provide potentials that can be used in Monte Carlo calculations or computer simulation is obviously pointing the way in which calculations must develop. On the other hand the gravity of the problem of producing the energy hypersurface for $(H_2O)_2$ must cause scepticism about some of the published attempts to deal with the more complex situations encountered in bulk water or with ions in solution.

4. POLYMERS OF WATER

The fact that *ab initio* calculations did not produce perfect predictions about the structure of the spectroscopic entity $(H_2O)_2$, much less produce a totally convincing potential function, suggests caution in extending the study. Indeed, many questions about trimeric $(H_2O)_3$ remain unresolved despite a number of published calculations.[232,776] Because of the number of degrees of freedom, searching the potential hypersurface for the most stable arrangement of atoms is no mean task. Consequently it is by no means certain whether the trimer is an open structure with two hydrogen bonds or a cyclic entity with three.

In using calculations to study liquid water, on the other hand, it is not the structure of the most stable form of the constituents that is of prime importance. Thermodynamically, it is the distribution of conformations that are adopted at a given temperature that determines properties. Thus, ideally calculations should be used to study the whole range of statistically accessible distributions. To do this entirely quantum mechanically is not feasible, but as with the dimer, calculations can provide an analytical potential by fitting a function to the numerical potential energy surface.

Clementi[184] has followed this approach using the pair potential derived from his $(H_2O)_2$ studies and assuming pairwise additivity. Figs. 3 and 4 present Clementi's conclusions about the most stable configurations of water polymers. The calculations give some idea of the stabilization energy per molecule. Clementi concluded that no particular individual cluster is especially favored, in contrast to mixture models of liquid water. Although impressive in many respects the work of Clementi and his colleagues is at a stage where it suggests the basis of an approach rather than answers to questions about water polymers.

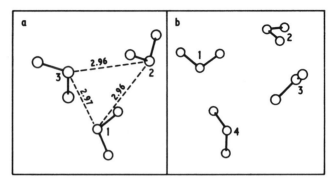

Fig. 3. Most stable configuration for (a) the water dimer and (b) the tetramer at $T = 0$ K. Distances are in angstrom units. (Reproduced with permission from Ref. 184.)

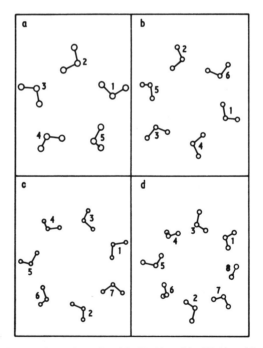

Fig. 4. Most stable configurations for (a) $(H_2O)_5$, (b) $(H_2O)_6$, (c) $(H_2O)_7$, and (d) $(H_2O)_8$, subject to the constraint that all the oxygen nuclei lie on a circle, but the protons are free to assume any orientation. (Reproduced with permission from Ref. 184.)

Straightforward calculations by molecular orbital methods on the higher polymers provide no serious information.

5. SOLVATED ION PAIRS

Calculations very much in the spirit of those described for water polymers have again been performed by Clementi.[184,341] In particular the system Li^+–F^-–H_2O has been the basis of an extensive study involving a cluster of 200 water molecules as well as the ion pair. The ions were held at fixed separations of 6, 8, and 10 Å and the temperature taken as 298 K and higher.[341]

The three interaction potentials were taken from Hartree–Fock calculations and a separate *ab initio* study provided details of the H_2O–Li^+–F^- three-body interaction.[517] The potentials were then utilized on a Monte Carlo program. The conclusions as to the disposition of water molecules were taken from averages of over 10^6 configurations. It was found that the cluster did not disperse and video displays suggest that filaments of linearly linked water molecules extend from the ions towards the surface of the cluster.

The computer experiments using theoretically calculable potentials can also give details about the coordination numbers of the ions. This rather imprecise characteristic is reproduced by the calculations in reasonable agreement with the scatter of experimental estimates, which themselves generally depend on some model of water structure surrounding the ion.

6. SOLVATION OF MOLECULES AND IONS

Ab initio calculations can provide results that are more accurate than experiment but only for the lone isolated molecule. Although this is appropriate for studies of interstellar molecules and perhaps gas-phase spectroscopic entities, the fact remains that the preponderance of experimental investigations are undertaken in solution. It is thus a clear goal of many research groups to attempt to incorporate the effects of solvent into theoretical calculations.

In view of the difficulties mentioned in the study of small polymers of water, it is clear that a more modest level of success is to be expected when one is dealing with a molecule or ion in an aqueous medium. Three independent lines of approach have been followed. The so-called "supermolecule" method incorporates tightly bound water molecules in the molecular orbital calculation. At the other extreme, in the "continuum method" the effect of solvent water is approximated by treating the surrounding medium as a continuous dielectric. Finally some tentative steps have been made towards treating solvent water by statistical mechanical methods based on solute–solvent and solute–solute potentials derived from *ab initio* quantum mechanical calculations. We will consider each of the approaches in turn.

6.1. The Supermolecule Approach

The first stage of the supermolecule approach is to find the sites and geometrical arrangements of the tightly bound water molecules to the solute. This is done by discrete calculations usually of an *ab initio* quality on solute plus a single water molecule whose position can be varied. These *ab initio* calculations are in general of a minimal basis set standard and with such limitations overestimate the strengths of hydrogen bonds. Since, however, the relative strengths of the hydrogen bonds to different quarters on the solute are the matters of interest, this defect is probably not crucial.

One of the earliest such calculations was concerned with formamide[16] and the method has been reviewed by the Pullmans.[724] In the case of formamide, six minima on the potential surface were located and correspond with chemical intuition. The calculations suggest that water molecules would act as proton donors to the carbonyl oxygen; as proton acceptors from the N–H bond, with less strongly bound water molecules accepting a proton from –C–H and donating yet another proton to the amine nitrogen. When binding to the carbonyl, the water is preferentially in the formamide plane, whereas accepting a proton it orients itself perpendicular to the plane so as to be bisected by it, as summarized in Fig. 5.

The study has been conducted to a level of fine detail, including effects such as rotating the water molecules.

A similar investigations has been made[722] of the tightly bound water to *N*-methylacetamide, which provides a model for the peptide bond. Three monohydrates are found to be particularly stable but the stabilization energies are, as might be expected, slightly decreased with respect to formamide. The same research group has considered in the identical way the solvation of the hydrogen-bonded peptide linkage of proteins and amino acid side chains;[716] purine and pyrimidine bases of nucleic acids[714] and ammonium and alkylammonium groups.[715,723] The results of all these investigations and a very full discussion of details of preferred geometrical

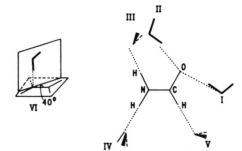

Fig. 5. Hydration of formamides, showing the six most favored regions.

arrangements are incorporated in the review by the Pullmans.[724] In general the results are in accord with intuition but are also quantitative, at least in a relative sense.

Although interesting and a definite contribution to the understanding of solvation, the results discussed earlier on statistical studies of water polymers lead to the conclusion that knowledge of preferred hydration sites is only a small part of the understanding of solvation. Indeed the most strongly bound water molecules may contribute statistically much less than the myriad of loosely bound alternatives. Contributions to the properties of solvated species will come from all those configurations that are likely at a given temperature. The configurations found to be most stable in the supermolecule of solute plus a single water molecule may represent the minima on the energy hypersurface. All other configurations within an energy of approximately kT of the minimum will exist, and if the minimum is very deep then its effects will be swamped by contributions from the flatter low-lying parts of the energy surface. Thus, as far as thermodynamic properties of solvated species are concerned, the supermolecule approach has little to offer, save as a starting point for finding the energy surface.

In fact the origins and utilizations of the supermolecule technique have been more modest and more specific. It has been largely with the question of the effect of aqueous solvent upon the conformation of flexible biologically active molecules that the method has been developed.

A major category of active biological molecules which incorporated neurotransmitters and many drugs has a flexible conformation which can be specified by two torsion angles, as indicated in Fig. 6. The great interest in the conformational properties of these molecules and ions, for example acetylcholine, adrenaline, and histamine, has resulted in many theoretical calculations of conformational preference.[748] The calculations have normally been performed on the isolated molecule whereas experimental

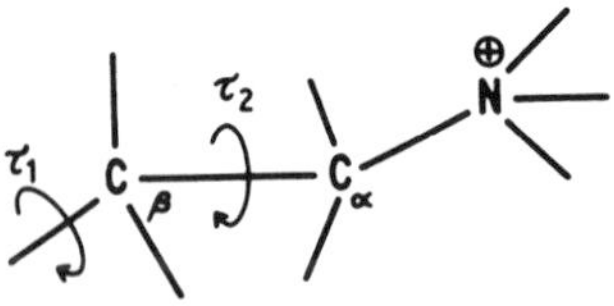

Fig. 6. Torsion angles specifying the conformation
of biologically active amines.

measurements of the ratio of *gauche* to *trans* forms (more properly *syn-clinal* to *anti*-periplanar ratios) have been by nuclear magnetic resonance (nmr) spectroscopy. Since the active environment of the small amines may well be aqueous to a great extent, there is an obvious need to incorporate solvation effects into the theoretical calculations, which are themselves necessary because the active conformation may well be distorted from the equilibrium form.

The precise locations of tightly bound water molecules can be found from *ab initio* calculations on the supermolecule consisting of amine and solvent water molecule. The conformation of the entire supermolecule may then be studied by systematically varying the angles τ_1 and τ_2 (Fig. 6) but keeping other bond lengths and angles and the bound water constant. The results may be presented as contour diagrams.

A case of particular interest is the histamine mono-cation (Fig. 7), on whose conformation solvent water is known to have a profound effect. In Fig. 8 the potential energy surface of the unsolvated cation is shown. The stable form suggested by these calculations is a *gauche* conformer with the side chain nitrogen atom coupled to the nitrogen of the imidazole ring by a hydrogen bond. This type of result is suggested both by *ab initio* calculations and by the more reliable variants of semiempirical molecular orbital study.

In contrast Fig. 9 shows the potential surface for histamine with the tightly bound water molecules included as part of the supermolecule. The entire qualitative result now changes and it is the extended *trans* form that is predicted as being the preponderant species. This result accords very well with experiment, and similar success has been found for acetyl-

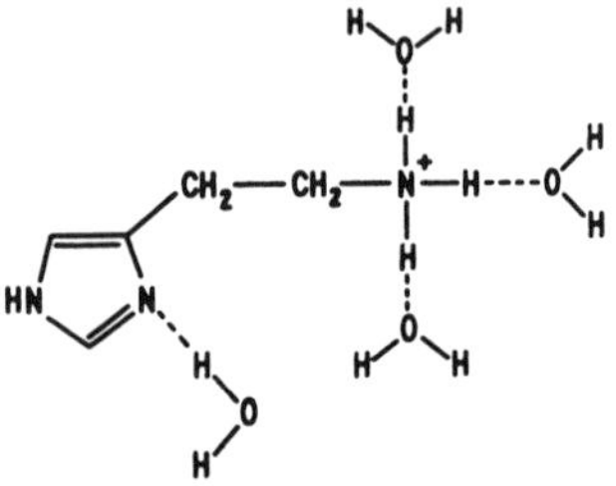

Fig. 7. The histamine monocation with the strongly
bound water molecules.

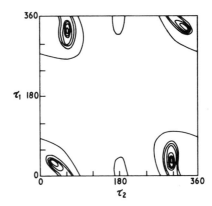

Fig. 8. Histamine monocation potential energy surface; contours in kcal mol^{-1}. (Reproduced with permission from Ref. 748.)

choline, γ-amino butyric acid (GABA), and a number of other cases, a summary of which can be found in a recent book.[748] In most instances in this type of molecule the inclusion of water molecules bound to the side-chain nitrogen atom results in a greater percentage of *trans* conformers, but a particularly interesting case is that of GABA, where the supermolecule calculations[725] predict the coexistence of several conformers in solution and nmr studies confirm that at least five conformational species are present in water.

As these successes indicate, the supermolecule calculations have produced results that alter lone molecule predictions in a manner consistent with experiment. On the other hand the triumphs ought not to overwhelm us, as the calculations are at best only a first stage in incorporating solvent effects. A major factor in increasing the percentage of *trans* forms of biologically active amines in solution could be merely the bulk of the solvated ammonium group which prevents folding back into *gauche* forms. Any bulky substitution such as methyl for hydrogen would similarly favor

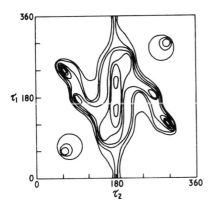

Fig. 9. Histamine monocation supermolecule potential energy surface; contours in kcal mol^{-1}. (Reproduced with permission from Ref. 748.)

trans conformations. Thus it is hard to be sure that the results are genuinely based upon the physics of the situation.

It must also be remembered with regard to the specific problem of the conformation of biological amines that they are mostly in the monocationic form with presumably a negatively charged ion in the vicinity to preserve local neutrality. Attempts to incorporate a chloride or fluoride ion in the calculated supermolecule suggest that these too could alter conformational preference,[7] although there is again a statistical problem in deciding upon the location of the anion.

With this type of reservation about the generality of the supermolecule approach it is probably best used with caution and considered as a starting point for solvation studies rather than a complete technique.

6.2. The Continuum Approach

The extreme alternative to treating solvent water in terms of individual molecules is to account for the bulk effects by means of a structureless dielectric continuum. Noteworthy in this respect are the contributions of Sinanoglu,[810] Beveridge,[105,106] Hopfinger[418] and Hylton, and Christoffersen and Hull.[430] The latter authors go to the heart of the physics of the problem and applications have been restricted to such systems as helium in carbon tetrachloride. The method is based on the classical electrostatic treatments of interacting systems due to Born,[132] Onsager,[672] and Kirkwood.[487]

As with the supermolecule model the main goal of the users of this approach has been to quantify the effects of solvent water on conformational properties of flexible molecules.

The total energy of a molecule or ion with a defined geometry and conformation can be written as

$$E_{\text{total}} = E_{\text{solute}} + E_{\text{solvation}}$$

with E_{solute} being the energy found in an isolated molecule *ab initio* molecular orbital computation.

The solvation energy, $E_{\text{solvation}}$, is approximately made up of three contributions,

$$E_{\text{solvation}} = E_{\text{es}} + E_{\text{dis}} + E_{\text{cav}}$$

Here E_{es} is the electrostatic solvent–solute binding energy arising from the interaction of permanent and induced electric moments; E_{dis} is the interaction energy due to dispersion forces; both E_{es} and E_{dis} are negative; E_{cav}, which is positive in sign, is the energy required to form a cavity in the solvent to accommodate the solute molecule. Each of these terms may depend parametrically on quantities that vary as functions of molecular conformation or geometry, and may be estimated from theoretical calculations.

For the calculation of E_{es} the solute is treated as a point dipolar ion of charge Q and total dipole moment $\mathbf{m}$ at the center of a sphere of effective radius α imbedded in the solvent. The solvent is represented as a polarizable dielectric continuum of dielectric constant E. The solute induces a reaction field $\mathbf{E}_R$ in the solvent which then acts back on the solute system. If differences in the monopolar term as a function of conformational change are ignored, the following expression for E_{es} is given:

$$E_{es} = -\frac{1}{2}\,\mathbf{m}\cdot\mathbf{E}_R \qquad \text{with } \mathbf{E}_R = \frac{2(\varepsilon - 1)}{2\varepsilon + 1}\,\frac{\mathbf{m}}{\alpha^3}$$

Both $\mathbf{m}$ and α depend on conformation; $\mathbf{m}$ can be computed from the wave function and α is the molecular cavity radius defined by equating the molecular volume with the volume of a sphere:

$$\alpha = (3V/4\pi)^{1/3}$$

The molecular volume can be estimated from the Cartesian coordinates of the atoms used as data in the isolated molecule calculation.

E_{dis} is estimated using the expression

$$E_{dis} = \frac{\varrho}{2}\int_0^\infty v^{eff}(r)g^{(2)}(r)4\pi r^2\,dr$$

where ϱ is the number density of the solvent, v^{eff} is the effective pairwise potential function for solute–solvent interaction, and $g^{(2)}(r)$ is a radial distribution function. For simplicity $g^{(2)}(r)$ is taken as zero for $r < \alpha$ and unity for $r > \alpha$. The function v^{eff} may be a modified Kihara potential.

E_{cav} is estimated from the cavity area A and the solvent surface tension γ as

$$E_{cav} = f\cdot 4\pi\alpha^2\gamma$$

where f is a factor that relates macroscopic to microscopic dimensions.[810]

Both E_{dis} and E_{cav} vary parametrically with conformation through the cavity size α.

It must be clear from this summary that the approximations that are necessary are inevitably very crude, but the parameters that enter the calculations are based on physicochemical properties of the solvent and do not obviously result in changing conformational preference to achieve a known result. It is thus impressive that the calculated effect on conformation is in accord with experiment. Most striking perhaps has been Beveridge's work on acetylcholine.[106] When the solvent effect is calculated by the continuum method the experimentally observed *gauche* (*syn*-clinal, *anti*-periplanar) conformation is predicted as being the most stable.

The obvious weakness of the continuum model is the strength of the supermolecule method: lack of explicit consideration of solute–solvent hydrogen bonding effects.

To overcome this defect a combined approach has been suggested by Beveridge and Schnuelle.[107] This extension does come back to the heart of the solvation problem by accepting that a satisfactory theory must have a statistical aspect.

6.3. The Statistical Approach Using Solvation Potentials

The inevitable direction in which one is forced on examining the solvation problem from a standpoint of quantum mechanical calculations is towards a statistical approach. A solvated molecule is not a supermolecule and the solvent is far more than a continuous medium. Thus if one disregards the amount of computation implied, a logical approach could be described as a three-stage process:

  (i) Compute the interactions of the solute with water molecules using reliable quantum mechanical methods.
 (ii) Fit the interaction energy to an analytical expression.
(iii) Use the potential in a Monte Carlo calculation to determine the organization of the solvent around the solute at a specified temperature.

This head-on attack on the problem has been the basis of the most recent work of Clementi and his collaborators,[184] but even to within this idealized scheme approximations have proved necessary, particularly as the systems studied have included amino acids.[598]

For stage one the quantum mechanical calculations have been less accurate than those used in the study of the water dimer, and the basis set used is likely to overstress the strength of hydrogen bonds. The interaction with one single water molecule placed at different distances around the solvent is also a serious restriction. The potential would no doubt alter were further solvent molecules to be incorporated. Only very limited variation of the orientation of the lone solvent water has been possible. Nonetheless even with these limitations over 2000 separate quantum mechanical calculations have been performed.

Stage two requires an analytical expression to which the quantum mechanical data are to be fitted. In the published work the Lennard-Jones 6–12 potential has been used. This choice is based on convenience and rapidity of use in the subsequent Monte Carlo calculations rather than for convincing physical reasons. It must be remembered that this potential is not notably successful in much simpler liquid state problems such as that of liquid rare gases, and its use for non-spherically-symmetrical partners is a gross extension. Nonpairwise additivity is again perforce neglected.

The actual analytical expression used to fit the interaction energy I is

$$I = \sum_i \sum_{i \neq j} \left(\frac{A_{ij}^{ab}}{r_{ij}^6} + \frac{B_{ij}^{ab}}{r^{12}} + \frac{C_{ij}^{ab} q_i q_j}{r_{ij}} \right)$$

The third term in the expression allows for direct charge–charge interaction and may be either positive or negative depending on the signs of the q_i and q_j. Indices i and j refer to atoms in solute and solvent respectively and q_i and q_j are the net changes on those atoms resulting from the quantum mechanical calculations. The indices a and b differentiate between atoms of differing atomic number and also between those with the same nuclear charge but differing environments. In all, six classes of hydrogen atom were considered as separate categories and for the main atoms, ten for carbon, two for oxygen, four for nitrogen, and one for sulfur. As a result, stage two yields 23 sets of interaction constants with the oxygen of H_2O and 23 sets of interactions with the hydrogen of H_2O. The quality of the fit of these expressions to the original numerical data provided by the quantum mechanical calculations is indicated by a standard deviation of about 2.5 kJ mol^{-1}.

The analytical potentials provide the basis for the third stage, the statistical mechanical calculation. Work is currently in progress on the solvation of amino acids and, perhaps optimistically, on proteins.

7. CONCLUSIONS

Review articles are almost invariably written by enthusiasts describing a subject in which their own contributions have advanced the topic. The present case is an exception, the author is an enthusiast for *ab initio* calculations which, despite many boring and trivial applications, can illuminate experiment and occasionally even surpass experimental accuracy or produce answers to questions that are beyond the capacity of experimentalists. The author is also well aware of the importance of an understanding of solvation and conscious of the defects in his own work consequent upon ignorance of how water influences an isolated molecule. Nevertheless, being an outsider it is possible to give a more objective assessment than is often the case at the end of a chapter. That considered view, not taken on a Monday morning nor after a heavy night, is pessimistic.

The problems of the water dimer and small polymers are within the scope of *ab initio* molecular orbital calculations if one only seeks to predict the structure of the spectroscopic entities found in beam experiments. Even here, however, the dreaded theoreticians' hedging phrase "in principle" must be used. The predictions on the dimer, while being accurate in many respects, were only moderately successful in others. A lot of work still needs to be done before one can be confident about the predictions for the trimer which will undoubtedly succumb to spectroscopic investigation in the near future.

To many the obvious role of theoretical calculations is to provide potentials for statistical thermodynamic studies. However, although in principle it is possible to compute the energies of a very large number of solute-plus-solvent molecule configurations, the number of degrees of freedom is such that astronomic amounts of computer time would be required before a satisfactory interaction energy hypersurface could be prepared. Then additionally the effects of water–water interactions and nonpairwise additivity should be considered.

It is hard to be confident enough to predict any success in the understanding either of water or of solvated species resulting from molecular orbital studies, except perhaps that the very nature of the failures may be illuminating.

The highest form of goodness is the water, water knows how to
benefit all things without striving with them.   *Lao-Tzu*

CHAPTER 4

Mixed Aqueous Solvent Effects on Kinetics and Mechanisms of Organic Reactions

J. B. F. N. Engberts

Department of Organic Chemistry
The University, Zernikelaan
Groningen, The Netherlands

1. INTRODUCTION

Water and mixtures of water with organic solvents are commonly
used as solvents for a large variety of organic reactions. Over the last two
decades, the mechanisms of many of these processes have been investigated
in some detail, usually by kinetic techniques. Comparison of these results
with those obtained for the same reactions in nonaqueous media have
impressively illustrated that water and highly aqueous mixed solvents
display remarkable and often amazing solvent effects on the rates and
thermodynamic activation parameters of nucleophilic displacement reactions, acid–base catalyzed processes, and many other important reactions
in organic chemistry. An interesting example of a large solvent rate effect
is provided by the uncatalyzed decarboxylation of 6-nitrobenzisoxazole-3-
carboxylate (1) studied by Kemp and Paul[476] and also discussed by

Jencks.[446] Upon changing the reaction medium from hexamethylphosphoramide (HMPA), a dipolar aprotic solvent, to water the rate is decreased by a factor of $\sim 10^8$. In ethanol the rate retardation is about 10^6. Apparently, delocalization of the negative charge in the transition state for decarboxylation (2) is highly unfavorable in hydrogen bonding solvents and especially in water. On the other hand, numerous reactions are known that involve either the creation or localization of charge in the transition state of the rate-limiting step and which are markedly *accelerated* if an aprotic solvent medium is replaced by water. Although the overwhelming importance of hydrogen bonding interactions in aqueous solvents is well established experimentally, the solvent effects on kinetic parameters cannot be exclusively rationalized simply in terms of the better hydrogen bonding properties of water as compared with, for instance, simple aliphatic alcohols or dipolar aprotic solvents. For many highly aqueous solvent mixtures, a complex and often intriguing behavior of kinetic parameters as a function of water concentration is observed. One example is furnished by the water-catalyzed (i.e., pH-independent) detritiation of *t*-butylmalononitrile (3) in dimethylsulfoxide (DMSO)–H_2O, which involves rate-limiting transfer of tritium from the substrate to water (see Section 5.2)[403]:

$$\underset{\mathbf{3}}{t\text{-Bu}\overset{\displaystyle CN}{\underset{\displaystyle CN}{\overset{|}{\underset{|}{C}}}}\text{—T}} \;+\; H_2O \;\longrightarrow\; t\text{-Bu}\overset{\displaystyle CN}{\underset{\displaystyle CN}{\overset{|}{\underset{|}{C}}}}\text{—H} \;+\; HTO$$

A plot of $\log k/k_w$ vs. n_{H_2O} (mole fraction of water) is shown in Fig. 1 (k and k_w are the pseudo-first-order rate constants in DMSO–H_2O and pure H_2O, respectively). Interestingly, a *maximum* in the kinetic basicity of the solvent mixture is observed at $n_{H_2O} \sim 0.80$. The expected rate decrease upon addition of the less polar cosolvent DMSO ($\varepsilon = 46$) is only observed below $n_{H_2O} = 0.40$. The authors have proposed[403] that the marked rate acceleration observed upon the initial addition of DMSO to water reflects an increase in water structure, which would stabilize the polar transition state relative to the apolar substrate, but an alternative explanation may also be offered (Section 5.2). At this point it should be noted that such extrema in $\Delta G^{\ddagger}$ are relatively rare and that the behavior of the Gibbs free energy parameter as a function of n_{H_2O} in aqueous binaries often conceals interesting and large solvation effects. This is, in a final introductory example, shown in Fig. 2, which displays $\Delta G^{\ddagger}$, $\Delta H^{\ddagger}$, and $-T\Delta S^{\ddagger}$ as a function of n_{H_2O} for the water-catalyzed hydrolysis of *p*-methoxyphenyl dichloroacetate (4) in *t*-BuOH–H_2O.[281] This reaction, which will be

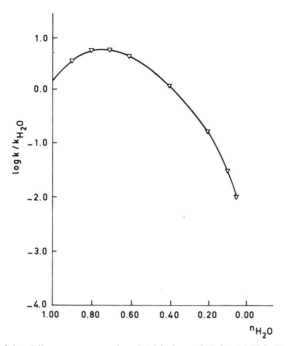

Fig. 1. Plot of $\log k/k_{H_2O}$ vs. n_{H_2O} for detritiation of **3** in DMSO–H_2O containing $10^{-3}\,M$ HCl (25°C).[403]

further discussed in Section 5.2, most probably occurs via water-catalyzed nucleophilic attack of water on the ester carbonyl with proton transfer from water to water in the rate-limiting step. As $\Delta G^{\ddagger}$ increases smoothly in the solvent composition range $n_{H_2O} = 1.00$–0.90, $\Delta H^{\ddagger}$ and $-T\Delta S^{\ddagger}$ show dramatic variations and pass through extrema at $n_{H_2O} = 0.95$, the "magic mole fraction" of maximum solvent structural integrity.

$$HCCl_2CO_2C_6H_4OCH_3\text{-}p + H_2O \longrightarrow HCCl_2CO_2H + p\text{-}CH_3OC_6H_4OH$$

4

Mixed aqueous binaries offer the possibility of continuously changing the three-dimensional hydrogen-bond network of water by gradually increasing the mole fraction (n_s) of the organic component. If the organic cosolvent is completely miscible with water, large changes in solvent properties may be effected. In dioxane–water, for instance, the dielectric constant may be varied from 78 (H_2O) to 2.2 (dioxane) at 25°C. Since the change in free energy associated with the change of the cybotactic region (Section 2.2.1) of the substrate as induced by the nonaqueous solvent com-

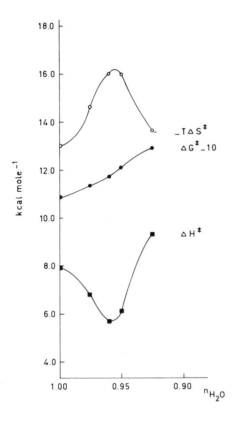

Fig. 2. Plot of $\Delta G^{\ddagger}$, $\Delta H^{\ddagger}$, and $-T\Delta S^{\ddagger}$ vs. n_{H_2O} (or n_{D_2O}) for the neutral hydrolysis of **4** in t-BuOH–H_2O at 25°C.[281]

ponent will be different from that of the transition state, a change in the rate and thermodynamic activation parameters of a reaction of the solute ensues. The continuous change in microenvironment of the reactant (change of n_{H_2O}) offers a powerful tool to assess the importance of solvent effects on organic reactions, and this method may be superior to that involving the study of solvent effects by changing from one pure solvent to another. Further we note that interpretation of solvent effects on reactions in aqueous binaries is often less complex than that for solvent effects in aqueous electrolyte solutions because of the absence of long-range Coulombic interactions of the reactants with cationic and anionic species which would modulate solvation changes.

Although we will be largely concerned with mixed aqueous solvent effects on reaction rates, it will be obvious that solvation effects may also greatly affect chemical equilibria, especially when charged particles are involved.[41,60] For example, gas-phase acidities[585,611] and basicities,[44,398] which have recently come available, have not only demonstrated the dominant role of solute–solvent interactions in determining solution acidities and basicities[37] but have also shed new light on the nature of substituent effects on the *intrinsic* acidity and basicity of a large variety of compounds.[143,847]

For instance, it was found that alkyl groups increase both acidity and basicity in the gas phase,[143] apparently due primarily to polarization of the alkyl moiety by an ionic center in its proximity. The solvent may also determine the *site* of protonation. A spectacular example involves the protonation of phenol which occurs predominantly on oxygen in aqueous solution but in the gas phase is by some 15 kcal mole^{-1} more favorable for protonation on the aromatic ring.[230] Despite the dramatic solvation effects on kinetics and equilibria, a few cases have been reported in which gas-phase properties have been correlated successfully with kinetic data obtained in solution.[618]

It will be clear that any enquiry into mixed aqueous solvent effects on kinetic parameters of organic substrates should quantitatively consider the interactions between substrate and solvent molecules and the effect of solvent reorganization as a result of transferring the reactant(s) into the transition state. Fortunately, one is usually only interested in the latter *change* of solvation, which leaves out the consideration of all those hydrophilic and hydrophobic parts of the substrates that are not directly involved in the reaction under study.

Apart from their own intrinsic interest, the study of solvent effects on organic reactions in aqueous binaries has relevance for our understanding of microenvironmental effects in biological processes. Life processes generally occur in the presence of water, which provides a unique medium for the solvation, interactions, and reactions of biopolymers.[55] Water not only serves as a transport medium for neutral and charged molecules in living organisms, it also determines to a large extent the conformation of peptides, proteins,[269,330,391] lipids,[391] carbohydrates,[838] and other biochemical molecules by interacting specifically with polar as well as nonpolar sites in these compounds.[329,388] At the active site of enzymes, water acts as an acid–base catalyst or solvating agent in a large variety of chemical transformations and also as an entropy-controlling medium in displacement reactions of water[553] in that region of the protein. Usually one can discern both hydrophilic and hydrophobic regions at or near the active sites, and X-ray analysis*[265] has revealed that binding of the substrate may be accompanied by removal of a number of strongly hydrogen-bonded water

---

* See Chapter 2 of this volume.

molecules. Therefore, the overall catalytic process will be strongly affected by the presence of organic cosolvents, as has been demonstrated in some detail for the acylation of α-chymotrypsin.[76] In attempts to assess the solvent effects in biochemical environments, particular aspects of solvation phenomena may be studied using model reactions which mimic the solvation behavior in the enzymic process. Much progress has been made, for instance, in the broad area of catalysis by micelles of surfactants in water[201,298] and by reversed micelles in aprotic solvents which contain water molecules solubilized in the polar, hydrophilic cavity.[297]

It has become clear that for any detailed interpretation of solvent effects in highly aqueous binaries, the solvent structural properties should be taken into account. However, the complexities associated with these factors, coupled with the question of the nature of the immediate environment of the reacting species and the intricacy of the reaction mechanisms themselves, have all frustrated a detailed approach in terms of quantitative theories. Nevertheless, the peculiar properties of water as a reaction medium as well as a reactant (acid, base, electrophile, or nucleophile) are well recognized and much current research is devoted to the quantitative elucidation of rates and mechanisms of chemical and biological processes in aqueous environments.

It is impossible to review here the thousands of papers on solvent effects on organic reactions in mixed aqueous solvents, some of which stretch back in time for some 80 years. Mixed aqueous solvent effects have been reviewed previously,[118,432,755] and particularly solvent effects of one-component solvents on rates of organic reactions have been comprehensively discussed[1,31,509] in attempts to reconcile these data with the known properties of the solvents and of the substrates and their transition states. We will here concentrate on a selection of recent studies of organic reactions in water perturbed by the presence of an organic cosolvent and which especially serve to demonstrate the interesting and unique solvation effects in these media.

2. MIXED AQUEOUS SOLUTIONS

2.1. Classification

The physical and thermodynamic properties of aqueous binaries strongly depend on the nature of the nonaqueous component. Deviation from ideal behavior is quite a common phenomenon and may lead to eccentric solvent properties. The wealth of data has been extensively reviewed in Volume 2 of this Treatise, but no general and comprehensive theory has been developed that can rationalize the experimental data in a quantitative fashion. As argued by Franks,[326] it is convenient to divide aqueous binaries into two classes.

I. Typically Aqueous (TA) Solutions. In these solutions the apolar, hydrocarbonlike organic cosolvent initially tends to occupy voids in the hydrogen-bond network of water, thereby promoting the formation of hydrogen bonds between the surrounding water molecules in a cooperative manner. This leads to lower rates of reorientation and diffusion of these water molecules. The formation of these *hydrophobic hydration spheres* reaches a maximum in the water-rich region at a mol fraction of water that is determined by the nature of the organic component. In some cases, this value for n_{H_2O} has been referred to as the "magic mole fraction" of a maximum in "water structure."[580] The most peculiar solvation behavior is usually observed around this n_{H_2O}, and is associated with pronounced concentration fluctuations (microheterogeneity). At lower n_{H_2O} the solvent properties tend to resemble more and more those of nonaqueous mixtures of polar compounds. Some examples of TA solutions are listed in Table I.

II. Typically Nonaqueous (TNA) Solutions. These solvent systems are usually confined to aqueous solutions of hydrophilic solutes (Table I). The solvent properties are now determined by specific, orientation-dependent, short-range interactions between water and the polar, hydrophilic

TABLE I. Mixed Aqueous Solutions

Typically aqueous solutions: Organic component	Typically nonaqueous solutions: Organic component
Alkanes	H_2O_2
Aromatic hydrocarbons	Acetonitrile
Alcohols (monofunctional)	Urea
Ethers	DMSO
Diethers	CCl_3CH_2OH, $CBr_3CH_2OH^a$
Ketones	Carbohydrates
Amines (monofunctional)	Polyalcohols
CF_3CH_2OH, $(CF_3)_2CHOH^a$	Amides
	Sulfolane

^{a} Reference 582.

sites of the organic component.[155,333–334] Consequently, the degree of long-range order in highly aqueous TNA solvents is often diminished relative to water, and, furthermore, TNA solvents hamper the formation of hydrophobic hydration spheres around hydrophobic solutes.[240,597] TNA solvents usually show only relatively small deviations from ideal behavior, and there is no evidence for pronounced microheterogeneities.

It is fairly obvious that mixed aqueous solvent effects on organic reactions, which directly reflect solute–solvent interactions, will be critically dependent on the intermolecular interaction between the solvent molecules themselves. As a consequence, they are expected to be essentially different for TA and TNA solvent systems. A proper analysis then requires the quantitative dissection of solvent effects on rates and thermodynamic activation parameters into initial-state and transition-state contributions (Section 4). In a final analysis, one may then venture to rationalize these data in terms of solute–solvent interaction mechanisms (see Section 3). But now the discussion is perforce qualitative in view of the absence of adequate theories, and, at best, an intellectual account can be given of trends in the thermodynamic parameters of initial and transition state as a function of solvent composition.

One more simple approach in treating solvent effects on organic reactions is to try to correlate kinetic parameters (usually $\log k$ or $\Delta G^{\ddagger}$) with a solvent parameter, selected to reflect the characteristic solvation requirements of the particular chemical transformation. Such attempts have met with only limited success, especially for organic reactions in highly aqueous binaries, for a number of reasons. One particular problem is found in the fact that kinetic solvent effects reveal changes of solvation of both the initial and transition state as a function of solvent composition. The solute–solvent interaction is governed by several interaction mechanisms, and in mixed aqueous solutions this may well lead to preferential solvation.[475,535] by one component of the solvent mixture. For instance, for hydrolysis of a relatively nonpolar organic substrate in a mixed aqueous solvent, the probability of encounters between reactant and water will be dependent on the possible, and in most cases likely, preferential solvation of the organic solute by the unreactive organic cosolvent. This problem has been studied mainly by nmr methods[324,468,535] in terms of solute chemical shifts and solvent relaxation times (T_2) as a function of solvent composition.[535] Therefore, it is not, a priori, expected that a solvent parameter shows a simple correlation with the observed kinetic solvent effects. Nevertheless, some useful correlations have been discovered, and, in addition, these studies have the fundamental merit that they have contributed considerably to our thinking about environmental effects on chemical processes. Accordingly, some of the most popular solvent parameters will now be briefly examined.

2.2. Solvent Parameters

2.2.1. Dielectric Constant and Functions Thereof

The dielectric constant (ε) is a macroscopic property defined for a continuous medium containing no microscopic structure. Solvent systems may be classified as polar ($\varepsilon > 15$) or nonpolar ($\varepsilon < 15$). Generally, the polarity of bulk solvent will be different from that in the solvation shell of a solute (i.e., in the "cybotactic region," being the volume around the solute in which the order of the solvent molecules has been affected by the solute molecule). A useful function of the dielectric constant is represented by the Kirkwood function[486] $(\varepsilon - 1)/(2\varepsilon + 1)$, which may correlate dipole interactions for dipolar substrates or transition states (indicated by ‡). For a bimolecular reaction involving the substrates A and B, the following equation has been proposed:

$$\ln \frac{k}{k_0} = -(kT)^{-1}\left[\frac{(\varepsilon - 1)}{(2\varepsilon + 1)}\right]\left(\frac{\mu_A^2}{r_A^3} + \frac{\mu_B^2}{r_B^3} - \frac{\mu_‡^2}{r_‡^3}\right) \qquad (1)$$

In this equation, k and k_0 are the rate constants in media of dielectric constants ε and unity, respectively, and μ_A, μ_B, and $\mu_‡$ are the dipole moments of the spherical species A, B, and the transition state, of radius r. This function has been successfully applied for a variety of reactions and useful quantitative information has been obtained about the nature of the transition state of these processes. Unfortunately, hydroxylic solvents and mixed aqueous solvents usually afford anomalous results,[1] although approximately linear correlations were found upon plotting $\log k$ vs. $(\varepsilon - 1)/(2\varepsilon + 1)$ for the solvolysis of t-BuCl in aqueous MeOH, EtOH, acetone, and dioxane.[835] Since $(\varepsilon - 1)/(2\varepsilon + 1)$ is essentially a linear function[835] of ε^{-1} at $\varepsilon > 5$, $\log k$ may also be plotted vs. ε^{-1}.

2.2.2. Solvatochromism Scales and Related Polarity Scales

A quite useful solvent parameter has been proposed by Kosower.[514−515] He has measured the maximum of the longest-wavelength intramolecular charge-transfer absorption band (λ_m) of 1-ethyl-4-carbomethoxypyridinium

iodide (**5**) in a large number of solvents to set up an empirical solvato-
chromism scale of solvent polarity. The Z value is defined as Z = transition
energy in kcal mol^{-1} = $2.859 \times 10^5 \lambda_m^{-1}$ (λ in Å). Now the solvent parameter reflects a model reaction and may be called a model, cybotactic solvent

5

polarity parameter.[501] Z values for some aqueous binaries are given in
Table II. Similarly, the Dimroth–Reichardt E_T value measures the solvent-
induced frequency shift of the electronic absorption maximum of the
betaine **6**.[255,739] There is a fairly linear correlation between Z and E_T

6

values for single solvents as well as for some aqueous binaries. Both the
Z and E_T values have been extensively used for the correlation of kinetic
data but the best results are often obtained by correlating solvent effects on spectroscopic parameters. For mixed aqueous solvents, linear
plots of $\log k$ vs. E_T were found for some $S_N 1$-type solvolysis reactions, but the slopes of these lines have as yet not been given a definite
mechanistic interpretation.[740] The optical absorption spectra of N-alkyl-
pyridinium iodides have also been employed to probe into the polarity of
the Stern layer of micelles and surfactant-solubilized waterpools.[299,637,837]
Several other empirical solvent parameters have been developed, among
them Berson's Ω values,[102] Brooker's χ_R solvent polarity scale,[147] Brownstein's S scale,[149] and a solvent polarity scale based on the tautomeriza-
tion equilibrium of pyridoxal 5'-phosphonate Schiff's bases.[568] Recently,
Kamlet and Taft[461,462] have constructed an α scale of solvent hydrogen-
bond donor ability and a β scale of solvent hydrogen-bond acceptor ability
using solvatochromic comparisons of uv–visible spectral data in a large
number of pure solvents including water.

    Katritzky[316] has utilized multiparameter equations containing linear
combinations of existing empirical solvent parameters for the correlation

TABLE II. Z Values of Some Aqueous Binaries

Solvent	Solvent composition in percentage by volume of the organic component	Z value (kcal mol^{-1})
H_2O	—	94.6
MeOH	—	83.6
MeOH–H_2O	97.5	84.1
MeOH–H_2O	95	84.5
MeOH–H_2O	92.5	84.9
MeOH–H_2O	90	85.5
MeOH–H_2O	87.5	85.8
EtOH	—	79.6
EtOH–H_2O	98	80.2
EtOH–H_2O	96	80.8
EtOH–H_2O	95	81.2
EtOH–H_2O	92	82.0
EtOH–H_2O	90	82.5
EtOH–H_2O	85	83.8
EtOH–H_2O	80	84.8
EtOH–H_2O	75	85.7
EtOH–H_2O	70	88.6
t-BuOH	—	71.3
t-BuOH–H_2O	95	76.5
t-BuOH–H_2O	90	80.4
t-BuOH–H_2O	80	83.3
Acetone	—	65.7
Acetone–H_2O	99	68.1
Acetone–H_2O	95	72.9
Acetone–H_2O	93	74.8
Acetone–H_2O	90	76.6
Acetone–H_2O	85	78.7
Acetone–H_2O	80	80.7
Acetone–H_2O	75	82.1
Acetone–H_2O	70	83.2
Acetone–H_2O	65	84.3
Acetone–H_2O	60	85.5

of a variety of chemical phenomena. In the most successful treatment, E_T values were combined with dielectric and refractive index functions.

Finally, we note that the solvent dependence of the epr nitrogen hyperfine splitting constants (A_N) of persistent nitroxide free radicals like 7–9 can also be employed as an empirical measure of solvent polarity.[501] These A_N values are also cybotactic probes but do not involve a model reaction. They can be measured in low concentration and are in several cases more easily determined than Z or E_T values. Interestingly, the A_N values show a linear correlation with Z, E_T, and Ω, which suggests that it is more important that a solvent polarity parameter be a cybotactic probe

than it is that it involve a model reaction.[501] By contrast, A_N values measured for t-butyl nitroxide in several mixed aqueous solvents do not correlate with Z, but instead show a linear correlation with ε, provided that ε is not too low.[834] However, the sensitivity of both A_N and A_H to ε is dependent on the nature of the organic cosolvent, presumably because of preferential solvation of the nitroxide by the nonaqueous solvent component.

2.2.3. Dipole Moment and Polarizability

Polar solvent molecules may be characterized by their permanent dipole moment μ, which measures the internal separation of charge within a molecule of a particular geometry and which will determine the proclivity of the dipolar species towards interaction with another dipolar solvent or solute molecule. The polarizability determines the response of the molecule to external electric fields in terms of polarization of its electron distribution. This response may be expressed as a polarizability tensor α. The mean molar polarizability $\bar{\alpha}$ can be determined from the refractive index n by the equation[626]

$$\bar{\alpha} = \frac{n^2 - 1}{n^2 + 2} \frac{M}{d} \frac{3}{4\pi N_0} \tag{2}$$

In eqn. (2) M is the molecular weight, d the density, and N_0 Avogadro's number. At high frequencies (visible light), the molar refraction R_m [eqn. (3)] is also a measure for the polarizability, which includes only the contribution from induced dipoles:

$$R_m = \frac{\varepsilon - 1}{\varepsilon + 2} \frac{M}{d} \tag{3}$$

If the orientation polarizability due to the permanent dipole moment is added, the total molar polarizability P_M is obtained[630]:

$$P_M = \frac{\varepsilon - 1}{\varepsilon + 2} \frac{M}{d} = \frac{N_0}{3\varepsilon_0} \left(\alpha + \frac{\mu^2}{3kT} \right) \qquad (4)$$

The high ε of water may now be understood on the basis of mutual polarization between neighboring water molecules in an external field and the high propensity to align the permanent dipole moments to form a space-filling network of hydrogen bonds. In Table III values of μ, ε, and $\bar{\alpha}$ are given for water and a number of organic solvent molecules frequently used as cosolvent in aqueous binaries.[574] The dramatic decrease of ε upon introduction of the second N-methyl substituent in formamide also demonstrates the pronounced effect of intermolecular hydrogen bond interactions in determining the value of ε.

In water and aqueous mixtures, both dipolar and dispersion interactions are dependent on the quantities μ and $\bar{\alpha}$ of the interacting species. It has been known for a long time that the dipole moment of an isolated

TABLE III. Dipole Moments (μ), Dielectric Constants (ε), and Polarizabilities ($\bar{\alpha}$) of Water and Some Organic Solvents (25°C)[574]

Solvent	μ (Debye)	ε	$\bar{\alpha} \times 10^{24}$ (cm^3)
H$_2$O	1.855	78.5	1.444
MeOH	2.87	32.70	3.26
EtOH	1.66	24.55	5.13
n-C$_6$H$_{13}$OH	1.55	13.3	12.46
t-BuOH	1.66	12.47	8.82
Dioxane	0.45	2.21	8.60
Acetone	2.69	20.70	6.41
HCOOH	1.82	58.5	3.39
MeCN	3.44	37.5[b]	4.41
DMSO	3.9	46.7	7.99
HCONH$_2$	3.37	109.5	4.22
HCONHMe	3.86	184.5	
HCONMe$_2$	3.86	36.71	7.90
CH$_3$NO$_2$	3.56	35.87[c]	4.85
HMPA[a]	5.54	30[b]	18.90

[a] HMPA = hexamethylphosphoramide.
[b] At 20°C.
[c] At 30°C.

water molecule is much smaller than that of a water molecule in the liquid.[674] Now the general problem is that upon formation of a hydrogen bond X–H $\cdots$ S the polarity of the bond X–H will be enlarged and this will lead to an increase of the dipole moment. For bulk water, molecular dynamics simulation[832] studies (Berendsen, personal communication) have indicated that it is not possible to account for the electrostatic interactions between water molecules just in terms of alignment of hydrogen bonds of "monomeric" water molecules but that an increase in dipole moment must be included to obtain a fit with the experimental μ.* Mutual polarization of the hydrogen bond donor and acceptor molecule will also be a general phenomenon in mixed aqueous solutions employed in kinetic studies and will markedly affect solvation forces in these media. A quantitative treatment must await further theoretical studies of chemical reactions in water and aqueous mixtures.

2.2.4. Viscosity, Internal Pressure, Cohesive Energy Density

In solutions showing small deviations from ideality, the macroscopic solvent parameter viscosity (η) has been utilized in quantitative theories to explain medium effects on reaction rates. As an example, the rate constant for diffusional separation of caged radical pairs to form free radicals in solution has been correlated with some power of the fluidity ($1/\eta$) of the solvent medium.[31,256] For water and aqueous mixtures containing three-dimensional hydrogen-bonded networks these correlations break down.[170,658] Viscosity studies of aqueous solutions of alcohols, ureas, and amides have revealed the deviant behavior of these solvents from the Einstein and Simha theories, and cosolvent-induced changes in water structure have been invoked to account for these deviations.[402] Cases where diffusion does not bear a simple relationship to viscosity are discussed in Section 5.5.

In several studies the term "internal volume factor" has been employed in order to account for the effect of internal pressure or solvent cohesion on reaction rates and product composition.[668] The internal pressure of a solvent (P_i) is a measure of the susceptibility of the internal energy to a very small isothermal volume change and is represented by[220]

$$P_i = \left(\frac{\partial E}{\partial V}\right)_T = T\left(\frac{\partial P}{\partial T}\right)_V - P \approx \frac{T\alpha}{\beta} - P \qquad (5)$$

---

* Computer simulation techniques, as applied to water, are discussed in Chapter 6 of this volume.

in which α is the coefficient of thermal expansion and β the isothermal compressibility. As argued by Dack,[220,221] a very small isothermal expansion does not disrupt all intermolecular interactions associated with one mole of the solvent, but rather reflects contributions from dispersion, repulsion, and weak dipole–dipole interactions, which vary most rapidly near the equilibrium separation of the solvent.

The cohesive energy density (c.e.d. or D_{ce}) of a solvent measures the total molecular cohesion per volume of solvent and is given by[220]

$$D_{ce} = \frac{\Delta E_{vap}}{V_m} = \frac{\Delta H_{vap} - RT}{V_m} \qquad (6)$$

in which ΔE_{vap} and ΔH_{vap} are the energy and latent heat of vaporization of the solvent, respectively, and V_m the molar volume. Since *all* intermolecular interactions between solvent molecules are destroyed upon vaporization, D_{ce} reflects intermolecular hydrogen-bonding interactions in addition to the interactions that are measured by P_i. Consequently, P_i and D_{ce} values will be approximately equal for solvent molecules having $\mu < 2D$.

Table IV contains D_{ce} and P_i values for a number of solvents. It has been proposed[220,221] that $(D_{ce} - P_i)$ will represent the total hydrogen-bonding energy (E^{HB}) within a solvent and which may overwhelm the energy due to all other interactions within the solvent (E^0, as measured by D_{ce} and increasing with increasing molecular polarizability of the solvent) only for protic solvents of considerable hydrogen-bonding ability (water, alcohols, etc.). Presumably, P_i and E^0 for highly polar solvents like DMF and DMSO contain appreciable contributions due to dispersion and dipole–dipole interactions. E^{HB} is seen to increase with increasing solvent dipole moment and apparently also reflects some part of *strong* dipole–dipole interactions in dipolar aprotic solvents as well as hydrogen bonding interactions in protic solvents. Most remarkable is the very low E^0 for water, indicating that more than 90% of the total cohesion of water at 25°C is due to intermolecular hydrogen bonding. In contrast to most other solvents, E^0 for water increases with increasing temperature, which may be explained by invoking disruption of the water structure, which brings water molecules in closer contact for attractive interactions other than hydrogen bonding. Internal pressures have been measured for aqueous solvents and electrolyte solutions[220,221,584,607] and the results have been explained largely in terms of changes in water structure.

Solvent effects on reaction rates have been correlated with solvent cohesion.[221] Media of high cohesion are expected to enhance rates of

TABLE IV. Internal Pressures, Cohesive Energy Densities, and E^0 and E^{HB} Values for Water and Some Organic Solvents[a]

Solvent	D_{ce} (cal cm^{-3})	P_i (cal cm^{-3})	E^0 (kcal mol^{-1})	E^{HB} (kcal mol^{-1})
Water	550.2	41.0	0.67	9.2
CH$_3$OH	208.8	70.9	3.0	5.6
C$_2$H$_5$OH	161.3	70.0	4.1	5.7
t-BuOH	110.3	81[b]	10.5	2.8
CH$_2$OHCH$_2$OH	213.2	128	6.7	5.2
THF	86.9			
Dioxane	94.7	119.3[b]	10.1	0
Acetone	94.3	79.5	6.0	1.0
CH$_3$CN	139.2	96	5.1	2.3
DMF	139.2	114	8.7	2.1
Formamide	376.4	131	5.2	9.8
DMSO	168.6	123.7	8.6	3.3
Propylene carbonate	182	129	11.0	~4.5
CCl$_4$	73.6	80.6	8.0	0
C$_6$H$_6$	83.9	88.4	8.0	0
Cyclohexane	66.9	77.8[b]		0
n-C$_6$H$_{14}$	52.4	57.1[b]		0

[a] Data from Ref. 221.
[b] Calculated from $T\alpha/\beta$.

reactions that lead to products of higher cohesion than the reactants and to diminish rates of reactions that yield products of lower cohesion than the reactants.[182,750] Although the theory is well established for electrolyte effects on reaction rates, the quantitative usefulness of this approach to solvent effects has been questioned.[221] If the cohesive energy density of a solvent system is viewed as a measure of the "freedom" of solvent molecules to be involved in solvation of a substrate and transition state of a reaction, then a relation between D_{ce} and the entropy of activation might be anticipated. Such a correlation has recently been found.[618]

2.2.5. Solubility Parameter

The solubility parameter δ (cal$^{1/2}$ ml$^{-1/2}$) of regular solution theory is somewhat different in nature from the solvent parameters discussed so far. It relates the activity coefficient (f_i) of a nonpolar nonelectrolyte to its molar volume V_i by eqn. (7), in which δ_i and δ are the solubility parameters of the solute and solvent, respectively[404]:

$$RT \ln f_i = V_i(\delta_i - \delta)^2 \tag{7}$$

The solubility parameter is defined by

$$\delta = \left(\frac{\Delta H_{vap} - RT}{V} \right)^{1/2} \tag{8}$$

where ΔH_{vap} is the latent heat of vaporization. Equations (7) and (8) can be combined with eqn. (9), which provides the relationship between the rate constant k in a particular solvent and the rate constant k_0 for standard conditions ($f_i = 1$) for a reaction between two substrates A and B:

$$k/k_0 = f_A f_B / f_{\ddagger} \tag{9}$$

Thus,

$$RT \ln k/k_0 = V_A(\delta_A - \delta)^2 + V_B(\delta_B - \delta)^2 - V_{\ddagger}(\delta_{\ddagger} - \delta)^2 \tag{10}$$

The utilization of eqn. (10) has been discussed by Abraham,[1] and he showed that either a plot of $\ln k/k_0$ vs. δ^2 or a plot of $\ln k/k_0$ vs. δ may yield a straight line depending on the polarity of the transition state relative to that of the reactants. The intercept and slope of these plots give valuable information about transition-state properties. In the past, solvent effects on the solvolysis of t-BuCl and t-BuBr in protic and in aprotic solvents have been analyzed successfully using solubility parameters.[1] However, the solubility parameter approach to mixed aqueous solvent effects is hampered by both theoretical and practical difficulties. As noted by Abraham (personal communication), for many solvents the internal pressures fall into roughly the same sequences as solubility parameters.[220] But this does not hold for hydrogen-bonding solvents such as water and alcohols: these solvents possess large solubility parameters (as measured from heats of vaporization) and very low internal pressures. Therefore, the applicability of solubility parameters to polar hydrogen-bonding solvents is uncertain at the moment.

2.2.6. Nucleophilicity and Ionizing Power

Solvent effects on S_N1 solvolyses (also referred to as k_c solvolyses) have frequently been correlated by a linear free-energy relationship originally proposed by Grunwald and Winstein.[369,377] If k and k_0 are the rate constants for solvolysis of a given substrate in a particular solvent and in 80%

(v/v) EtOH–H$_2$O, respectively, the equation (11) may be defined, in which the parameter Y is called the "ionizing power" of the medium:

$$\log(k/k_0) = mY \tag{11}$$

The slope m of the linear plot of $\log(k/k_0)$ vs. Y represents the sensitivity of $\log k$ to Y. The Y values[292] are determined from measurements of k for solvolysis of t-BuCl ($m = 1$, by definition) in a solvent relative to the standard solvent 80% (v/v) EtOH–H$_2$O. The m value is indicative of the degree of charge separation in the transition state and represents a useful mechanistic criterion.[2,387] However, the limitation of eqn. (11) is illustrated by the fact that quite often slightly different slopes m are obtained if for a given substrate $\log k$ is plotted vs. Y for different solvent systems.

In order to account for nucleophilic solvent assistance in solvolysis processes of primary and secondary substrates, eqn. (11) has been extended[915,916] by an additional term to give eqn. (12), in which N denotes the nucleophilicity of the solvent and l the sensitivity of $\log k$ to N:

$$\log(k/k_0) = m^*Y + lN \tag{12}$$

This equation has recently been evaluated thoroughly by Schleyer and his associates.[787,788] They defined N values for the solvolysis of methyl tosylate which were employed in connection with Y values based on solvolysis of 2-adamantyl tosylate in order to retain the same leaving group and eliminate complications from ion pairing effects:

$$N = \log(k/k_0)_{\text{CH}_3\text{OTs}} - 0.3Y_{2-\text{AdOTs}} \tag{13}$$

The two sets of (Y, N) values are compared for aqueous EtOH in Table V. 2-Adamantyl tosylate was selected as a model substrate because its solvolysis shows all characteristics of a limiting S_N1-type process. Moreover, it is a suitable compound with which to compare the solvolysis of primary and other secondary substrates. Schleyer et $al.$[787,788] found that for a wide range of primary and secondary substrates the sensitivity to solvent ionizing power (reflected in m^*) decreases proportionately as the sensitivity to solvent nucleophilicity increases. Varying amounts of nucleophilic solvent assistance could be detected, depending largely on steric factors. Employing eqn. (12), previous theories about solvolysis processes in the S_N1–S_N2 spectrum could be critically evaluated and some solvolyses involving transition states in between those of nucleophilic assisted (S_N2) and unassisted (S_N1) processes were proposed to occur via nucleophilically solvated ion pairs ["S_N2 (intermediate) mechanism"].

TABLE V. Comparison of the Parameters (Y, N) with $(Y_{2\text{-AdOTs}}, N_{\text{CH}_3\text{OTs}})$ for EtOH–H$_2$O^a

Solvent	Y^b	N^b	$Y_{2\text{-AdOTs}}{}^c$	$N_{\text{CH}_3\text{OTs}}{}^c$
EtOH	−2.03	0.09	−1.75	0
90% (v/v) EtOH–H$_2$O	−0.75	0.05	−0.58	0.01
80% (v/v) EtOH–H$_2$O	0	0	0	0
70% (v/v) EtOH–H$_2$O	0.60	−0.09	0.47	−0.05
60% (v/v) EtOH–H$_2$O	1.12	−0.14	0.92	−0.08
50% (v/v) EtOH–H$_2$O	1.66	−0.20	1.29	−0.09

^{a} Data from Ref. 787, 788.
^{b} t-BuCl scale.
^{c} 2-AdOTs scale.

Although eqn. (12) is always superior to eqn. (11) unless the solvolysis is of the limiting S_N1 type, several authors have noted that the simple eqn. (11) is sometimes surprisingly effective in correlating solvolysis rates of even primary and secondary substrates. This situation has now been analyzed in some detail by Kaspi and Rappoport,[471] who could show that in several cases the solvents compared differ in polarity and nucleophilicity to an equal extent.

Several other linear free-energy relationships[840–842] have been proposed for the correlation of reaction rates with solvent nucleophilicity, electrophilicity, and polarity, but these equations seem to provide a mechanistically less useful approach than the extended Grunwald–Winstein relation given in eqn. (12). We like to emphasize, however, that the Grunwald–Winstein equation provides "kinetic" solvent parameters Y and, despite its mechanistic significance, contributes but little to our understanding of solvation behavior in aqueous media.

2.2.7. Water Structure

Although the problem of hydrogen-bonding interactions in aqueous solutions will be discussed in Section 3.1, we note here that the often eccentric behavior of the kinetic parameters of organic reactions in aqueous binaries has frequently been rationalized in a qualitative way by invoking solvation effects originating from changes in water structure. However, this is a controversial topic (Section 5). As argued in Section 3.1, the precise characterization of the three-dimensional hydrogen-bond patterns in water and aqueous binaries awaits further efforts along theoretical and experimental lines. The unique three-dimensional hydrogen-bond structure of liquid water originates from the fact that the water molecule possesses two hydrogen-bond donor and two hydrogen-bond acceptor sites. This gives rise to a unique type of solvent association, which may be compared with that of simple, monofunctional alcohols. Alcohols do not form three-dimensional hydrogen-bond networks but rather associate via short-lived polymeric chains of hydrogen-bonded molecules.[366]

Many theories on water structure have evolved,[319,904] and still in 1968 water was described as an "exceedingly complicated pudding."[420] However, during the last decade progress in our "understanding" of water has been fast and recent computer simulations of water have revealed that the hydrogen-bond network topology of liquid water may be best viewed as a "defective, strained, random network."[676,828]* For our purposes, a more simplified concept of water structure is desirable. This has been provided by Ben-Naim,[83] who introduced a simple definition for the stiffness of the fluctuating network of distorted hydrogen bonds in bulk water. He defined the "degree of water structure" as the average number of hydrogen bonds per water molecule as

$$\frac{1}{2} \sum_k \frac{kN_k}{N} \tag{14}$$

in which N_k is the number of water molecules participating in k hydrogen bonds, and N is the total number of water molecules. In an equally simplified fashion, water structure can be defined in terms of a two-state equilibrium between "dense" and "bulky" water[318]:

$$(H_2O)_d \rightleftarrows (H_2O)_b$$

Dense water is less structured and has a higher density, bulky water is highly structured and has a low density. The lifetime of the two states (10^{-10}–10^{-11} sec) is much shorter than the half-lives of the reactions to be discussed in Section 5. The organic cosolvent as well as any other species present in aqueous solution will affect the water structure unless it would fit perfectly into the hydrogen-bond network characteristic of pure water.[84] A CNDO study[414] has offered evidence that even air in water might have a stabilizing effect on the water structure. It is clear, however, that the hydrogen bond régime in the cybotactic region of the solute will differ

---

* Computer simulation of water and aqueous solutions is reviewed in detail in Chapter 6 of this volume.

from bulk solvent and will specifically reflect the influence of the solute. In kinetic studies, solvation effects of this type can only be investigated by dissecting solvent effects on initial and transition states, but the complexities of intermolecular interactions in aqueous media usually prevent quantitative formulation. We wish to emphasize that explanations of irregularities of kinetic parameters at high n_{H_2O} values in terms of changes in water structure very often suffer from vagueness and ambiguity. Anyone willing to invoke water structure effects as an explanation for solvent effects on kinetic parameters of reactions in aqueous binaries is recommended to consult the recent fundamental paper by Roseman and Jencks.[766]

3. SOLUTE–SOLVENT INTERACTIONS IN MIXED AQUEOUS SOLUTIONS

There is surprisingly little quantitative information about the forces that determine solute–solvent and solute–solute interactions in aqueous solutions. That these forces are strong and dominantly influence reaction rates and reactivity patterns in water and aqueous mixtures is illustrated by the rapidly growing information about rates and substituent effects of comparable chemical reactions in the absence of solvation forces, i.e., in the gas phase (compare Section 1). It goes without saying that for a proper understanding of solvent effects on reaction rates, a discussion of the experimental data in terms of interaction mechanisms is highly desirable. Since intermolecular interactions and general concepts of solvation of organic solutes in aqueous media have been reviewed authoritatively,[445] we will restrict ourselves to affording only a list of the most important forces in aqueous environments and a few comments which may be appropriate in view of the discussion of solvent effects in Section 5.* Solvation effects for polar and ionic solutes will be mainly discussed in terms of the cosphere models introduced by Frank,[317] Gurney,[371] and Friedman.[338] Cospheres may be defined as layers of water molecules around the solute and which are perturbed by the presence of the solute particle. They represent arbitrarily defined sectors of the cybotactic region of the solute. For ionic species Gurney[371] describes three cospheres, their nature and extent being characteristic of the specific ion, and the temperature and pressure of the solution. In the innermost cosphere A, the water molecules are

---

* Current views on intermolecular forces with special emphasis on aqueous solutions are found in Chapters 3 and 5 of this volume.

strongly oriented and largely immobilized owing to the electric field of the ion. The second cosphere B, adjacent to A, contains water molecules that are more randomly organized than in bulk water owing to opposing orienting influences of the ion and of bulk solvent. In bulk solvent, sometimes referred to as cosphere C, the effect of the electric field of the ion is too weak to disturb the structural network to a significant degree. According to Friedman and Krishnan[338] hydration in cosphere A arises from ionic or polar field effects due to the presence of the solute. These inner cospheres may be further classified.[338] Hydration in cosphere B involves water perturbed by the proximity of a solute particle but which is not affected by directional solute–solvent forces. These states of water can also be further classified.[338] At finite concentration of organic solutes in water, or when an organic cosolvent is added to the aqueous solution of an organic solute, the organic species in solution will interact with each other. When hydration cospheres overlap, the water molecules in the overlapping region undergo specific rearrangements. The degree of rearrangement will then depend on the degree of dissimilarity of the order–disorder characteristics of the overlapping cospheres of the solutes.[239,243a,922] In addition, in mixed aqueous solvents specific solute–water and solute–cosolvent interactions may lead to preferential solvation by one component of the solvent mixture (see Section 2.1).

3.1. Hydrogen Bonding

Solvent effects in (mixed) aqueous solutions are dominated by hydrogen-bonding interactions. The hydrogen bond structure of water (mean hydrogen bond energy per OH group $\sim 4\,\text{kcal mol}^{-1}$)[577] provides the matrix for hydrogen-bond interactions between water and hydrophilic sites in organic solutes.[506] For instance, immersing a carboxylic group in water is about 3 kcal mole^{-1} more exothermic in the enthalpic sense than immersing a methyl group.[145] In addition, water–water interactions provide the driving force for hydrophobic interactions between hydrophobic sites of organic solutes of which one may belong to the organic component of the aqueous binary (Section 3.2). However, in aqueous media the hydrogen-bonding phenomenon is connected with a continuous variation in interaction energies determined by the hydrogen bond lengths and angles. Now a definition of a hydrogen bond can be given either in terms of the configuration of the pair of particles which interact with each other[82] or in terms of the strength of the interaction between the pair of particles.[87] Following the second approach, Stillinger,[828] in his molecular dynamics studies of water employing a model using an effective pair potential v, has taken a negative cutoff energy E_{HB} of -1.7 kcal mol^{-1}. Thus, in deciding whether or not a pair (i, j) of water molecules is hydrogen bonded, one can take

$$v_{(x_i, x_j)} \leq E_{HB}, \qquad i \text{ and } j \text{ hydrogen bonded}$$
$$> E_{HB}, \qquad i \text{ and } j \text{ not hydrogen bonded}$$

Rahman and Stillinger[733] found that the conventional hydrogen-bond pattern of ice can be successfully reproduced if v_{HB} is taken between -1.7 and -4.5 kcal mol^{-1}.

The number of hydrogen bonds in which a water molecule or a solute simultaneously participates can now be defined using the selected E_{HB}. This approach also allows for a definition of water structure, although the parameter concerned will be difficult to quantify. Solute–water hydrogen bonding[506] may be similarly defined but an element of arbitrariness will be involved in the selected magnitude of E_{HB}. This also applies to any other type of interaction in solution and precludes an exact separation of all contributing interaction mechanisms in solvation processes.

Another problem of considerable interest and controversy involves the question of the existence of hydrogen-bonding interactions between two nonionic solute molecules or between a solute and the nonaqueous component (e.g., an alcohol or urea) of a water-rich mixed aqueous solvent system. In a much-cited paper, Klotz and Franzen[499] have proposed that intermolecular hydrogen bonding between N-methylacetamide molecules in aqueous media is weaker than the N-methylacetamide–water hydrogen-bond interaction. Ever since this work, it has often been suggested that water is the dominant hydrogen-bond donor in aqueous solutions, leaving little or no chance for other hydrogen-bond donors to compete with water for available acceptor sites at solute molecules. Considerable interest has been shown in this subject[243,330,348,766] since direct protein–urea hydrogen-bond interaction could provide a substantial driving force in protein denaturation in aqueous urea solutions. The extensive literature has been critically reviewed recently,[330] and the conclusion is that there exists hardly any reliable evidence for direct binding of urea or other highly polar organic cosolvents to proteins and a concomitant effect on protein stability. Of course many model systems of proteins have been examined by a variety of techniques. Interestingly, Jencks[761,766] has proposed the formation of bifunctional hydrogen bonds between urea or guanidine hydrochloride and acetyltetraglycine ethyl ester. Favorable polar interaction effects between urea and glycylglycine have also been claimed on the basis of nmr measurements[845] and activity coefficient studies.[875] Recently, Bonner et al.[131] determined the heat capacities of ureas and water in water and N,N-dimethylformamide. From these results the authors conclude that the relative hydrogen bonding strengths of protons to the peptide carbonyl oxygens are in the order $NH_2 > OH_2 > NH$. This would indicate that specific binding of urea to peptide moieties might contribute to protein destabilization in aqueous urea solutions. As argued by Savage and Wood,[782] the low heats of dimerization found by Klotz and Franzen[499] are the result of efficient cancellation of the negative enthalpy of amide–amide interaction via CONH $\cdots$ CONH hydrogen bonding by a positive enthalpy due to $CH_3 \cdots CH_3$ hydrophobic interaction. These authors have performed studies of enthalpies of dilution of a wide variety of molecules containing alkyl, amide, and hydroxyl groups, which reveal that in the enthalpic sense, CONH $\cdots$ CONH interactions may well compete with $CH_2 \cdots CH_2$ interactions but that $-CH(OH) \cdots CH(OH)$ association is definitely weaker. It is clear, however, that the problem of the strength and nature of polar solute–solute and solute–cosolvent interactions in aqueous binaries is by no means settled at the moment.

Finally, we would like to comment briefly on the possibility that kinetic solvent effects find their origin in cosolvent-induced changes in the water structure. At the moment, the quantitative importance of these effects is hard to assess, although theoretical models may predict substrate–cosolvent structural interactions. Usually, water structure effects on reactions in aqueous media will be revealed by often large and mutually compensating changes in $\Delta H^{\ddagger}$ and $\Delta S^{\ddagger}$ upon variation of n_{H_2O} (see Section 5) and which will have little or no effect on $\Delta G^{\ddagger}$. However, the properties of initial state and transition state usually differ substantially with respect to their size, polarity, and polarizability, and their degree of hydration will not only depend on their polarity and hydrogen-bonding properties but also on their steric compatibility with bulk solvent structure. Juillard et al.[45,257,710] have systematically studied transfer enthalpies ($\Delta H_{tr}^{\ominus}$, see Section 3.5) from water to the TA mixed solvent t-BuOH–H_2O for polar and ionic solutes. They find that $\Delta H_{tr}^{\ominus}$ is governed both by *specific solvation effects* (for instance, charge–dipole interactions), which depend on n_{H_2O} and reflect the solvation of polar and ionic functional groups, and by *hydrophobic hydration effects*, largely depending on the shape and size of the hydrocarbon part of the solute. Quite generally $\Delta H_{tr}^{\ominus}$ passes through endothermic extrema in the region $n_{H_2O} = 0.95$–0.90, that is within the solvent composition range of enhanced structuredness. The hydrophobic hydration effect shifts the maximum in $\Delta H_{tr}^{\ominus}$ in the opposite direction as the specific solvation effect, i.e., to higher values of n_{H_2O} and also increases the magnitude of $\Delta H_{tr}^{\ominus}$. When a solute is dissolved in a mixed aqueous solvent, a cavity must be created to accommodate the molecule. The thermodynamics of this cavity-formation process (see also Section 3.5) will depend on the degree of solvent structure and will be subject to enthalpy–entropy compensation. These compensatory effects are caused by the fact that formation of a cavity in water containing a structure-making cosolvent (e.g., in t-BuOH–H_2O, $n_{H_2O} = 0.95$) will be associated with an unfavorable enthalpy term since more and/or stronger hydrogen bonds must be broken but, on the other hand, entropy will be gained since strongly hydrogen-bonded water molecules will be released. The latter effect is consistent with the occurrence of hydrophobic contacts between the solute and the cosolvent. Usually, the entropy term is dominating leading to solubilization of nonpolar solutes by the addition of the organic cosolvent to form a TA solution (Section 2.1). But analysis of thermodynamic quantities purely in terms of solvent structure is hazardous, since these parameters are composites of contributions of several interaction mechanisms. One way partly to avoid these complexities is to study the thermodynamic parameters of transfer of a solute from H_2O to D_2O taking into account the larger structural order in D_2O as compared with H_2O at a given temperature.[39,222,454,882]

Finally, one has to consider the possibility that solvent structure affects the kinetic basicity and acidity of water. Cooperativity in H $\cdots$ O hydrogen bonding has long been recognized as one of the essential features of intermolecular interactions between hydroxylic species[755] and especially in water. Calculations on water dimers by Stillinger[60,379] $et\ al.$ and by Kollman and Allen[506–508] indicate an altered charge distribution upon association via hydrogen bonding. They showed an increase in net negative charge on the "free" oxygen atom and an increase in net positive charge on the unbonded hydrogen atoms, both relative to one isolated water molecule.

Nmr chemical shifts measurements[357] on $CHCl_3$ in t-BuOH–H_2O seem to support the greater dynamic basicity of more ordered water. Furthermore, we submit that organic cosolvents which are good hydrogen bond acceptors (DMSO, dioxane, etc.) will form hydrogen bonds with water and thereby induce an enhanced electron density on the water–oxygen atom as compared with water–water complexes. This cosolvent–water association will then lead to an enhanced kinetic basicity of water (see Section 5.2). Khoo and

Chee-Yan[482] in their discussion of the transport numbers of hydrochloric acid in DMSO–H_2O, also assume increased solvation of a proton by increased ion–dipole interactions upon accommodation of DMSO in the secondary and subsequent solvation shells of the proton. Previously, electrolyte effects on the kinetic basicity and acidity of water have in a similar way been rationalized in terms of polarization of water molecules in the solvation shells of the ions.[386,670,671,833]

3.2. Hydrophobic Interactions*

In aqueous solution, nonpolar, hydrophobic molecules or hydrophobic sites in molecules have been found to associate with each other. These so-called hydrophobic interactions in water are supposed to be appreciably stronger than solvophobic interactions in nonaqueous solvents like alcohols, 1,4-dioxane, and cyclohexane.[925] The mechanism of the hydrophobic interaction is the subject of considerable debate[327,497] and even the concept of the hydrophobic effect has been severely attacked.[211] A considerable stimulus for active investigation of hydrophobic interactions is found in the fact that "hydrophobic forces" may play an important role in the stabilization of protein structures in aqueous solution[330,526] and in enzyme–substrate interactions. Furthermore they provide the major driving force for the formation of micelles in aqueous solutions of surfactant molecules.[298] The subject has been reviewed several times,[213,327,553,855] and is treated in detail in Chapter 5 of this volume. It is therefore only necessary to summarize the salient features of hydrophobic effects. Thus it is known that the free energy of mixing of water and "hydrophobic" molecules is dominated by a large negative entropic contribution, believed to originate from a redistribution of water–water hydrogen bonds in the vicinity of the solute molecule.

The hydrophobic interaction between nonpolar residues in aqueous solution is considered to be a partial reversal of this effect and manifests itself as an apparent attractive contribution to the forces between such molecules.

Such hydrophobic interactions may occur between two nonpolar reactants in an aqueous medium or between a nonpolar reactant and the organic cosolvent in an aqueous binary in the water-rich region. Although these types of interaction do not imply that the hydrophobic molecules come into

---

* The concept of solvent structure contributions to intermolecular forces is dealt with in Chapter 5 of this volume.

contact,[332] it should be emphasized, however, that direct "contact interactions" may also occur and may contain contributions from attractive dispersion forces if the interacting components are larger and more polarizable than water.[327]

Hydrophobic interactions between neutral organic solutes and between a neutral solute and structure-promoting ions like tetraalkylammonium ions have been mainly studied by thermochemical methods and the thermodynamics seem to be dependent on the solvent structural integrity.[83,90, 91,146,243,244] Several other potentially useful methods for the study of hydrophobic interactions have been developed including the measurements of nmr relaxation rates $(1/T_1$ or $1/T_2)$ in the presence of paramagnetic solutes.[452,453] Using the latter method, Jolicoeur et al.[452] could show that with neutral solutes like t-BuOH and t-BuNH$_2$ the proton relaxation induced by a persistent nitroxide free radical is much greater in water than in other solvents of comparable viscosities, presumably as a result of hydrophobic interactions. Cosolvents like urea and formamide were found to decrease these interactions, which is consistent with other thermodynamic evidence.[240]

The strength of hydrophobic interactions in H$_2$O as compared with those in D$_2$O is a matter of considerable debate,[82,89,90,665,667,924] and the matter is still more complicated in mixed aqueous solvents.[665,666,924] Based on conductance measurements and kinetic experiments, Oakenfull[665] reached the conclusion that the free energy of hydrophobic interaction between two methylene groups of a hydrocarbon chain is more favorable in D$_2$O than in H$_2$O by 86 cal mol^{-1}. Consistent with this conclusion, the critical micelle concentration of some ionic surfactants has been found to be lower in D$_2$O than in H$_2$O.[636] Isotope effects in hydrophobic binding have recently been investigated by high-pressure liquid chromatography, and definite CH(CD) isotope effects were found for aliphatic as well as aromatic hydrocarbons on a μ-Bondapak C$_{18}$ column.[853,854] The effect of hydrophobic interactions between two reactants on reaction rates has been studied quite extensively.[123,372,663,665,675,876] and interesting enzyme models based upon this concept have been developed.[373] Guthrie,[372] Jones and Gorden,[455] and Oakenfull[665] have shown that in studies of bimolecular reactions between two substrates containing hydrophobic alkyl chains it should be carefully checked that the substrates themselves do not aggregate,[123,663] a process which may occur at extremely low concentrations in aqueous solution. For reactions of p-nitrophenyl alkanoates in aqueous solutions,[665] the critical concentration above which aggregation takes place has been suggested to be the solubility limit of the ester.[372]

The effects of hydrophobic interactions on solvolysis reactions persist to some extent in highly aqueous mixed solvents but finally disappear completely if n_{H_2O} is further reduced.[665]

3.3. van der Waals Interactions

Intermolecular dipolar interactions in solution may involve forces between species containing either a permanent or an induced dipole moment.[626,630] For most neutral hydrocarbonlike organic compounds, these interactions are primarily due to van der Waals forces which include the sum of all interactions between uncharged molecules. For two uncharged molecules A and B with dipole moments μ_A and μ_B, the net attractive energy for the dipoles in rapid thermal motion is given by Keesom's formula, in which k is the Boltzmann constant and r the distance between the dipoles:

$$E_d = -\frac{2\mu_A^2\mu_B^2}{3kTr^6} \tag{15}$$

In aqueous solvent mixtures of not too low n_{H_2O}, it is expected that dipole–dipole interactions between water and a nonaqueous component will dominate over dipolar interactions between two nonaqueous components because of the high dipole moment and hydrogen-bond ability of water.

If a permanent dipole of molecule A induces a dipole in molecule B, the net attraction will be dependent on the polarizabilities α_A and α_B as has been shown by Debye:

$$E_i = -\frac{\alpha_B\mu_A^2 + \alpha_A\mu_B^2}{r^6} \tag{16}$$

Pure dispersion forces between two neutral apolar molecules A and B are governed by London forces. The attractive contribution is often simply represented by

$$E_L = -\frac{3I_AI_B}{2(I_A + I_B)}\frac{\alpha_A\alpha_B}{r^6} \tag{17}$$

in which I_A and I_B are the first ionization potentials of the molecules A and B. Although other expressions for E_L have been proposed and the dependence of E_L on r is partly determined by the distance of separation between the molecules, London forces are also short range in nature and are controlled by the polarizabilities of the interacting molecules. In fact the repulsive contribution to the dispersion forces is of very much shorter range still.

Consideration of the surface tension of saturated hydrocarbons in comparison with that of water leads to the conclusion that the contribution of dispersion energy per unit of volume of water is rather similar to that for hydrocarbons.[803] Thus, the enthalpy of mixing of hydrocarbons with water will predominantly reflect changes in hydrogen-bonding interaction of water molecules surrounding the hydrocarbon. However, if an organic cosolvent is added to water, the solvation change for a relatively nonpolar organic solute may contain a contribution of London forces, either as part of hydrophobic interaction with the cosolvent in the water-rich region or via London forces originating from (preferential) solvation of the solute by the nonaqueous solvent at lower n_{H_2O}.[778]

The above discussion of van der Waals forces only includes the very simple elements, to show that they are essentially short- (or very-short-) range effects and depend on physical properties of the bulk phases. Although some very refined treatments have been developed[587] for their evaluation, it is not yet immediately clear how such calculations can be applied to characterize short-range molecular interactions of the kind that result in chemical reactions.

Studies aimed at the quantitative assessment of the contribution of dispersion forces to solvent effects are few[313,365,367,368] and at the moment theoretical predictions are also difficult. There is little doubt, however, that these short-range forces may affect microenvironmental effects on reaction rates. In the formation of enzyme–substrate complexes, the contribution of London forces[899] may be appreciably enhanced as compared with, for instance, solute–solvent interactions by this mechanism.[680] This is caused by the fact that the interaction takes place with closely packed and covalently bound atoms of the polypeptide chains. In this connection, it has been noted that in protein molecules the fraction of space occupied by atoms may be as large as 0.76, which may be compared with upper limits of this value for cyclohexane (0.44) and water (0.36).[221]

3.4. Electrostatic Interactions

Although the solvent is of paramount importance in determining electrostatic interactions between charged solutes[486,489,606] and also between ionic sites within the same molecule,[181] discussion of this topic is less relevant for the purpose of the present review. The subject has been treated by Jencks,[445] Amis and Hinton,[31,509] Harned and Owen,[382] and Gordon,[358] who also cite many pertinent references.

3.5. Treatments of Solution Processes in Aqueous Media

The multitude of interaction mechanisms available for intermolecular interactions in (mixed) aqueous solvents as well as the absence of a viable quantitative theory for these solvent systems throw up a high barrier for a quantitative analysis of solvent effects on rate processes in these media. Any single solvent parameter (see Section 2.2) will be able to correlate only a limited set of kinetic data, and even if a significant correlation of this type is found, its physical significance will usually be far from obvious. The complexity of the problem at present also strongly hampers theoretical treatments of aqueous binary solutions although significant progress has recently been made.[82,321,506,809] In the analysis of dissolution processes much current attention is paid to the scaled-particle theory,[706,741] which assumes that fluid mixtures may be treated as hard sphere systems perturbed by intermolecular interactions. In other words, the formation of a solution is described as a packing phenomenon and thermodynamic quantities are expressed as functions of the measured density of the fluid. This approximation in terms of van der Waals fluids seems at first sight hardly appropriate for hydroxylic solvents and, in particular, water. Nevertheless, the scaled-particle approach, especially in terms of Stillinger's modification,[827] has been applied with remarkable success to dilute solutions of nonpolar molecules in water and has been advocated as a challenge to those theories that emphasize solvent structure perturbation around the hydrophobic solute as a significant factor in the dissolution process.

Klapper[497] has elaborated on the merits of the scaled-particle theory for aqueous solutions and has tried to reconcile some of the controversies by arguing that the high solute vaporization enthalpies and entropies do not reflect water–water reorientations around the solute but rather imply van der Waals interactions between lattice water and the solute. It has then been emphasized that the uniqueness of water as a solvent is found in the structure of the liquid and not in a unique dissolution process for apolar compounds. However, we note that despite the success of the scaled-particle theory in reproducing the thermodynamic properties of dilute aqueous solutions,[575,576,706] there is a wealth of evidence, other than in terms of thermodynamic data, that water structure perturbation accompanies the dissolution of nonpolar solutes. Even for an interstitial type of solution, the rotational diffusion rate of the solvent molecules has been experimentally shown to be seriously decreased although the freedom of rotation of the solute molecules is almost unrestricted.[326,929] Therefore it is highly significant to note that the molar structural entropy change, introduced by DeVoe,[245] which measures the increase in the degree of structural ordering in water upon dissolution of nonpolar solute molecules can in fact be calculated using scaled-particle theory. These values were shown to be substantially larger than the corresponding values for non-aqueous solvents.

A particular efficient and relatively unambiguous approach to rationalize intermolecular interactions in aqueous media has recently been employed by Jolicoeur[454] and by Roseman and Jencks.[766] In a beautifully detailed study, the latter authors have investigated the effects of urea and other cosolvents on the solubilities and activity coefficients of uric acid (**10**) and naphthalene (**11**) in aqueous solution. These molecules were

chosen because they are approximately of equal size but differ greatly in polarity and in their ability to participate in hydrogen-bonding interactions. Solute–solvent interactions were rationalized by considering two thermodynamic quantities. One term included the sum of Gibbs free-energy terms reflecting the formation of a cavity in the solvent to accommodate the solute (ΔG^{cav}) and the development of relatively nonpolar solute–solvent interactions (ΔG^{nonpol}) due to dispersion, dipole-induced dipole, and quadrupole forces. The other term represented the Gibbs free energy for polar solute–solvent interactions (ΔG^{pol}) between the solvent and a polar solute, which are primarily due to hydrogen bonding forces. The free energy for the transfer of a solute from water to an aqueous binary (ΔG_{tr}) may then be expressed as

$$\Delta G_{tr} = (\delta\Delta G^{cav} + \delta\Delta G^{nonpol}) + \delta\Delta G^{pol} \tag{18}$$

Of course, analogous expressions may be written for ΔH_{tr} and ΔS_{tr}. These are frequently obtained from the temperature dependence of the relative solubilities in the solvent systems.

In view of the high cohesive energy density (see Section 2.2.4) and surface tension of water, the contribution of ΔG^{cav} will be an important one and will depend, at a given temperature, on solvent structure (determining intermolecular solvent–solvent interactions and the number of voids in the solvent) and on the size and shape of the solute. In terms of the scaled-particle theory[321,708,741,870] the standard free energy of solution is usually written as

$$\Delta G^{\ominus} = \Delta G_{cav}^{\ominus} + \Delta G_{int}^{\ominus} + RT \log \frac{RT}{V} \qquad (19)$$

where $\Delta G_{int}^{\ominus}$ measures the solute–solvent interaction and V is the molar volume of the solvent. Calculations using this theory require the effective hard-sphere diameter of the solvent. Recently, an empirical relationship between this diameter and the surface tension of the liquid could be established.[653]

If an organic cosolvent is added to water, the solubility of *both* polar and nonpolar solutes is usually increased, indicating that the term $(\Delta G^{cav} + \Delta G^{nonpol})$ becomes more favorable. Unfortunately, the contributions of ΔG^{cav} and ΔG^{nonpol} are exceedingly difficult to dissect. Furthermore, an analysis of $\delta \Delta G_{tr}^{\ominus}$ in terms of $\delta \Delta H_{tr}^{\ominus}$ and $\delta \Delta S_{tr}^{\ominus}$ values is also complicated. Thus one can envisage that the $\Delta H_{tr}^{\ominus}$ term for a relatively apolar organic solute will be a composite of terms due to solute–solvent hydrogen bonding (if a hydrophilic site is present in the solute), attractive van der Waals interactions between solute and solvent, and reorientation of water molecules in the cybotactic region of the solute. Of special interest is Roseman and Jenck's observation[766] that the polar uric acid molecule in aqueous media can interact favorably with polar cosolvents like urea. As expected for a polar interaction, this effect decreases with increasing temperature. It was suggested that hydrogen bonding could not provide a driving force for the polar interaction effect, but in the light of more recent results (Section 3.1) a definite conclusion about the nature of the polar interaction forces cannot be drawn at the moment.

Finally we note that Guillet et al.[370] have recently determined the thermodynamic parameters for transfer of water and the [$H^{\oplus}OH^{\ominus}$] couple from water to aqueous mixtures of EtOH, t-BuOH, acetone, DMSO, and urea in concentrations from 0 to 40%. These measurements quantitatively illustrate the specific solvation properties of TA and TNA solvent systems and are highly useful in conjunction with transfer parameters of organic substrates for the elucidation of transition-state solvation behavior. Trends in the thermodynamic parameters for initial state and transition state as a function of solvent composition may then serve to illustrate the different solvation requirements of both species in terms of different intermolecular interaction mechanisms. This approach will be further outlined in Section 4.3.

4. MIXED AQUEOUS SOLVENT EFFECTS AND TRANSITION-STATE THEORY

4.1. Transition-State Theory

Although alternative theories have been developed,[312] transition-state theory[286,290,534] remains the most popular and convenient vehicle for the analysis of solvent effects on the kinetics of organic reactions that are not extremely fast. The applicability of the theory to reactions between solvated reactants in mixed aqueous solvents has been amply demonstrated.[613,683] Assuming a pseudothermodynamic equilibrium situation between reactants and transition state (or activated complex), the rate constant of a reaction (k) is determined by the difference in Gibbs free energy ($\Delta G^{\ddagger}$) between the transition state and the reactants:

$$k = k_B T/h \, \exp(-\Delta G^{\ddagger}/RT) = k_B T/h \, \exp(-\Delta H^{\ddagger}/RT) \exp(\Delta S^{\ddagger}/R)$$

where $\Delta G^{\ddagger}$, $\Delta H^{\ddagger}$, and $\Delta S^{\ddagger}$ are the Gibbs free energy, enthalpy, and entropy of activation, k_B the Boltzmann constant, h Planck's constant, and T the absolute temperature. In terms of activity coefficients (f_i), the rate constant for a bimolecular reaction between the substrates A and B may be expressed as

$$k = \varkappa_x K_x \frac{k_B T}{h} \frac{f_A f_B}{f^{\ddagger}}$$

in which $\varkappa_x$ is the transmission coefficient and K_x the equilibrium constant for the formation of the transition state from the reactants. Now the rate constant in a solution (k) can be related to that in an ideal solution ($k_0; f_A = f_B = f_{\ddagger} = 1$) by

$$k = k_0 \frac{f_A f_B}{f^{\ddagger}}$$

The solvent effect on a rate constant is then determined by the ratio of the activity coefficients, which in turn depends on the specific solvation behavior of the transition state and reactants. A thorough discussion of solvent effects based on "solvent activity coefficients" has been given by Parker.[682] However, in our discussion of kinetic solvent effects we will concentrate on the thermodynamic quantities of transfer (see Section 4.3) rather than on activity coefficient behavior.

4.2. Activation Parameters

Free-energy–reaction-coordinate diagrams are highly useful for the interpretation of kinetic parameters for reactions in aqueous media.[445] For a reaction carried out in two solvents S_1 and S_2, the difference in rate constants may be expressed as a difference in $\Delta G^{\ddagger}$:

$$\delta \Delta G^{\ddagger} = \Delta G^{\ddagger}_{S_2} - \Delta G^{\ddagger}_{S_1}$$

In the analysis it may reasonably be assumed that the transition state is in thermal equilibrium with the solvent.[1] Ever since the pioneering work of Winstein,[914] Hyne and Robertson,[434] and their co-workers, the isobaric thermodynamic activation parameters $\Delta H^{\ddagger}$ and $\Delta S^{\ddagger}$ have frequently been employed as sensitive probes for the study of mixed aqueous solvent effects. The solvent dependence of these thermodynamic quantities of activation is usually much more pronounced and complex than that of $\Delta G^{\ddagger}$, mainly because in aqueous binaries $\Delta H^{\ddagger}$ and $\Delta S^{\ddagger}$ vary in a compensating fashion with changing water concentration:

$$\Delta G^{\ddagger} = \Delta H^{\ddagger} - T \Delta S^{\ddagger}$$

The enthalpy of activation reflects the height of the energy barrier over which the reacting species must pass in the form of an activated complex. This barrier is dependent not only on the intrinsic energies of the bond-breaking and bond-making processes but also on the relative solvation of initial and transition state. In this context we note that there may be marked changes in activation parameters between aqueous "hydrophilic" solvent media and nonaqueous "hydrophobic" environments. For the latter type of solvent, even substantial negative $\Delta H^{\ddagger}$ values have occasionally been observed.[814]

The entropy of activation measures the orientational requirements inherent in the formation of the transition state and, among other factors, depends on whether the reaction is mono- or bimolecular and on the loss or gain of translational and rotational degrees of freedom of solvent molecules upon formation of the transition state.[789] For instance, if a highly charged transition state is formed from essentially neutral reactants, the freedom of motion of polar solvent molecules will be restricted during the activation process and this loss of solvent entropy will be reflected in a strongly negative contribution to $\Delta S^{\ddagger}$.

Especially in aqueous binaries, solvent effects on rate constants may be dominated by entropy effects (Section 5). The same is true for substituent effects, as demonstrated by the activation parameters for the neutral

TABLE VI. Activation Parameters for the Neutral Hydrolysis of $Cl_2CHCO_2C_6H_4X$-p in H_2O at $25°C^a$

X	$\Delta G^{\ddagger}$ (kcal mol^{-1})	$\Delta H^{\ddagger}$ (kcal mol^{-1})	$\Delta S^{\ddagger}$ (e.u.)
NO_2	19.03	8.6	-35
OCH_3	20.88	7.9	-44

^{a} The data for X = NO_2 are from Ref. 281; those for X = OCH_3 have recently been re-investigated by H. A. J. Holberman and J. B. F. N. Engberts (to be published).

hydrolysis of p-nitro- and p-methoxyphenyl dichloroacetates (Table VI).[281] The larger $\Delta G^{\ddagger}$ for hydrolysis of the p-methoxy derivative is primarily due to the more strongly negative $\Delta S^{\ddagger}$. It has been suggested that the partly charged p-methoxyphenolate group in the transition state for hydrolysis is more effective in orienting neighboring water molecules relative to the initial state than is the p-nitrophenolate group. This stronger orienting effectiveness results from a greater localization of the negative charge on the phenolic oxygen compared to the p-nitrophenolate leaving group. This argument is in agreement with the conventional ideas about the small tendency of the methoxy substituent to take part in resonance interaction with the phenolic function and is consistent with the observed more strongly negative entropy of ionization for p-methoxyphenol than for p-nitro-phenol.[622,762] In this connection, it should be mentioned that the difference in acidity by more than one pK_A unit between m- and p-nitro-phenol is dominantly determined by the standard entropies of ionization ($\Delta \bar{S}_i^{\ominus} = -21.8$ e.u. and -17.1 e.u. for the *meta* and *para* isomer, respectively).[301,537] Hepler[537] has rationalized this observation by assuming that in the m-nitrophenolate ion the negative charge is more localized on oxygen and, hence, is more effective in orienting solvent molecules in its cybotactic region.[686] Recently, volumes of ionization ($\Delta \bar{V}_i^{\ominus}$) have become available which reveal that $\Delta V_i^{\ominus}$ is more negative for the *meta*-substituted isomer, as expected.[563] But it was surprising to see that the origin of the $\Delta V_i^{\ominus}$ difference lies exclusively with the un-ionized phenols rather than with the conjugate bases. Subsequent studies[419] of heat capacity changes (*vide infra*) for the ionization of aqueous phenols revealed only small differences for m- and p-nitro and -cyano substituents. A more definite analysis of the heats of ionization of phenols has been given by Arnett,[42] who compared ionization properties in the gas phase, water, and DMSO. As expected, p-nitrophenolate ion is much less exothermically solvated both in water and in DMSO than is the unsubstituted phenolate ion. In addition to the deprotonation equilibria of phenols, the thermodynamic parameters for protonation equilibria of organic compounds in water also exhibit highly interesting features.[291,693]

A highly valuable activation parameter for studying the specific role of solvent reorganization in rate processes is the heat capacity of activation, $\Delta C_P^{\ddagger}$, which represents the second temperature differential of the rate constant:

$$\Delta C_P^{\ddagger} = \left(\frac{\delta \Delta H^{\ddagger}}{\delta T}\right)_P$$

As argued by Robertson,[755] the rationalization of solvent effects on hydrolytic processes in terms of $\Delta C_P^{\ddagger}$ rather than $\Delta S^{\ddagger}$ has some advantages. Among them is the fact that $\Delta S^{\ddagger}$ contains contributions from steric factors involved in the activation process, not easily separable from contributions due to solvent reorganization, and which do not influence $\Delta C_P^{\ddagger}$.

Accurate measurements[1,42,432] of $\Delta C_P^{\ddagger}$ can only be accomplished if sufficiently refined kinetic equipment is available allowing the determination of rate constants with an accuracy of 0.2% or better, implying temperature stability within $\sim 0.001\,°C$. The solvent dependence of $\Delta C_P^{\ddagger}$ has been analyzed for several hydrolytic reactions in both TA and TNA solvents and plots of $\Delta C_P^{\ddagger}$ vs. n_{H_2O} have been found to pass through extrema at characteristic values of n_{H_2O}.[1,505]

The interpretation of $\Delta C_P^{\ddagger}$ and its solvent dependence is still under active discussion. It is a useful mechanistic probe and its magnitude has been related to the degree of charge development on the anion at the transition state of nucleophilic displacement reactions.[1,758] Furthermore, the solvent dependence of $\Delta C_P^{\ddagger}$ seems to be sensitive to changes in the water structure. The solvent dependence of $\Delta C_V^{\ddagger}$ has not been studied in any detail, but it has been suggested[48] that the function $\delta_m \Delta C_V^{\ddagger}$ will be easier to understand than $\delta_m \Delta C_P^{\ddagger}$.

The volume of activation $(\Delta V^{\ddagger})$ is another useful quantity in the study of solvent effects on reaction rates.[432] This quantity is obtained from the dependence of $\ln k$ on the external pressure (P) and measures the change in partial molal volume on transforming the reactants into the transition state:

$$\left(\frac{\partial \ln k}{\partial P}\right)_T = \frac{-\Delta V^{\ddagger}}{RT}$$

The difference in $\Delta V^{\ddagger}$ for a reaction in two solvents S_1 and S_2 is given by

$$-\Delta(\Delta V^{\ddagger}) = RT\left(\frac{\partial \ln(k_1/k_2)}{\partial P}\right)_T$$

in which k_1 and k_2 are the rate constants at the same pressure in the solvents S_1 and S_2. The slope (at zero pressure) of a plot of $\ln(k_1/k_2)$ vs. P then gives $\Delta(\Delta V^\ddagger)/RT$.

In aqueous binaries $\Delta V^\ddagger$ is usually strongly dependent on solvent composition and on the nature of the nonaqueous cosolvent. In TA solvent mixtures, $\Delta V^\ddagger$ usually exhibits extrema in the highly aqueous region, which have been explained in terms of varying water ordering effects.

The isothermal compressibility coefficient of activation $\partial \Delta V^\ddagger/\partial P$, which is the second derivative of rate with respect to pressure, has also been determined for reactions in aqueous binaries and, of course, requires highly accurate kinetic techniques.[889,432]

Quite often, mixed aqueous solvents display large deviations from ideal thermodynamic behavior and activation parameters of organic reactions in these media show a complex variation with n_{H_2O}. In a thoughtful paper, Grunwald and Effio[363] have argued that the complexity of the thermodynamic functions measured at constant solvent composition reflects changes in the relative partial molal functions of the solvent components. If the thermodynamic quantities are measured in such a way that the ratio a_1/a_2 of the activities of the solvent components remains constant (i.e., under endostatic conditions) the complexities can be avoided because the endostatic functions bear an exact analogy to corresponding functions in one-component solvents. The authors have demonstrated the transformation for solvent effects on kinetic data for the $S_N 1$ solvolysis of t-BuCl in EtOH–H_2O and the results reveal that the consequences of the transformation are different for the solvent dependence of $\Delta G^\ddagger$ and $\Delta H^\ddagger$. The method has recently been criticized[104] and clearly further development of the theory awaits the availability of more accurate data for the required transformations.

In contrast to the isobaric activation parameters $\Delta H_P^\ddagger$ and $\Delta S_P^\ddagger$, the isochoric parameters of activation $\Delta E_V^\ddagger$ and $\Delta S_V^\ddagger$ have hardly been studied as a function of solvent composition in mixed aqueous solvents.[46–48]

The difference between the enthalpy of activation at constant pressure and at constant volume is given by[48]

$$\Delta E_V^\ddagger = \Delta H_P^\ddagger - RT^2 \left(\frac{dV}{dT}\right)_P \left(\frac{dP}{dV}\right)_T \left(\frac{d\ln k}{dP}\right)_T$$

$$= \Delta H_P^\ddagger + \frac{RT\alpha}{\beta} \left(\frac{d\ln k}{dP}\right)_T = \Delta H_P^\ddagger - \frac{T\alpha \Delta V^\ddagger}{\beta}$$

Analogously,

$$\Delta S_V^{\ddagger} = \Delta S_P^{\ddagger} - \frac{\alpha \Delta V^{\ddagger}}{\beta}$$

In these equations α and β are the thermal expansivity and compressibility of the reaction medium. The important observation was made that the extrema in the isobaric parameters observed for several hydrolysis reactions in mixed aqueous solvents of varying n_{H_2O} disappeared or were greatly reduced when transformed to constant volume.[46–48] Whalley has suggested[48] that minima in $\delta_m \Delta H_P^{\ddagger}$ and $T\delta_m \Delta S_P^{\ddagger}$ must, at least in part, be ascribed to the abnormally low thermal expansivity of water near room temperature as compared with that of organic solvents. Therefore the isochoric activation parameters may perhaps be preferred over those at constant pressure because they appear to be conceptually more simple since they do not contain contributions of terms due to work of expansion against the external pressure and the internal pressure of the solution.[48,911] However, this argument has been criticized,[48] but Whalley himself has pointed out that—although $\Delta E_V^{\ddagger}$ and $\Delta S_V^{\ddagger}$ presumably better reflect fundamental processes that occur in aqueous binaries—a thermodynamic analysis cannot tell a priori whether the set $(\Delta H_P^{\ddagger}, \Delta S_P^{\ddagger})$ or $(\Delta E_V^{\ddagger}, \Delta S_V^{\ddagger})$ is inherently simpler to interpret. Unfortunately, the mechanistic relevance of isochoric activation parameters is only partly assessed at the moment. It is expected that further studies of the effect of pressure on a large variety of organic reactions in aqueous organic solvents may be highly rewarding.

4.3. Solvent Effects on Initial State and Transition State

The variation in the thermodynamic quantities of activation as a function of solvent composition is composed of initial-state and transition-state contributions. Following Leffler and Grunwald,[542] a solvent operator δ_m may be defined which determines the change in, say, $\Delta G^{\ddagger}$ on changing the reaction medium from pure water ($\Delta G_{H_2O}^{\ddagger}$) to a mixed aqueous solvent S_1 of mol fraction of water n_{H_2O}. Thus,

$$\delta_m \Delta G^{\ddagger} = \Delta G_{S_1}^{\ddagger} - \Delta G_{H_2O}^{\ddagger}$$

For a bimolecular reaction between two reactants A and B in water, $\Delta G_{H_2O}^{\ddagger}$ is defined as

$$\Delta G^{\ddagger} = G_{H_2O}^{\ddagger} - G_{H_2O}^{A} - G_{H_2O}^{B}$$

and

$$\delta_m \Delta G^{\ddagger} = (G_{S_1}^{\ddagger} - G_{H_2O}^{\ddagger}) - (G_{S_1}^A - G_{H_2O}^A) - (G_{S_1}^B - G_{H_2O}^B)$$

$$= \delta_m' G_{\ddagger} - \delta_m'' G_A - \delta_m''' G_B$$

Quite generally, $\delta_m \Delta G^{\ddagger} \neq 0$ since the free energy of the transition state and of the reactants will respond differently towards variation of n_{H_2O}. The terms $\delta_m'' G_A$ and $\delta_m''' G_B$ are the free energies of transfer for A and B ($\Delta G_{tr,A}$) resp. ($\Delta G_{tr,B}$), from water to the mixed aqueous solvent S_1. These transfer functions may be compared with the corresponding excess functions of the mixed solvent.[432] The thermodynamic quantities $\delta_m'' G_A$ and $\delta_m''' G_B$ are directly accessible (in contrast to $\delta_m' G_{\ddagger}$) by several techniques including calorimetric measurements, vapor-pressure determinations, or solubility experiments.[34,347] The crucial factor here is often the stability of A and B towards solvolysis, but sophisticated methods have been developed in order to overcome this difficulty.[34] For further analysis, several approaches are possible. First $\delta_m' G_{\ddagger}$ may be obtained from the experimentally determined $\delta_m \Delta G^{\ddagger}$, $\delta_m'' G_A$, and $\delta_m''' G_B$. The function $\delta_m' G_{\ddagger}$ may yield valuable information about the nature of the transition state if it is assumed that the position of the transition state along the reaction coordinate does not vary appreciably with n_{H_2O}. The function $\delta_m' G_{\ddagger}$ may then be compared with $\delta_m G_M$ values measured in the same solvent system for a compound M which is thought to be a "transition state model." If $\delta_m' G_{\ddagger}$ and $\delta_m G_M$ show closely similar behavior, the solute M may be held to resemble the transition state. As an example of this approach, the elegant work by Abraham[1] may be cited, in which it was shown that the transition-state behavior of several nucleophilic substitution reactions is very close to $\delta_m G_M$ for M = tetraalkylammonium halide ion pairs. At this point it should be noted that $\delta_m G_A$ values often conceal much larger variations in $\delta_m H_A$ and $\delta_m S_A$ owing to compensatory behavior of the enthalpy and entropy of transfer upon variation of the solvent composition.[205] Consequently, $\delta_m H_A$ and $\delta_m S_A$ data provide more sensitive probes for solvation behavior of solutes. The heats of transfer may be obtained from the appropriate van 't Hoff plots:

$$\delta_m H_A = (H_{S_1,A} - H_{H_2O,A}) = \frac{\delta(\delta_m G_A)/T}{\delta(1/T)}$$

Enthalpy–entropy compensation phenomena occur frequently for the transfer functions of both reactants and transition state, leading to similar enthalpy–entropy compensation phenomena for the enthalpy and entropy of activation (Section 4.4).

In the case that one (or both) of the reactants is too reactive to allow the measurement of the thermodynamic transfer parameters, less reactive model compounds may be employed. Of course, a factor of uncertainty is introduced here, but it may be noted that for structurally related non-electrolytes the gross features of the thermodynamic transfer quantities as a function of solvent composition are often quite similar.[38]

The heat capacity of activation $\Delta C_P{}^\ddagger$ measures the difference in heat capacity between the initial state and the transition state:

$$\Delta C_P{}^\ddagger = C_{P,\ddagger} - C_{P,A} - C_{P,B}$$

Partial heat capacities at infinite dilution ($\bar{C}_P{}^0$) are usually obtained from specific heat measurements at different solute concentrations. They are sensitive probes for the temperature dependence of solute–solvent interactions. Measurements of this type have provided important information on the effects of solutes on the water structure[802] and on the hydrophobic hydration of α-amino acids[15,719] and proteins.[839] From the results it has been concluded that the hydrogen-bonded structure near apolar solutes must be more temperature sensitive (i.e., melt with a larger ΔH) than that of pure water.

Recently, heat capacities were determined[243a] for a series of ternary aqueous systems at 25°C. The results obtained in the water-rich region are largely determined by strong solute–water interactions. At the nonaqueous end the heat capacities are reflecting mostly solute–solvent interactions. The concentration dependence of the thermodynamic function in both solvent composition ranges may differ not only in magnitude but even in sign, leading to inflection points or extrema in these functions in the intermediate solvent composition range. In another interesting but more limited study, heat capacities of transfer ($\Delta C_{P,\mathrm{tr}}^\ominus$) have been determined[406] for some simple monofunctional alcohols from a solution in H_2O to a solution in D_2O. These $\Delta C_{P,\mathrm{tr}}^\ominus$ are positive and are consistent with slightly enhanced structure making in D_2O as compared with H_2O by the hydrophobic part of the alcohol molecule and with progressive breakdown of this structure upon increasing temperature.

The volume of activation $\Delta V^\ddagger$ can also be dissected into its component initial and transition-state volume contribution. This has been accomplished by measuring the "instantaneous" change in volume as a result of dissolution of the substrate in the appropriate solvent mixture.[354] The behavior of $\delta_m V_\ddagger$ may also be compared with that of a suitable transition-state model. Finally we note here that $\Delta V^\ddagger$, just as $\Delta H^\ddagger$, often shows extrema when plotted as a function of n_{H_2O}, although usually not at the same n_{H_2O}. The exact origin of these extrema is not well understood but is usually attributed to some aspect of solvent structure. Usually, a contraction in volume is observed when a hydrocarbonlike solute is transferred from a nonaqueous phase to water or a water-rich solvent system, in agreement with the concept that hydrophobic hydration occurs with an economy of space.[243,339]

The study of solvent effects based on the analysis of the solvent operator δ_m for reactants and the transition state is a fundamental advance over many previous frameworks towards the understanding of microenvironmental effects on reaction rates. Taking into account the proper solution standard states,[1,446] this effective approach allows the investigation of solvent effects on rates in aqueous binaries in terms of intermolecular interactions. It does not rely on unwarranted assumptions inherent in correlation of rates with ion-solvating power[425] of the medium or other solvent parameters nor does it consider only transition-state solvation. As argued before (see also Section 2), the main barrier for a quantitative analysis of kinetic solvent effects is provided by the complexities of water and its mixtures with organic solvents. Since the understanding of the solvent properties of these systems is steadily increasing, at present especially via molecular dynamics computer simulations, it may be foreseen that the coming years may witness the development of relations between kinetic data and molecular theories of aqueous solvent structure. This would then lead to the possibility of successfully predicting optimum reaction conditions for synthetic reactions and, furthermore, it would set the stage for a proper understanding of the medium effects operating on biological processes in the living cell.

4.4. Isokinetic Relationships

Many chemical processes in mixed aqueous solution exhibit enthalpy–entropy compensation effects upon variation of n_{H_2O}.[287,377,542,580] As a result, large changes in enthalpy and entropy are frequently accompanied by much more moderate variations in free energy. It is quite plausible that the enthalpy–entropy compensatory behavior is the thermodynamic manifestation of changes in hydrogen-bonding interactions.[580,690] Thus, in kinetic studies one often finds a linear relationship between $\Delta H^{\ddagger}$ and $\Delta S^{\ddagger}$ upon changing n_{H_2O} in specific ranges of n_{H_2O}. The slope of the plot of $\Delta H^{\ddagger}$ vs. $\Delta S^{\ddagger}$ has the unit of absolute temperature and is called the isokinetic or compensation temperature $(\Delta H^{\ddagger} = \alpha + T_c \Delta S^{\ddagger})$. The $\Delta H/\Delta S$

compensation phenomena occurring in aqueous media have been extensively reviewed by Lumry and Rajender.[580] These authors propose that the linear $\Delta H/\Delta S$ relationship is a ubiquitous property of aqueous solvent systems and may even serve as a diagnostic test for the participation of water in protein processes. Nevertheless, the origin of the isokinetic relationships and the mechanistic significance of T_c values for kinetic processes are not well understood. Recently, Melander[614] has suggested that the thermal dependence of the dielectric constant, influencing the energetics of reactions proceeding via ionic mechanisms, could be the "macroscopic source" for the compensation phenomena. During the last decade, the usual procedure for determining linear compensation behavior from ΔH and ΔS values has been strongly criticized. Since ΔS is usually calculated from ΔG and ΔH, or ΔH from ΔG and ΔS, errors in ΔH (or ΔS) are compensated in ΔS (or ΔH), leading to a statistical compensation pattern. Therefore different tests for linear compensation behavior have been developed.[288,289,918] These methods rely on the use of the original independent experimental observations (reaction rates and temperatures) rather than on the interdependent parameters $\Delta H^{\ddagger}$ and $\Delta S^{\ddagger}$. According to Exner, linear compensation behavior exists if the plot of $\ln k_1$ at T_1 vs. $\ln k_2$ at T_2 yields a straight line.[287] The use of plots of $\ln(k \cdot T^{-1})$ vs. T^{-1} has been proposed by Petersen.[701] A real linear compensation behavior is claimed if the lines intersect at a single point, representing the compensation temperature. Linear plots of $\Delta G^{\ddagger}$ vs. $\Delta H^{\ddagger}$ have also been taken as indications for real compensation phenomena.[522,579,580] Despite extensive previous analysis, the thorough statistical treatment by Krug et al.[522] has indicated that extrathermodynamic enthalpy–entropy effects are rare, many earlier claims for the phenomenon being the result of random errors in the parameters. Quite generally it is observed that the less accurately $\Delta H^{\ddagger}$ and $\Delta S^{\ddagger}$ are measured, the greater the likelihood that their changes as a function of solvent composition will be proportional to one another. A statistical compensation pattern that arises solely from experimental error usually has an "isokinetic temperature" near the experimental temperature. Therefore, the crucial problem is to establish which part of an observed linear $\Delta H^{\ddagger}/\Delta S^{\ddagger}$ compensation pattern is the consequence of experimental errors and which part is due to a real isokinetic relationship. There is no doubt that solvation changes in mixed aqueous solvents are very often accompanied by mutually compensating changes in $\Delta H^{\ddagger}$ and $\Delta S^{\ddagger}$. But there are so many uncertainties in the reliability of the tests for *linear* enthalpy–entropy compensation as well as in the mechanistic meaning of an isokinetic temperature that at present we shall refrain from further discussion.

5. KINETIC SOLVENT EFFECTS ON ORGANIC REACTIONS IN MIXED AQUEOUS SOLVENTS

5.1. Introduction

It is the present author's belief that solvent effects on reaction rates in solution may be best analyzed by employing a thermodynamic approach, that is, by considering the thermodynamic quantities of transfer of the reactants which in combination with the relevant changes in the activation parameters yield the transfer functions of the transition state (Section 4.3). This type of analysis has been applied frequently for reactions in sets of pure solvents and the field has recently been reviewed authoritatively by Abraham.[1] A similar treatment is also the most straightforward and efficient one for rates of organic reactions in mixed aqueous solvents, usually taking the hypothetical ideal solution of unit molar concentrations as the standard state (Section 4.3). An alternative approach involves an analysis of activity coefficient behavior as a function of n_{H_2O} but suffers from the disadvantage that the experimental data are less easily rationalized in terms of solute–solvent interaction mechanisms.

Because of our incomplete understanding of water and mixed aqueous solutions, a quantitative interpretation of the thermodynamic quantities of transfer and of activation is not feasible at the moment. Therefore, *trends* in the changes of the thermodynamic data as a function of n_{H_2O} are usually explained by considering the contributions of different types of solute–solvent interactions to the solvation changes as the initial state is transferred into the transition state. Needless to say, the most successful results have been obtained for the most simple kinetic processes involving the smallest number of species in the transition state of the slow step.

5.2. Water-Catalyzed Reactions

Water-catalyzed hydrolysis reactions are pH-independent hydrolytic processes involving proton transfer to or from water molecules in the transition state of the rate-determining step. Activated esters and amides may hydrolyze via this mechanism, which is characterized by solvent deuterium isotope effects (k_{H_2O}/k_{D_2O}) in the region 2–4 and by large and negative entropies of activation (often < -30 e.u.).[284,451,755] Several transition states are possible depending on the extent of bond breaking and bond making between the molecules in the transition state.[528–530] These "water reactions," in which water is both the *reactant* and the *solvent* have the advantage of being relatively simple processes for quantitative studies of solvent effects. In pure water, the reaction is best described as pseudo-first-order in water since the formation of a bimolecular encounter complex has no well-defined meaning as the substrate is completely surrounded by water molecules.[535] Furthermore, there is no partial desolvation of both reactants upon going to the transition state, which partly explains the strongly negative activation entropies. In mixed aqueous solvents, the usually unknown order in water is no serious drawback since the interpretation of solvent effects usually concentrates on $\delta_m \Delta H_{tr}$ values, which are largely independent of the molecularity of the reaction. Finally, the magnitudes of the thermodynamic functions of transfer for *water* from water to aqueous binaries are usually small compared to those of the substrate and transition state, so that at sufficiently low substrate concentration initial-state solvation effects need only be considered for one reactant.

The first example involves the neutral, pH-independent hydrolysis of *p*-nitrophenyl dichloroacetate $\mathbf{12}$[302,447] in *t*-BuOH–H$_2$O, already referred to in Sections 1 and 4.2[280,281]:

$$\text{HCCl}_2\text{CO}_2\text{C}_6\text{H}_4\text{NO}_2\text{-}p \;+\; \text{H}_2\text{O} \;\longrightarrow\; \left[\begin{array}{c} \text{O} \\ \overset{\|}{\text{C}}{}^{\delta\ominus} \\ \text{HCCl}_2\text{—C}\cdots\text{OC}_6\text{H}_4\text{NO}_2\text{-}p \\ \text{O—H} \\ \text{H}\;{}^{\delta\oplus} \\ \text{OH}_2 \end{array} \right]^{\ddagger}$$

$$\mathbf{12} \qquad\qquad\qquad\qquad \mathbf{13}$$

$$\longrightarrow\quad \text{HCCl}_2\text{CO}_2\text{H} \;+\; \text{HOC}_6\text{H}_4\text{NO}_2\text{-}p$$

In water, $k_{H_2O}/k_{D_2O} = 3.1$ and $\Delta S^{\ddagger} = -35$ e.u. These data as well as the effect of *para*-substituents in the phenyl ring (linear correlation of log k_{obsd} with σ^-)[280] are best reconciled with partial proton transfer and considerable solvent immobilization in the transition state of the slow step. It has been suggested that the hydrolysis process occurs via a transition state in which the leaving group is already partially expelled ($\mathbf{13}$) although a tetrahedral addition intermediate is not ruled out.

Plots of $\Delta G^{\ddagger}$, $\Delta H^{\ddagger}$, and $-T\Delta S^{\ddagger}$ vs. n_{H_2O} for *t*-BuOH–H$_2$O and *t*-BuOD–D$_2$O (at 25°C) are shown in Fig. 3. The pseudo-first-order rate constants (k_{obsd}) used in the calculation of the activation parameters pertain only to water reactions since solvolysis of $\mathbf{12}$ in pure *t*-BuOH is at least 10^4 times slower than in water. There is no simple correlation of log k_{obsd} with ε, $(\varepsilon - 1)/(2\varepsilon + 1)$, or with Z or E_T values (Fig. 4). In order to separate initial-state and transition-state effects, thermodynamic quantities of transfer

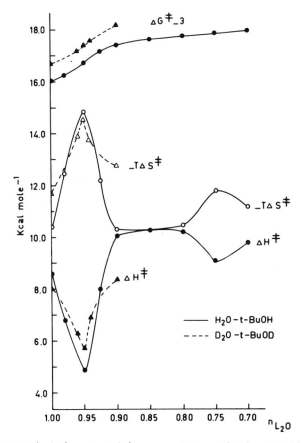

Fig. 3. Plot of $\Delta G^{\ddagger}$, $\Delta H^{\ddagger}$, and $-T\Delta S^{\ddagger}$ vs. n_{H_2O} (or n_{D_2O}) for the neutral hydrolysis of **12** in t-BuOH–H$_2$O and t-BuOD–D$_2$O at 25°C.[280,281]

from H$_2$O and D$_2$O to the binary mixtures were determined for p-nitrophenyl acetate (**14**) as a model for **12**. Trends in the changes in the free energy of the transition state ($\delta_m \Delta G^{\ominus}_{tr,\ddagger}$) as a function of n_{H_2O} or n_{D_2O} are shown in Fig. 5. As expected, addition of t-BuOH or t-BuOD induces a stabilization of the *reactant*, which is primarily responsible for the increase of $\Delta G^{\ddagger}$ upon decreasing n_{H_2O} or n_{D_2O}. The variation of $\Delta G^{\ominus}_{tr}$ for the initial state (i) with solvent composition may be analyzed by using eqn. (18) (Section 3.5):

$$\Delta G^{\ominus}_{tr,i} = (\delta \Delta G^{cav} + \delta \Delta G^{nonpol}) + \delta \Delta G^{pol}$$

The favorable effect of the addition of t-BuOH on $\Delta G^{\ominus}_{tr,i}$ must be due to the sum of the first two terms. The contribution of $\delta \Delta G^{pol}$ will be relatively small since **12** is only a weak hydrogen-bond acceptor.[280] The decrease of $\Delta G_{tr,i}^{\ominus}$ may then be the result of several factors. First, the addition of the organic cosolvent leads to a smaller average density of hydrogen bonds per unit volume and area and a concomitant reduction in the free energy necessary for cavity formation. However, it is likely that this favorable effect on $\Delta G_{tr,i}^{\ominus}$ is partly offset in the water-rich region by a decrease of natural cavities in the water hydrogen bond network with decreasing n_{H_2O} owing to accommodation of cosolvent molecules in the void space. Second, interaction of the hydration sphere of **12** with the hydrophobic hydration sphere of the cosolvent will provide a favorable contribution to $\Delta G_{tr,i}^{\ominus}$ due to the gain in entropy associated with the liberation of water molecules from the hydrophobic hydration envelope around t-BuOH.

The solvation changes of the polar transition state in the water-rich region will be largely determined by hydrogen-bonding interactions with the solvent ($\delta \Delta G^{pol}$ term). For many polar solutes, the favorable enthalpy dominates the unfavorable entropy of hydration, providing the driving force for dissolution of these solutes in water. The small variation of $\Delta G_{tr,\ddagger}^{\ominus}$ in the region $n_{H_2O} = 1.00$–0.85 is consistent with this theory and

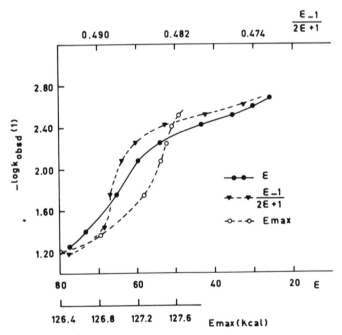

Fig. 4. Plot of $\log k_{obsd}$ for neutral hydrolysis of **12**, vs. ε, $(\varepsilon - 1)/(2\varepsilon + 1)$, and E_T in t-BuOH–H$_2$O at 25°C.[280]

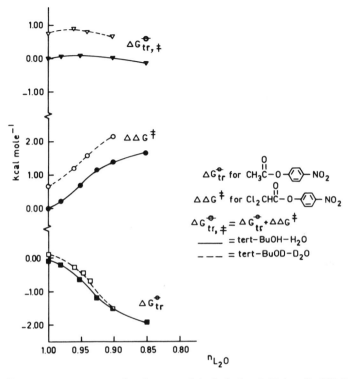

Fig. 5. Free-energy parameters for the neutral hydrolysis of **12** in t-BuOH–H$_2$O and t-BuOD–D$_2$O at 25°C.[280]

will be the outcome of mutually compensating contributions of ($\delta\Delta G^{\text{cav}}$ + $\delta\Delta G^{\text{nonpol}}$) and $\delta\Delta G^{\text{pol}}$.

The free energy of transfer of **14** from H$_2$O to D$_2$O is only 0.07 kcal mol^{-1}. Quite generally, $\Delta G_{\text{tr}}^{\ominus}$(H$_2$O → D$_2$O) values for nonelectrolytes are small and variable in sign (Section 3.1).

We now turn to the dramatic solvent effects on $\Delta H^{\ddagger}$ and $\Delta S^{\ddagger}$ in the range $n_{\text{H}_2\text{O}}$ = 1.00–0.90 (Fig. 3). Whereas the contributions of $\Delta H^{\ddagger}$ and $-T\Delta S^{\ddagger}$ to $\Delta G^{\ddagger}$ differ by only 1.8 kcal mol^{-1} in water and are almost equal in the solvent composition range 0.90 > $n_{\text{H}_2\text{O}}$ > 0.80, there exists a difference of 10 kcal mol^{-1} between these two quantities at $n_{\text{H}_2\text{O}}$ = 0.95! Extrema in $\Delta H^{\ddagger}$ and $\Delta S^{\ddagger}$ are located at $n_{\text{H}_2\text{O}}$ = 0.95 for both t-BuOH–H$_2$O and t-BuOD–D$_2$O. This is exactly the "magic mole fraction" of maximum solvent structural integrity. Again using **14** as a model for **12**, $\delta_m\Delta H^{\ddagger}$ and $\delta_m\Delta S^{\ddagger}$ can be separated into initial- and transition-state contributions (Fig. 6). As shown in Fig. 6, $\Delta H_{\text{tr},i}^{\ominus}$ exhibits a large increase upon initial addition of t-BuOH and $\delta_m \Delta H^{\ominus}_{\mathrm{tr},i}$ reaches a maximum value of 7.1 kcal mol^{-1} at $n_{\mathrm{H_2O}} = 0.95$ and of 8.2 kcal mol^{-1} at $n_{\mathrm{D_2O}} = 0.95$. Similar extrema in $\Delta H^{\ominus}_{\mathrm{tr},i}$ at $n_{\mathrm{H_2O}} \sim 0.95$ have been found for other nonelectrolytes, but the parameter δ_m is strongly dependent on the solvation properties of the solute (vide infra). For the polar transition state the maximum in $\Delta H^{\ominus}_{\mathrm{tr},\ddagger}$ is somewhat less pronounced and occurs at a lower $n_{\mathrm{H_2O}}$, as expected for polar or ionic solutes.[257] In more general terms, we conclude: (i) the location and strength of the extreme in $\Delta H^{\ddagger}$ will be determined by the difference in δ_m for initial and transition state, which reflects differences in specific and hydrophobic hydration effects for both species; (ii) the specific $\Delta H^{\ddagger}/T\Delta S^{\ddagger}$ mirror image behavior determines whether or not an extreme in $\Delta G^{\ddagger}$ will be observed.

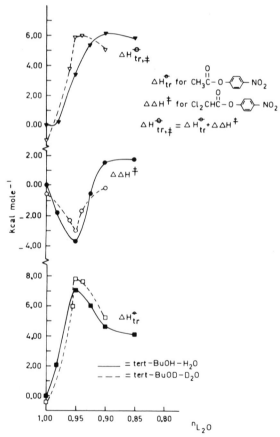

Fig. 6. Enthalpy parameters for the neutral hydrolysis of **12** in t-BuOH–H$_2$O and t-BuOD–D$_2$O at 25°C.[280]

Because $\Delta G^{\ominus}_{\text{tr},i}$ is decreasing upon addition of t-BuOH to water (Fig. 5), the large increase in $\Delta H^{\ominus}_{\text{tr},i}$ is clearly overcompensated by a still larger contribution from $-T\Delta S^{\ominus}_{\text{tr},i}$. Therefore, the entropy term governs the sign of $\Delta G^{\ominus}_{\text{tr},i}$ in the water-rich region, which is often a characteristic of processes related with hydrophobic hydration effects in aqueous solutions.

Following a similar treatment as for $\Delta G^{\ominus}_{\text{tr},i}$, $\Delta H^{\ominus}_{\text{tr},i}$ may be written as

$$\Delta H^{\ominus}_{\text{tr},i} = (\delta\Delta H^{\text{cav}} + \delta\Delta H^{\text{nonpol}}) + \delta\Delta H^{\text{pol}}$$

Since $\delta\Delta H^{\text{pol}}$ will provide a minor contribution to $\Delta H^{\ominus}_{\text{tr},i}$, the behavior of $\Delta H^{\ominus}_{\text{tr},i}$ in the range $n_{\text{H}_2\text{O}} = 1.00\text{--}0.85$ is largely dominated by $\delta\Delta H^{\text{cav}}$. At low concentration of t-BuOH, the cosolvent will initially occupy the holes in the water structure and the nonpolar part of the alcohol will be surrounded by a hydrophobic hydration sphere. Dissolution of **14** will then require more enthalpy for cavity formation. At higher mole fractions of cosolvent, the decrease in the average density of hydrogen bonds per unit volume and area (evidenced by a decrease in D_{ce}) will attenuate the further increase in $\delta\Delta H^{\text{cav}}$. Furthermore, interactions between the solvation sphere of **14** and the hydrophobic hydration sphere of t-BuOH will become more and more unfavorable in the enthalpic sense since hydrogen bonds will be broken (compare Section 3). If direct contacts between **14** and t-BuOH would occur in this solvent composition range, some enthalpy may be gained from the $\delta\Delta H^{\text{nonpol}}$ term. Below $n_{\text{H}_2\text{O}} = 0.95$, the hydrophobic hydration spheres around t-BuOH gradually collapse, resulting in a decrease in ΔH^{cav} for **14**. It is also anticipated that at lower $n_{\text{H}_2\text{O}}$ the ΔH^{nonpol} term will acquire more importance and the overall result will be a decrease of $\Delta H^{\ominus}_{\text{tr},i}$ (Fig. 6). Since the polar transition state is extensively hydrogen bonded to water molecules, there will be a reduced tendency for making hydrophobic contacts with t-BuOH as compared with the nonpolar initial state. Consequently, $\Delta H^{\ominus}_{\text{tr},\ddagger}$ will increase less steeply than $\Delta H^{\ominus}_{\text{tr},i}$ upon addition of t-BuOH to water and owing to the specific solvation effects the maximum will be reached at a lower $n_{\text{H}_2\text{O}}$ (Fig. 6). The different response of $\Delta H^{\ominus}_{\text{tr},i}$ and $\Delta H^{\ominus}_{\text{tr},\ddagger}$ then leads to the observed minimum in $\Delta H^{\ddagger}$.

The deuterium isotope effects on the enthalpy terms (Fig. 6) are hard to evaluate quantitatively. Quite generally, structural changes induced by nonpolar solutes are more manifest in D_2O than in H_2O (Section 3.1). This may explain the larger $\delta_m\Delta H^{\ominus}_{\text{tr},i}$ in D_2O than in H_2O and the opposite behavior for $\delta_m\Delta H^{\ominus}_{\text{tr},\ddagger}$. The favorable $\Delta H^{\ominus}_{\text{tr},i}$ term for transfer of **14** from H_2O to D_2O may be a consequence of the larger void space assumed to be present in D_2O which initially renders cavity formation less difficult in this solvent. At higher mole fractions of *t*-BuOH, this void space is filled up by the cosolvent, and cavity formation will then become more difficult in D_2O than in H_2O because rupture of deuterium bonds costs more enthalpy than rupture of hydrogen bonds.

The neutral hydrolysis of **12** has also been investigated in the TNA solvent system acetonitrile–water and in dioxane–water (Fig. 7).[280] These organic cosolvents induce nearly equal rate retardations, but the behavior of $\Delta H^{\ddagger}$ and $\Delta S^{\ddagger}$ as a function of n_{H_2O} is somewhat different. Trends in the initial-state parameters were again obtained using **14** as a model compound. Unfortunately, transfer functions of **14** from H_2O to $MeCN$–H_2O could not be obtained for mixtures having $n_{H_2O} < 0.9$ because of phase separation upon addition of **14**.

The gradual increase of $\Delta G^{\ddagger}$ with decreasing n_{H_2O} is here also primarily due to stabilization of the initial state, dioxane being a little more effective than MeCN. This somewhat larger effect of dioxane has also been observed for the transfer functions of ethyl acetate from water to dioxane–water and

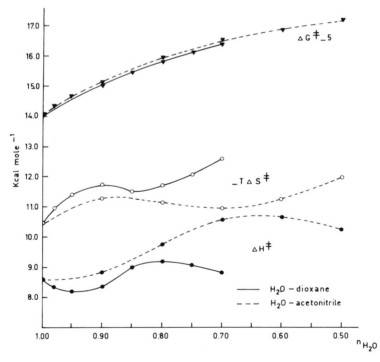

Fig. 7. Plot of $\Delta G^{\ddagger}$, $\Delta H^{\ddagger}$, and $-T\Delta S^{\ddagger}$ vs. n_{H_2O} for the neutral hydrolysis of **12** in dioxane–H_2O and MeCN–H_2O at 25°C.[280]

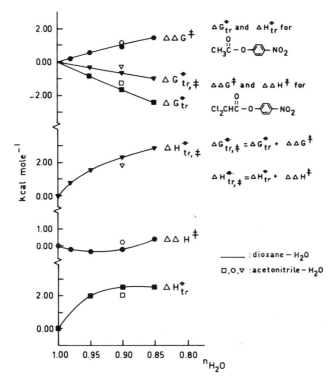

Fig. 8. Free-energy and -enthalpy data for the neutral hydrolysis of **12** in dioxane–H_2O and MeCN–H_2O at 25°C.[280]

MeCN–H_2O, respectively.[205] It is noteworthy that $\Delta G_{tr,i}^{\ominus}$ as well as $\Delta G_{tr,\ddagger}^{\ominus}$ are more sensitive to variation of n_{H_2O} in these TNA mixtures than in the TA solvent system t-BuOH–H_2O (Fig. 3). A similar difference has been found by Roseman and Jencks[766] for the corresponding transfer functions of naphthalene and uric acid (see also Section 3.5). Because the solvent structure around dioxane and MeCN is appreciably less ordered than around t-BuOH, the difference may well arise from specific interactions of the solute with dioxane and CH_3CN. The most striking difference between the TA solvent t-BuOH–H_2O and the TNA mixtures is exhibited by $\delta_m \Delta H^{\ddagger}$, $\delta_m \Delta H_{tr,i}^{\ominus}$, and $\delta_m \Delta H_{tr,\ddagger}^{\ominus}$ (Fig. 8). In contrast to t-BuOH–H_2O, $\Delta H^{\ddagger}$ in dioxane–H_2O varies only little (by ~ 1 kcal mol^{-1} between $n_{H_2O} = 1.0$ and 0.7), the rate retardation being primarily an entropy effect. In MeCN–H_2O $\Delta H^{\ddagger}$ exhibits a shallow maximum around $n_{H_2O} = 0.60$. In this medium, $\delta_m \Delta G^{\ddagger}$ is governed by $\delta_m \Delta H^{\ddagger}$ since $\Delta S^{\ddagger}$ remains almost constant in the region $0.90 \geq n_{H_2O} \geq 0.60$. The relatively small and smooth variation of

$\Delta H_{tr,i}^{\ominus}$ and $\Delta H_{tr,\ddagger}^{\ominus}$ with n_{H_2O} in the TNA mixtures are compatible with dominating specific solvation effects and a minor role played by hydrophobic hydration. The initial increase of $\Delta H_{tr,i}^{\ominus}$ upon addition of dioxane and MeCN may reflect the occupation of the void space in the water structure by the cosolvent, thereby reducing the possibilities for interstitial dissolution of the solute. As the concentration of cosolvent is further increased, cavity formation will require less and less breaking of hydrogen bonds, which will level off the increase of the enthalpy of transfer. Direct interactions between solute and cosolvent may also contribute to this decrease of the transfer enthalpy. As expected, the effect is seen to start at higher n_{H_2O} for the apolar (model) substrate than for the polar transition state.

Very extensive studies of the neutral hydrolysis of acyl activated carboxylic esters in both TA and TNA solvent systems have been made by the Finnish school of Euranto and Cleve.[189,191,284,285] For a variety of substrates, the solvent dependence of $\Delta H^{\ddagger}$, $\Delta S^{\ddagger}$, and $\Delta C_P^{\ddagger}$ was examined and characteristic extrema in these quantities of activation were noted in several cases. However, the main emphasis in this careful work has been on the recognition of specific patterns in the kinetic solvent effects as a function of the specific nature of the substrate rather than on a detailed analysis of interaction mechanisms in the different media.

The neutral hydrolysis of bis(4-nitrophenyl)carbonate (15) has been studied by Menger and Venkatasubban.[615] A proton inventory study revealed that only a single proton contributes to the solvent deuterium

$$(p\text{-}O_2NC_6H_4O)_2C{=}O$$

15

isotope effect ($k_{H_2O}/k_{D_2O} = 2.88$), which is best reconciled with a "two-water" general base mechanism and a transition state which is reached at an early stage on the reaction coordinate. The reaction rates (at 50°C) were only slightly affected by both TA (t-BuOH–H$_2$O) and TNA solvents

(MeCN–H$_2$O, DMSO–H$_2$O, dioxane–H$_2$O, urea–H$_2$O) in the water-rich region, once again demonstrating that $\Delta G^{\ddagger}$ values for the neutral hydrolysis of activated carboxylic esters are slightly or not at all influenced by changes in water structure.

TABLE VII. Activation Parameters for the Hydrolysis of Acetic Anhydride in
t-BuOH–H_2O and DMSO–H_2O[664]

Solvent	n_{H_2O}	$\Delta H^{\ddagger}$ (kcal mol^{-1})	$\Delta S^{\ddagger}$ (e.u.)
H_2O	1.00	10.8 ± 0.4	-41.8 ± 1
t-BuOH–H_2O	0.9	9.9 ± 0.4	-45.9 ± 1
	0.8	10.7 ± 0.6	-45.0 ± 2
	0.7	10.5 ± 0.6	-45.9 ± 2
DMSO–H_2O	0.9	9.8 ± 0.7	-47.1 ± 2
	0.8	12.3 ± 0.4	-40.0 ± 2
	0.7	15.4 ± 0.9	-29.0 ± 3

The rate of the neutral hydrolysis of acetic anhydride is also reduced upon addition of organic cosolvents like EtOH, t-BuOH, dioxane, and DMSO.[664] Since the rate constant for the reaction in water lies on the same Brönsted plot as the rate constants of the hydrolysis catalyzed by substituted pyridines ($\beta = 0.88$), water-catalyzed nucleophilic attack of water on the carbonyl group is here also the most likely mechanism. In view of the large β value, the hydrolysis involves a transition state with a considerable degree of charge separation; in DMSO–water the activity coefficient behavior of the transition state was found to follow a similar trend to that of N-phenylaniline. Activation parameters were determined in water and at $n_{H_2O} = 0.9$, 0.8, and 0.7 in t-BuOH–H_2O and in DMSO–H_2O (Table VII). The limited data, which show the expected $\Delta H^{\ddagger}/\Delta S^{\ddagger}$ compensatory behavior, do not allow any more detailed interpretation. Since no $\Delta H^{\ddagger}$ and $\Delta S^{\ddagger}$ values were measured at solvent compositions between $n_{H_2O} = 1.00$ and 0.90 in t-BuOH–H_2O, it may be that larger variations in these quantities of activation have been overlooked.

Kinetic solvent deuterium isotope effects on the neutral hydrolysis of carboxylic anhydrides[757,768] have been examined in some detail and have been compared with those for the water reaction of alkyl trifluoroacetates,[768,917] which also proceeds by a $B_{AC}2$ mechanism.* The observation that k_{H_2O}/k_{D_2O} values may increase or decrease with decreasing n_{H_2O}

---

* Definitions of common types of mechanisms like $B_{AC}2$, $A_{AC}2$, ElcB, etc., may be found in any good textbook on physical organic chemistry. See, for example, T. H. Lowry and K. S. Richardson, *Mechanism and Theory in Organic Chemistry*, Harper and Row, New York, 1976.

TABLE VIII. Differences^a between $\Delta H^{\ddagger}$, $\Delta S^{\ddagger}$, and $\Delta C_P^{\ddagger}$ for Solvolysis of Acyl Carbon (at 25°C) in H_2O and D_2O[768]

Substrate	$\delta\Delta H^{\ddagger}$ (cal mol^{-1})	$\delta\Delta S^{\ddagger}$ (cal deg^{-1} mol^{-1})	$\delta\Delta C_P^{\ddagger}$ (cal deg^{-1} mol^{-1})
Acetic anhydride	362 ± 64	-0.91 ± 0.22	42 ± 6
Propionic anhydride	179 ± 113	-1.40 ± 0.38	-63 ± 12
Succinic anhydride	-137 ± 100	-2.46 ± 0.34	21 ± 11
Benzoic anhydride	1256 ± 127	1.93 ± 0.42	-54 ± 8
Phthalic anhydride	1413 ± 365	3.17 ± 1.25	96 ± 23
Methyl trifluoroacetate	430 ± 192	-1.00 ± 0.66	2 ± 13

^{a} $\delta\Delta X^{\ddagger} = \Delta X^{\ddagger}_{D_2O} - \Delta X^{\ddagger}_{H_2O}$.

in mixed aqueous solutions[160] wrecks the hope for straightforward interpretation, and a comprehensive theory has not been advanced so far. Rossall and Robertson[768] determined $\Delta H^{\ddagger}$, $\Delta S^{\ddagger}$, and $\Delta C_P^{\ddagger}$ values for solvolysis of a series of anhydrides and for methyl trifluoroacetate in H_2O and D_2O (Table VIII). The differences between the activation parameters in H_2O and D_2O are markedly dependent on the nature of the substrate and will, among other factors, reflect differences in the detailed solvolysis mechanism. The changes in $\Delta C_P^{\ddagger}$, depending on solvent and electronic nature of the reactants, probably measure the temperature-dependent hydrogen-bonding interaction between the "ether" oxygen atom and the solvent in the activated acyl compound and in the transition state. Mixed aqueous solvent effects on kinetic solvent deuterium isotope effects have been more extensively studied for some acid- and base-catalyzed reactions (Section 5.4).

16a, R = CH_3, R' = H
16b, R = C_6H_5, R' = H
16c, R = p-$CH_3OC_6H_4$, R' = H
16d, R = p-$NO_2C_6H_4$, R' = H
16e, R = CH_3, R' = C_6H_5

The effect of substrate structure on the $\Delta H^{\ddagger}/\Delta S^{\ddagger}$ compensation behavior for hydrolysis in highly aqueous t-BuOH–H_2O has been investigated extensively by Karzijn and Engberts[469,470] for the pH-independent hydrolysis of the 1-acyltriazoles **16a–16e** (k_{H_2O}/k_{D_2O} = 2.4–3.4; $\Delta S^{\ddagger} \sim -30$ to -45 e.u.). This "water reaction" (see bottom of page 192) can be conveniently studied in the pH range 3.5–4.5 and involves the 1,2,4-triazole group ($pK_A = 10.3$) as the leaving group. Proton inventory experiments indicate that—in contrast to the hydrolysis of **15**—three protons in the transition state contribute to the solvent deuterium isotope effect. This implies a transition state which is reached relatively late on the reaction coordinate and considerable polarization of the OH bonds of the water molecule acting as a general base. The following mechanism is consistent with the available data and may be contrasted with the mechanism proposed by Staab[820] and later adopted by Potts:[717]

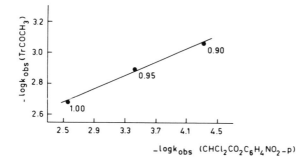

As shown in Fig. 9, the functions $\delta_m \Delta G^{\ddagger}$ for **16a** and **12** are linearly related for neutral hydrolysis in t-BuOH–H_2O. This suggests the operation of rather related mechanisms for both types of substrates. A similar plot for hydrolysis of **16a** and **16e** is shown in Fig. 10.

Figures 11–15 portray the variation of $\Delta G^{\ddagger}$, $\Delta H^{\ddagger}$, and $\Delta S^{\ddagger}$ as a function of n_{H_2O} for the neutral hydrolysis of **16a–16e** in highly aqueous t-BuOH–H_2O. Since **16a–16e** hydrolyze via a common mechanism, *it is clear that*

Fig. 9. Plot of log k_{obsd} for **16a** vs. log k_{obsd} for **12** for neutral hydrolysis in t-BuOH–H_2O in the range n_{H_2O} = 0.9–1.0 (25°C).[470]

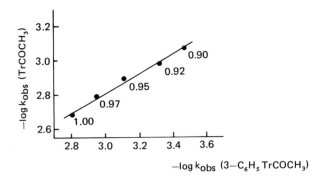

Fig. 10. Plot of log k_{obsd} for **16a** vs. log k_{obsd} for **16e** for neutral hydrolysis in t-BuOH–
H$_2$O in the range $n_{H_2O} = 0.9$–1.0 (25°C).[470]

the occurrence of extrema in $\Delta H^{\ddagger}$ and $\Delta S^{\ddagger}$ around $n_{H_2O} = 0.95$ is by no
means a general phenomenon but rather depends on the structure of the sub-
strate. Thermodynamic parameters of transfer from water to t-BuOH–H$_2$O
determined for the model substrate 1-benzoylmethyltriazole (**16f**) have
been combined with the activation parameters for hydrolysis of the isomer
16e in Figs. 16 and 17. Initial-state solvation is seen to be primarily responsible for the rate decrease induced by the organic cosolvent. The minimum
in $\Delta H^{\ddagger}$ at about $n_{H_2O} = 0.95$ is also primarily an initial-state effect (Fig. 17),
but comparison of the functions $\delta_m \Delta H^{\ominus}_{tr,i}$ for **16f** (Fig. 17) and for **14**
(Fig. 5) reveals that the variation of $\Delta H^{\ominus}_{tr,i}$ between $n_{H_2O} = 1.00$ and 0.90

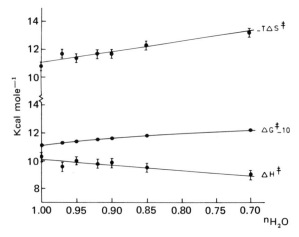

Fig. 11. Plot of $\Delta G^{\ddagger}$, $\Delta H^{\ddagger}$, and $-T\Delta S^{\ddagger}$ vs. n_{H_2O} for the neutral hydrolysis of **16a** in
t-BuOH–H$_2$O at 25°C.[470]

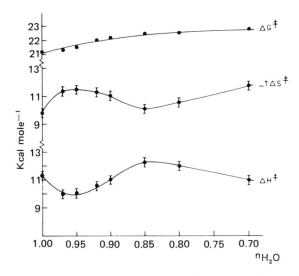

Fig. 12. Plot of $\Delta G^{\ddagger}$, $\Delta H^{\ddagger}$, and $-T\Delta S^{\ddagger}$ vs. n_{H_2O} for the neutral hydrolysis of **16b** in t-BuOH–H_2O at 25°C.[470]

is much smaller for the 1,2,4-triazole derivative. Presumably this difference reflects the presence of the hydrophilic triazole ring in **16f** and a contribution of polar interactions to the enthalpy of transfer. The parameter $\Delta H_{tr,\ddagger}^{\ominus}$ for **16e** is even less sensitive to variation of n_{H_2O} than the corresponding quantity for **12**, which most likely manifests strong hydrogen-bonding

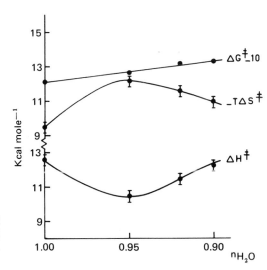

Fig. 13. Plot of $\Delta G^{\ddagger}$, $\Delta H^{\ddagger}$, and $-T\Delta S^{\ddagger}$ vs. n_{H_2O} for the neutral hydrolysis of **16c** in t-BuOH–H_2O at 25°C.[470]

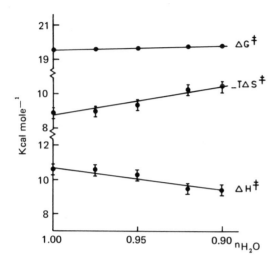

Fig. 14. Plot of $\Delta G^{\ddagger}$, $\Delta H^{\ddagger}$, and $-T\Delta S^{\ddagger}$ vs. n_{H_2O} for the neutral hydrolysis of **16d** in t-BuOH–H_2O at 25°C.[470]

interactions of the polarized water molecules in the transition state for hydrolysis of **16e**.

The presence of extrema in $\Delta H^{\ddagger}$ for hydrolysis of **16b** and **16e** in t-BuOH–H_2O and the absence of an extreme in $\Delta H^{\ddagger}$ for hydrolysis of **16a** may be taken as evidence that the size of the hydrophobic part in the substrate is one of the factors that determines whether or not extrema in $\Delta H^{\ddagger}$ are produced around $n_{H_2O} = 0.95$. If the solvent operator δ_m is sufficiently

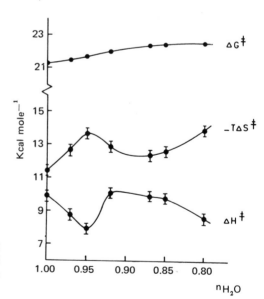

Fig. 15. Plot of $\Delta G^{\ddagger}$, $\Delta H^{\ddagger}$, and $-T\Delta S^{\ddagger}$ vs. n_{H_2O} for the neutral hydrolysis of **16e** in t-BuOH–H_2O at 25°C.[470]

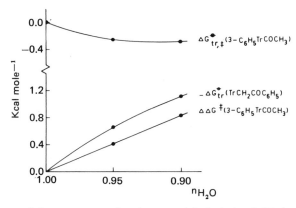

Fig. 16. Free enthalpy parameters for the neutral hydrolysis of **16e** in *t*-BuOH–H$_2$O at 25°C.[470]

different for the initial state and transition state as a result of markedly different contributions of hydrophobic and specific solvation effects for the two species, then the slopes of $\Delta H^\ominus_{tr,i}$ and $\Delta H^\ominus_{tr,\ddagger}$ vs. n_{H_2O} will be different and the functions will pass through extrema at significantly different values of n_{H_2O} thereby leading to extrema in $\Delta H^\ddagger$. Hence, the neutral hydrolysis of *p*-nitrophenyl dichloroacetate (**12**) exhibits more marked extrema in $\Delta H^\ddagger$ than the corresponding hydrolytic reaction of 1-acyl-1,2,4-triazoles

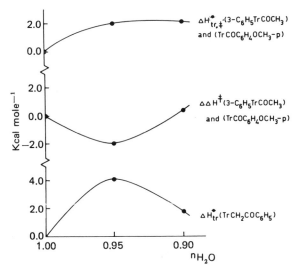

Fig. 17. Enthalpy parameters for the neutral hydrolysis of **16e** in *t*-BuOH–H$_2$O at 25°C.[470]

16a–16e because the activated esters are more hydrophobic than the acyl-triazoles as evidenced by their much lower solubility in water.

The different behavior of $\delta_m \Delta H^{\ddagger}$ for **16c** (Fig. 13) and **16d** (Fig. 14) is more difficult to pinpoint. The difference in solvation between the transition state and the initial state is smaller for hydrolysis of **16d** than for hydrolysis of **16c** since $\delta_m \Delta G^{\ddagger}$ is less for **16d** than for **16c**. This may then account for a smaller difference in the relative contributions of hydrophobic and specific hydration if **16d** is transferred into the transition state.

In an interesting study, Franks and his co-workers[567] have analyzed some mixed aqueous solvent effects on the water-catalyzed mutarotation of glucose. This reaction most likely proceeds via a concerted mechanism involving a cyclic, extensively hydrated transition state leading to ring opening to give the aldehyde form of the sugar.[169] It is proposed that the ease of formation of such a transition state will depend on the steric compatibility of the hydration sites at the C-1 OH group and the ring oxygen group with the intermolecular spacings and orientations of the water molecules involved in the molecular assembly. As indicated by the rate data, the formation of the transition state will be favored by aqueous solutions of the α-anomer. On the basis of model-building experiments, the authors conclude that a cyclic transition state containing four water molecules is the most favorable hydration geometry in aqueous solution, since it will lead to a minimum of perturbation in the hydrogen-bonding pattern of water. The participation of four water molecules in a cyclic transition state has been suggested for several other water reactions, notably the spontaneous hydration of carbonyl compounds like chloral.[819]

The rate of water-catalyzed mutarotation is decreased in both TA (t-BuOH–H_2O, THF–H_2O) and TNA solutions (DMSO–H_2O, DMF–H_2O), the organic cosolvent by itself not promoting mutarotation. However, $\delta_m \Delta G^{\ddagger}$ shows a quite different behavior for both types of aqueous binaries (Fig. 18). Figure 19 portrays the complex function $\delta_m \Delta H^{\ddagger}$ for t-BuOH–H_2O and THF–H_2O. Apparently, the increase of $\Delta G^{\ddagger}$ in the region $n_{H_2O} = 1.00$–0.95 is governed by the contribution of the $\Delta S^{\ddagger}$ term, which is consistent with a solvent effect originating from interference of the hydrophobic hydration spheres of t-BuOH and THF with the hydrophilic hydration spheres normally associated with the sugars.[162,334,850] This likely explanation also hinges on the observation of extrema in $\Delta H^{\ddagger}$ and $\Delta S^{\ddagger}$ and in the kinetic isotope factor at $n_{H_2O} = 0.95$ (Fig. 20), the solvent composition of maximum water ordering. Below $n_{H_2O} = 0.95$, the increase of $\Delta H^{\ddagger}$ becomes the dominating factor in reducing the rate, and below $n_{H_2O} = 0.88$ the solvent apparently loses its TA character (Fig. 19).

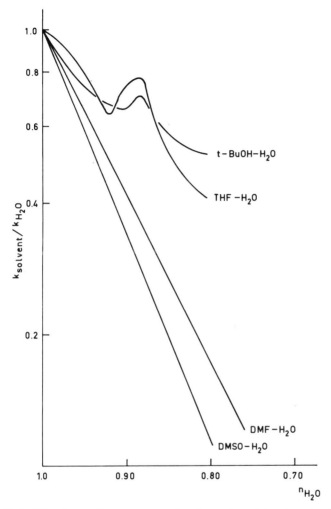

Fig. 18. Plot of $\log k_{solvent}/k_{H_2O}$ vs. n_{H_2O} for the water-catalyzed mutarotation of glucose in t-BuOH–H_2O, THF–H_2O, DMF–H_2O, and DMSO–H_2O at 25°C (except for DMF, 20°C).[567]

Water-catalyzed deprotonation and protonation of organic substrates constitute a second class of water reactions which have been quite extensively investigated in water perturbed by the presence of organic cosolvents. Thus, the water-induced detritiation of malononitrile **17** ($pK_A = 11.19$) and t-butylmalononitrile **3** ($pK_A = 13.10$) have been studied in EtOH–H_2O, dioxane–H_2O, and DMSO–$H_2O^{[403]}$ (see also Section 1). The reaction mechanism involves general base catalysis by the solvent. The Brönsted

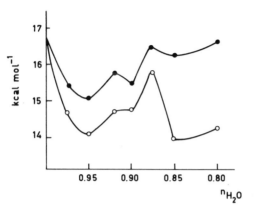

Fig. 19. Plot of $\Delta H^{\ddagger}$ vs. n_{H_2O} for the water-catalyzed muta- rotation of glucose in t-BuOH– H_2O (●) and THF–H_2O (○) (25°C).[567]

coefficient β (= 0.98 ± 0.02) is essentially unity and the rate of the reverse reaction is virtually diffusion-controlled ($k_{H_3O^\oplus} = 4 \times 10^9\ M^{-1}\ sec^{-1}$ at

$$RCT(CN)_2 + S \xrightarrow{\text{slow}} \left[\begin{array}{c} CN \\ | \\ R\overset{\ominus}{C}\cdots TS^\oplus \\ | \\ CN \end{array} \right]^{\ddagger} \longrightarrow R\overset{\ominus}{C}(CN)_2 + ST^\oplus$$

17, R = H

3, R = t-Bu

$$\downarrow \text{proton species (fast)}$$

$$RCH(CN)_2$$

25°C). Therefore, the transition state for proton transfer will resemble an ion pair in which the proton is almost completely transferred from the

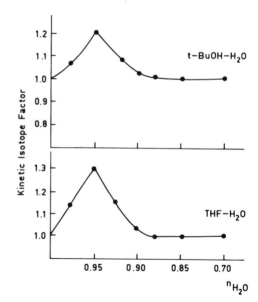

Fig. 20. The influence of solvent composition on the kinetic deu- terium isotope effect for the water- catalyzed mutarotation of glucose at 25°C. The results are nor- malized to $k_{H_2O}/2.9k_{D_2O} = 1$.[567]

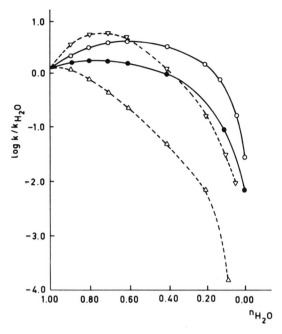

Fig. 21. Plots of log k/k_{H_2O} vs. n_{H_2O} for the detritiation of **17** (EtOH–H$_2$O, ●) and of **3** (dioxane–H$_2$O, △; EtOH–H$_2$O, ○; DMSO–H$_2$O, ▽) at 25°C. All solvents contain 10^{-3} M HCl.[403]

carbon acid to the general base. Extensive solvent immobilization in the transition state is revealed by the strongly negative $\Delta S^{\ddagger}$ (-21 e.u. in H$_2$O). The addition of all three cosolvents results in an initial rate acceleration (Fig. 21; see also Section 1).

The magnitudes of the rate maxima (DMSO > EtOH > dioxane) and the solvent composition at which they occur depend on the nature of the nonaqueous solvent component, and the authors suggest that the reaction could serve as a probe for *kinetic* solvent basicity. However, it is obvious that the process is not a general probe since the solvent operator δ_m working on $\Delta G^{\ddagger}$ will be substrate dependent (*vide infra*). Most interesting are the rate *maxima*, which are unexpected in view of both the low kinetic basicity of the pure cosolvents and the decreased polarity of the medium. In EtOH– H$_2$O, the rate of detritiation roughly follows the H$_-$ acidity function, but divergent behavior in this respect is observed for dioxane–H$_2$O. Solubility data indicate that, in contrast to most of the previous examples of water reactions, the kinetic solvent effects find their origin in transition-state solvation. The authors rationalize the rate data in terms of solvent structural properties, implying that increased water structure leads to enhanced hydro-gen-bonding stabilization[357] of the almost completely formed hydronium ion in the transition state. It should be noted, however, that other factors also play an important role since the TNA solvent DMSO–H_2O exhibits a more pronounced rate maximum than the TA solvent EtOH–H_2O. These factors are most likely specific hydrogen-bonding interactions which have been more explicitly considered in a related hydrolytic process, namely, the water-catalyzed deprotonation of covalent arylsulfonylmethyl per-chlorates **18**[151,282]: this reaction also exhibits general base catalysis

$$\text{ArSO}_2\text{CH}_2\text{OClO}_3 + \text{H}_2\text{O} \xrightarrow{\text{slow}} \text{ArSO}_2\overset{\ominus}{\text{CHOClO}}_3 + \text{H}_3\text{O}^\oplus$$

18a, Ar = p-$NO_2C_6H_4$

18b, Ar = p-$CH_3C_6H_4$

$$\text{H}_2\text{O} \downarrow \text{fast}$$

$$\text{RSO}_2\text{H} + \text{HCOOH} + \text{ClO}_3^\ominus$$

(primary kinetic deuterium isotope effect $k_H/k_D \sim 6$, $k_{H_2O}/k_{D_2O} = 1.7$, Hammett $\varrho = 0.79$).[151] However, the Brönsted β coefficient (~ 0.5) as well as the $\Delta S^\ddagger$ value (-9 e.u. in H_2O) are in this case indicative of only partial proton transfer to the general base in the transition state. The mechanism is either stepwise (E1cB) or involves an E2 pathway with the proton partly transferred and the O–Cl bond partly broken. However, this question is not a crucial one in the discussion of the kinetic solvent effects. Based on the above data and the solvolysis of **18a–18b** in pure organic solvents of sufficient kinetic basicity to deprotonate the substrate (mono-functional aliphatic alcohols, glycol, and a series of dipolar aprotic solvents), it is assumed that the transition state for deprotonation is characterized by (i) a partial and strongly delocalized negative charge which is only weakly involved in hydrogen bonding interactions with the solvent and (ii) a partial positive charge, residing mainly on the incompletely transferred proton and which interacts with the surrounding solvent molecules.[616,617] Pseudo-first-order rate constants as a function of n_{H_2O} are shown in Fig. 22 for solvolysis in t-BuOH–H_2O, dioxane–H_2O, MeCN–H_2O, and dioxane–EtOH. Two features of these plots are particularly noteworthy. First, addi-tion of the organic cosolvents leads to an initial *increase* in rate despite the fact that in dioxane–water and MeCN–H_2O the solvolysis is exclusively a water-induced process (**18a** and **18b** are stable in pure dioxane and MeCN). Second, addition of dioxane to EtOH leads to rate retardation at all values of n_{EtOH} for ethanolysis of **18a**. Furthermore, it was found that the rate-accelerating effect of t-BuOH is larger than that of EtOH (Fig. 23). Since t-BuOH is a more effective structure maker than EtOH, an interpretation

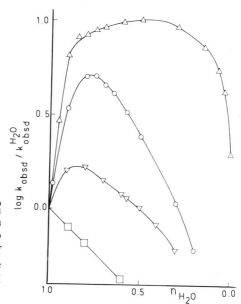

Fig. 22. Plots of $\log k_{obsd}/k_{obsd}^{H_2O}$ vs. n_{H_2O} for the solvolysis of **18a** in t-BuOH–H$_2$O ($\triangle$), dioxane–H$_2$O ($\bigcirc$), and MeCN–H$_2$O ($\triangledown$), and of $\log k_{obsd}/k_{obsd}^{EtOH}$ vs. n_{EtOH} for the solvolysis of **18a** in dioxane–EtOH ($\square$) (25°C).[617]

in terms of a relation between k_{obsd} and water structure suggests itself. But this interpretation is by no means conclusive for two main reasons. First of all it has been shown that the kinetic basicity of t-BuOH for deprotonation of covalent perchlorates **18** is greater than that of EtOH.[618] Second, the initial rate accelerating effect of MeCN in the TNA solvent system MeCN–H$_2$O is opposite to expectation on the basis of a water structure effect.

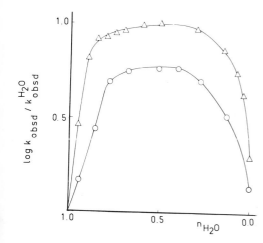

Fig. 23. Plots of $\log k_{obsd}/k_{obsd}^{H_2O}$ vs. n_{H_2O} for the solvolysis of **18a** in t-BuOH–H$_2$O ($\triangle$), and EtOH–H$_2$O ($\bigcirc$) (25°C).[617]

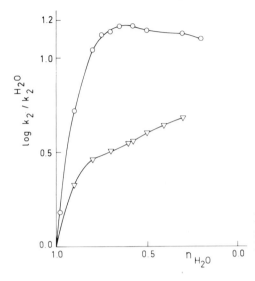

Fig. 24. Plots of $\log k_2/k_2^{H_2O}$ vs. n_{H_2O} for the hydrolysis of **18a** in dioxane–H_2O (○) and in MeCN–H_2O (▽) (25°C).[617]

Taking into account the effective water concentration in dioxane–H_2O and MeCN–H_2O, plots of "k_2" $= k_{obsd}[H_2O]^{-1}$ can be constructed,* which are shown in Fig. 24. Interestingly, the rate constants remain almost constant after a sharp increase with cosolvent concentration in the highly aqueous region. Since the solvation behavior of the transition state is predominantly determined by interactions between its cationic part and the solvent (*vide supra*), the rate accelerations found for the TNA mixtures are probably best explained by the concept of water polarization resulting from hydrogen bonding of water to the organic cosolvent in complexes like **19** (see Section 3.1).

$$S\overset{\frown}{\cdots}\underset{\delta\oplus}{H}-\overset{\overset{\delta\ominus}{\frown}}{O}\diagdown H$$

19, S $=$ dioxane, MeCN

Support of this idea is provided by the high intrinsic basicities of ethers[569,596] and acetonitrile[40] as compared with water and revealed by gas-phase proton affinities: $PA_{Et_2O} = 205$ kcal mol^{-1}, $PA_{Me_2O} = 190$ kcal mol^{-1}, $PA_{MeCN} = 186$ kcal mol^{-1}, $PA_{H_2O} = 165$ kcal mol^{-1}. It has also been demonstrated that $H_3O^{\oplus}$ interacts more favorably with three ether molecules than with three water molecules, owing to the greater

---

* For a discussion of the meaning of "second-order" rate constants for solvolysis reactions, see Ref. 535.

electron density on the ether–oxygen atom.[361,362] Now the low kinetic basicity in the liquid phase of pure dioxane and MeCN relative to water may be ascribed to the aprotic character of these solvents. The solvent-induced deprotonation of **18** requires a transition state of well-defined geometry in which at least one solvent molecule will be largely immobilized. This will cost more entropy for solvent molecules that are not inter-molecularly hydrogen bonded than for water, where the molecules form already part of an associated structure. This interpretation is supported by the $\Delta H^\ddagger/\Delta S^\ddagger$ data to be discussed below.

Assuming that the transition-state solvation dominantly affects[616,617] $\delta_m \Delta G^\ddagger$, the stabilizing effect of dioxane and MeCN may be schematically represented as follows:

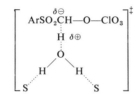

$$S = \text{dioxane, MeCN, } H_2O \text{ (for one S)}$$

The smooth decrease of k_{obsd} in *ethanol* upon addition of dioxane[617] is proportional to the decreasing ethanol concentration and reflects the relatively small difference between the gas-phase proton affinities of EtOH and ethers, and the absence of extensive three-dimensional hydrogen-bonded structures in liquid ethanol.

The observation that $k_{obsd}[H_2O]^{-1}$ values are practically constant below $n_{H_2O} = 0.6$ in dioxane–water may be the result of several factors. It may well be that the transition state as depicted above in which *both* molecules S are dioxane is unfavorable for entropic reasons. Furthermore, in this solvent composition range ε falls now below 30, which will discourage charge separation in the transition state. Thirdly, preferential solvation of the substrate by dioxane may now come into play (as suggested by nmr measurements for the disulfone $CH_3SO_2CH_2SO_2CH_3$).[468]

The behavior of $\Delta H^\ddagger$ and $\Delta S^\ddagger$ as a function of n_{H_2O} for hydrolysis of **18a** in dioxane–H_2O and MeCN–H_2O is displayed in Figs. 25 and 26, respectively. The extrema around $n_{H_2O} = 0.70$ in dioxane–H_2O are especially noteworthy. Upon the first addition of dioxane to water, the rate enhance-ment is governed by a decrease in $\Delta H^\ddagger$, which is only partly compensated by a decrease in $\Delta S^\ddagger$. Below $n_{H_2O} = 0.70$, the decrease of k_{obsd} is determined by a decrease in $\Delta S^\ddagger$. The pronounced variation of $\Delta H^\ddagger$ and $\Delta S^\ddagger$ with n_{H_2O}

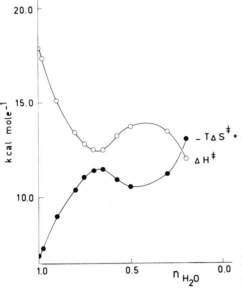

Fig. 25. Plots of $\Delta H^{\ddagger}$ and $-T\Delta S^{\ddagger}$ vs. n_{H_2O} for the hydrolysis of **18a** in dioxane–H_2O at 25°C.[617]

is typical for the aqueous mixed solvents and contrasts strongly with the small changes in these activation parameters for ethanolysis in dioxane–EtOH.[617] As outlined above, the proton-accepting water molecule is probably polarized in the transition state through hydrogen-bonding interactions to dioxane and MeCN. This is consistent with the observed decrease

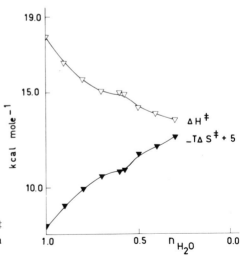

Fig. 26. Plots of $\Delta H^{\ddagger}$ and $-T\Delta S^{\ddagger}$ vs. n_{H_2O} for the hydrolysis of **18a** in MeCN–H_2O at 25°C.[617]

in $\Delta H^{\ddagger}$. However, if this were the only factor, a smooth and monotonic decrease of $\Delta H^{\ddagger}$ (and $\Delta S^{\ddagger}$) with n_{H_2O} would be expected. This behavior is, in fact, only observed for MeCN–H$_2$O. The extreme in $\Delta H^{\ddagger}$ in dioxane–H$_2$O then probably arises from effects due to cosolvent-induced changes in solvent structure. Unfortunately, there is considerable controversy about the solvent structural properties of dioxane–H$_2$O and, consequently, for a discussion of this factor we would do better to turn to solvolysis of the covalent perchlorates **18** in some representative TA solvent systems. Most experimental data are available for alcohol–water mixtures. The behavior of $\delta_m \Delta G^{\ddagger}$ may be summarized as follows: (i) rate maxima[617] (Fig. 22) are reached at lower n_{H_2O} when the hydrophobic part of the alcohol is smaller, (ii) the magnitude of the maximum rate enhancement becomes smaller in the series t-BuOH > EtOH > glycol, and (iii) most aliphatic alcohols cause rate enhancements except trifluoroethanol. The rate enhancements that occur upon initial addition of the alcohols are in line with the higher kinetic basicities of the alcohols towards the covalent perchlorates **18** as compared with water and which are related to their gas-phase proton affinities[63,569] ($PA_{MeOH} = 180$, $PA_{EtOH} = 187$, and $PA_{t\text{-}BuOH} = 198$ kcal mol^{-1}). In the transition state for proton transfer from **18** to water, the attacking water molecule may form part of a water–alcohol hydrogen-bonded complex of enhanced kinetic basicity similar to those described earlier for dioxane and MeCN. Support for this interpretation is provided by the decrease in rate of hydrolysis of **18a** upon addition of the weak hydrogen-bond acceptor TFE to water.[617] The inherent basicities of TFE and H$_2$O are comparable[596] ($PA_{TFE} = 168$, $PA_{H_2O} = 165$ kcal mol^{-1}), making TFE–H$_2$O complexes less favorable for acting as a general base than the other alcohol–water complexes. However, from the "plateau rates" in the plots of log k_{obsd}/log $k_{obsd}^{H_2O}$ vs. n_{H_2O} (Fig. 27) it is

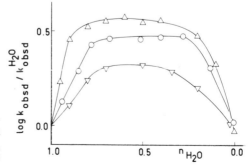

Fig. 27. Plots of log k_{obsd}/$k_{obsd}^{H_2O}$ vs. n_{H_2O} for the solvolysis of **18b** in t-BuOH–H$_2$O ($\triangle$), EtOH–H$_2$O ($\bigcirc$), and glycol–H$_2$O ($\triangledown$) at 25°C.[617]

obvious that the change in the kinetic basicity of the mixed solvent, based upon the kinetic basicities of the solvent components, is not the only factor that determines the kinetic solvent effect. Since the "plateau rate" is reached at higher n_{H_2O} when the alkyl part of the alcohol is more voluminous, the effect may be due to the decreasing ε of the medium with decreasing n_{H_2O}. The ε of the mixture at which the "plateau rate" is reached is approximately equal in t-BuOH–H_2O ($n_{H_2O}^{max} = 0.9$, $\varepsilon = 55$) and in EtOH–H_2O ($n_{H_2O}^{max} = 0.8$, $\varepsilon = 55$), but is higher in glycol–H_2O ($n_{H_2O}^{max} = 0.8$, $\varepsilon = 65$) and lower in dioxane–H_2O ($n_{H_2O}^{max} = 0.8$, $\varepsilon = 30$). A factor of further consideration may be alcohol-induced changes in water structure, the structure-promoting effect decreasing in the series t-BuOH > EtOH > glycol. The extrema in $\Delta H^{\ddagger}$ and $\Delta S^{\ddagger}$ at $n_{H_2O} = 0.85$ for solvolysis of **18a** in t-BuOH–H_2O are not unexpected in terms of this theory (Fig. 28) and the minimum in $\Delta H^{\ddagger}$ may be, at least in part, a consequence of the increase in the kinetic basicity of water with increasing degree of water structure.[617] Unfortunately, the situation is highly complicated, the more since both solvent components may act as general bases in the deprotonation of **18**. It seems likely, however, that water-structure effects modulate primary solvation changes induced by addition of the organic cosolvent to water and may be invoked to explain the extrema in $\Delta H^{\ddagger}$ and $\Delta S^{\ddagger}$ in the water-rich region.

Very recently, Symons[846] has interpreted the cosolvent-induced kinetic effects on the water-induced deprotonation of **18** by considering a

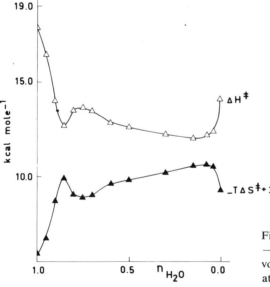

Fig. 28. Plots of $\Delta H^{\ddagger}$ and $-T\Delta S^{\ddagger}$ vs. n_{H_2O} for the solvolysis of **18a** in t-BuOH–H_2O at 25°C.[617]

simple equilibrium between completely hydrogen-bonded water molecules and water molecules containing "free" OH groups and "free" lone pairs:

$$(H_2O)_{bound} \rightleftarrows (OH)_{free} + (LP)_{free}$$

The physical significance of this model was sought in the observation of absorption bands in the overtone infrared region which are associated with $(OH)_{free}$ oscillators. It is proposed that aprotic solvents like dioxane scavenge $(OH)_{free}$ groups leading to a gain of $(LP)_{free}$ available for deprotonation of **18**. This implies that k_{obsd} will be proportional to the concentration of $(LP)_{free}$. Although the rate increase upon *initial* addition of dioxane may be accounted for by Symons' model, a quantitative test for the assumption that the *concentration* of $(LP)_{free}$ groups is the only parameter determining $\Delta G^{\ddagger}$ in dioxane–water has still to be given.

Another type of solvent-structure effect[843,913] may be responsible for the strong increase in $\Delta H^{\ddagger}$ and $\Delta S^{\ddagger}$ for the solvolysis of **18** in *t*-BuOH–H$_2$O below $n_{H_2O} = 0.1$. Some diffusion measurements on very dilute solutions of water in alcohols suggest that the addition of small quantities of water to *t*-BuOH leads to a more stable hydrogen-bonded structure (built around water molecules) as compared with that in the unperturbed alcohol. This may be reflected in the dramatic changes in $\Delta H^{\ddagger}$ and $\Delta S^{\ddagger}$. A similar phenomenon has been observed for some ligand substitution reactions.[163]

Rate-limiting proton transfer from water to a carbon *base* has also been investigated in mixed aqueous solvents. The neutral hydrolysis of ketene bis(2-methoxyethyl) acetal **20** is a representative example. As

found by Kankaanperä,[463,464] the Brönsted α (determined in bicarbonate–carbonate buffer solutions) of this A-S$_E$2 reaction is 0.54, and k_{H_2O}/k_{D_2O} amounts to 7.2. Both observations are indicative of a transition state in which the proton is only partly transferred from water to sp^2 carbon ($\Delta S^{\ddagger} = -20$ e.u. in water). Activation parameters have been determined by Huurdeman and Engberts[429] for solvolysis in *t*-BuOH–H$_2$O between $n_{H_2O} = 1.00$ and 0.85 and are graphically shown in Fig. 29. These data pertain only to water-induced reactions since solvolysis in pure *t*-BuOH is some orders of magnitude slower than in water. Rates of hydrolysis are

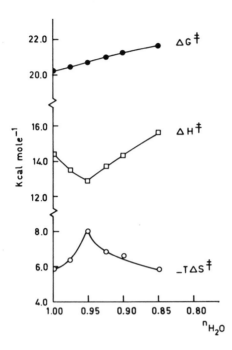

Fig. 29. Plots of $\Delta G^{\ddagger}$, $\Delta H^{\ddagger}$, and $-T\Delta S^{\ddagger}$ for hydrolysis of **20** in t-BuOH–H_2O at 25°C.[429]

clearly retarded upon addition of t-BuOH, as has also been found for dioxane as the cosolvent,[464] and the slope of a plot of log k_{obsd} vs. log c_{H_2O} is between 4 and 5. The smooth variation of $\Delta G^{\ddagger}$ with n_{H_2O} conceals mirror image behavior of $\Delta H^{\ddagger}$ and $\Delta S^{\ddagger}$. The minimum in $\Delta H^{\ddagger}$ at $n_{H_2O} = 0.95$ is in line with similar endothermic extrema in $\Delta H^{\ddagger}$ for several water-catalyzed reactions discussed above, and its rather modest magnitude possibly reflects the polar nature of the substrate due to the sizable contribution of resonance hybrid **20b**.

$$
\begin{array}{ccc}
\underset{\textstyle\text{H}}{\text{H}}\diagdown \hspace{-0.3em} \underset{}{\text{C=C}} \hspace{-0.3em} \diagup\underset{\textstyle\text{OCH}_2\text{CH}_2\text{OCH}_3}{\text{OCH}_2\text{CH}_2\text{OCH}_3} & \longleftrightarrow & \underset{\textstyle\text{H}}{\text{H}}\diagdown \hspace{-0.3em} \underset{}{\overset{\ominus}{\text{C}}-\overset{\oplus}{\text{C}}} \hspace{-0.3em} \diagup\underset{\textstyle\text{O}-\text{CH}_2\text{CH}_2\text{OCH}_3}{\text{O}-\text{CH}_2\text{CH}_2\text{OCH}_3}
\end{array}
$$

20a   **20b**

5.3. Acid- and Base-Catalyzed Reactions

Both general and specific acid- and base-catalyzed reactions have been frequently investigated in mixed aqueous solutions. The reaction mechanisms are more complex than those of water-catalyzed processes (Section 5.2) and in many cases involve one or even a series of charged intermediates.

Very often, hydroxylic solvent molecules will participate in the proton-transfer reactions.[18,364] The incorporation of two or more species other than the solvent in the transition state of the rate-determining step leads to kinetic solvent effects which critically depend on the mechanism of the reaction and on the position of the transition state on the reaction coordinate. These factors make the interpretation of solvent effects even more hazardous than for simpler processes. Nevertheless, the type of analysis outlined in Section 4.3 is often feasible. Additional complications are introduced for those systems in which the quantitative description of (de)-protonation equilibria necessitates the use of acidity functions. The limitations of this approach, which are more severe for mixed aqueous solutions (particularly those of low dielectric constant) than for water as the solvent, have been reviewed.[763]

In the following discussion we will consider some representative acid- and base-catalyzed processes for which the kinetic solvent effects have been rationalized in more depth than only in terms of some correlation with solvent polarity.

An early example involves the work of Patai et al.[687,688] on the condensation reaction of malononitrile and ethyl acetoacetate with benzaldehyde and substituted benzaldehydes. These reactions showed maxima in a plot of initial rate coefficients vs. solvent composition in EtOH–H_2O. The extrema were tentatively assigned to counteracting effects of the superior basicity of EtOH compared with water and the better carbanion solvating power and higher dielectric constant of water.

One category of simple base-catalyzed processes for which kinetic solvent effects have recently been analyzed in more detail comprise the hydroxide-ion-catalyzed deprotonation of carbon acids (i.e., species with relatively acidic C–H bonds) in TNA solvent systems like DMSO–H_2O and MeCN–H_2O.[343,456] Especially the addition of DMSO to water strongly decreases the overall hydrogen-bond donor ability of the solvent system and rates are often enhanced with increasing DMSO content as a result of decreased hydrogen-bond stabilization of the reactants and (or) enhanced solvation of the transition state. Interpretation of kinetic solvent effects in these media is facilitated by the availability of free energies, enthalpies, and entropies of transfer of a variety of organic solutes from water to DMSO–H_2O, MeCN–H_2O, and dioxane–H_2O.[205] The gross features of these thermodynamic transfer functions are rather similar, but DMSO shows particularly strong hydrogen-bond acceptor ability and polarizability.

The base-catalyzed deprotonation of tritium-labeled chloroform has been studied by several investigators. For this reaction, general base catalysis is not observable in aqueous buffer systems and hydroxide ion dominates the base catalysis (Brönsted $\beta = 0.98$, primary kinetic deuterium isotope effect $k_H/k_D = 1.42$ at 25°C).[594] The mechanistic implication of the large β is that proton (tritium) transfer from the substrate to $OH^\ominus$ will be virtually complete in the transition state of the slow step. Rate

$$TCCl_3 + OH^\ominus \xrightarrow{\text{slow}} {}^\ominus CCl_3 + TOH$$

$$^\ominus CCl_3 + HA \xrightarrow{\text{fast}} HCCl_3 + A^\ominus$$

constants and activation parameters for ionization in DMSO–H_2O are summarized in Table IX. Apart from the marked rate acceleration with decreasing n_{H_2O}, the most noticeable result is the decrease of $\Delta S^\ddagger$ by ~ 10 e.u. in the solvent composition range $n_{H_2O} = 1.00$–0.76. Long et al.[594] rationalized this result by assuming that in the transition-state desolvation of the heavily solvated $OH^\ominus$ leads to a release of water molecules, which is not compensated by an entropy loss, owing to hydration of the $^\ominus CCl_3$ ion, which is almost completely formed in the transition state. Of course, this desolvation becomes less the more DMSO is present in the medium. In a more recent analysis in terms of enthalpy changes, the change in transition state solvation with n_{H_2O} was obtained by Jones and Fuchs[457] by combining the $\Delta H^\ddagger$ values given in Table IX with enthalpies of solution of chloroform in water and DMSO–H_2O and enthalpies of transfer of $OH^\ominus$ from H_2O to DMSO–H_2O (Table IX). If it is assumed that $\delta_m \Delta G^\ddagger$ parallels

TABLE IX. Enthalpies of Activation and Enthalpies of Transfer of Reactants and Transition State in the $OH^\ominus$-Catalyzed Ionization of $CTCl_3$ in DMSO–H_2O

n_{H_2O}	$\Delta H^{\ddagger\,a}$ (kcal mol^{-1})	$\delta_m H_{OH^\ominus}{}^b$ (kcal mol^{-1})	$\Sigma \delta_m H_{\text{reactants}}{}^b$ (kcal mol^{-1})	$\delta_m H_\ddagger$ (kcal mol^{-1})
1.000	23.2	0.0	0.0	0.0
0.973	22.5	-0.2	-0.2	-0.9
0.941	21.6	-0.6	-0.3	-1.9
0.910	21.1	-0.8	-0.3	-2.4
0.870	20.2	0.0	0.7	-2.3
0.829	19.0	2.2	3.1	-1.1
0.797	18.0	4.0	4.8	-0.4
0.764	16.5	5.4	6.1	-0.6

[a] From Ref. 594.
[b] From Ref. 457.

$\delta_m \Delta H^{\ddagger}$, the picture emerges that the increase in rate induced by low concentrations of DMSO will be primarily due to increased transition state solvation. Higher concentrations of DMSO accelerate the ionization process mainly by a reduction in the solvation of $OH^{\ominus}$. The latter factor is also the major reason for the increase in the rate of $OH^{\ominus}$-catalyzed racemization of D-α-methyl-α-phenylacetophenone (21) in DMSO–H$_2$O between $n_{H_2O} = 0.278$ and 0.516.[441] In this solvent range the desolvation of $OH^{\ominus}$ occurs to the extent of 6.4 kcal mol^{-1}. However, an evaluation of transition state enthalpies of transfer revealed that desolvation of the transition state partly counteracts the DMSO-induced enhancement of the kinetic basicity of $OH^{\ominus}$.

$$C_6H_5\overset{O}{\overset{\|}{C}}\!-\!\overset{*}{C}H\!\underset{Ph}{\overset{Me}{<}} \quad \xrightarrow{OH^{\ominus}} \quad C_6H_5\overset{O^{\ominus}}{\overset{|}{C}}\!=\!C\!\underset{Ph}{\overset{Me}{<}} \quad + \text{ H}_2\text{O}$$

21

The base-induced deprotonation of carbon acids has also been employed as a probe for the study of the solvent dependence of primary kinetic deuterium isotope effects. Previously, it has been often assumed that for proton transfer from an acid HX to a base Y, a maximum isotope effect would occur in the vicinity of $\Delta pK = pK_{XH} - pK_{YH} = 0$. This was mainly based on a theory described by Westheimer,[910] which predicted a maximum isotope effect when the transition state for proton transfer is symmetrical. Consideration of quantum mechanical tunnel effects[73] also led to the same prediction. Now Bell, Cox, and their co-workers anticipated that mixed aqueous solvents could function excellently for continuously varying ΔpK through changing n_{H_2O} and keeping the substrate and catalyst constant. Indeed, k_H/k_D for the $OH^{\ominus}$-catalyzed inversion of (−)-menthone passes through a maximum ($k_H/k_D = 6.5$) at $\sim n_{H_2O} = 0.35$ in DMSO–H$_2$O, which approximately coincides with the solvent composition for which $OH^{\ominus}$ and the menthone anion have similar basicities.[74] Other examples for which maxima in k_H/k_D have been observed in DMSO–H$_2$O include the rate of ionization of nitroethane[75] and the $OH^{\ominus}$-induced β-elimination reaction of dimethyl-2-phenylsulfonium bromides.[193] Subsequent studies of rates and equilibria for acetate anion catalyzed proton transfer reactions from several carbon acids in DMSO–H$_2$O and CF$_3$CH$_2$OH–H$_2$O indicated however, that addition of the organic cosolvent to water may lead to large rate increases but relatively small changes in ΔpK and, consequently, that there is no simple relationship between rates and equilibria in these aqueous binaries. As argued by Cox and Gibson,[206,207] solvation effects on pK's of acids involved in proton transfer reactions will not generally be reflected in the transition states. This may result in anomalous Brönsted β values like $\beta = 1.4$ for the reaction of nitroalkanes with $OH^{\ominus}$ in water.[77] That these anomalous β values are a consequence of specific hydration effects on the acidity constants is consistent with the observation that the deviant β values disappear in $DMSO-H_2O$.

Recently it was found that for several simple proton-transfer reactions involving 2-nitropropane, nitroethane, and ethyl 2-methylacetoacetate, k_H/k_D exhibits no simple correlation with reaction rates.[208] In this work it was suggested that the magnitude of $\delta_m(k_H/k_D)$ would be sensitive to the involvement of the solvent in the proton transfer through its effect on the tunnel correlation.

In a recent study, Bunton and his co-workers[157] have measured solvent effects on the reaction of malachite green (22) with a series of nucleophiles. This work was carried out as part of a study aimed at the elucidation of the generality of Ritchie's N_+ scale of nucleophilicity.[751,752] In the N_+ scale, anionic nucleophilicity is related primarily to anionic desolvation.

$$(p\text{-Me}_2\text{NC}_6\text{H}_4)_2\overset{\oplus}{\text{C}}\text{---C}_6\text{H}_5$$

22

For $OH^{\ominus}$ as the nucleophile, the solvent isotope effects in the reaction of **22** are $k_{H_2O}/k_{D_2O} = 1.50$ and $k_{OH^{\ominus}}/k_{OD^{\ominus}} = 1.10$, which are consistent with general base catalysis by $OH^{\ominus}$ of the attack of water on the carbonium ion center. In $t\text{-BuOH-H}_2O$ and $MeCN\text{-H}_2O$, k_{rel} values ($k_{rel} = k/k_{n_{H_2O}=1}$) pass through minima, the one for the TA $t\text{-BuOH-H}_2O$ solvent at $n_{H_2O} = 0.95$ being the most pronounced (Fig. 30). The authors propose that

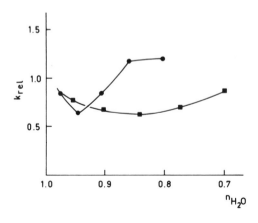

Fig. 30. Plot of k_{rel} vs. n_{H_2O} for the reaction of **22** with $OH^{\ominus}$ in $t\text{-BuOH-H}_2O$ (●) and MeCN–H_2O (■) at 25°C.[157]

these minima probably reflect the effect of the cosolvent on water structure. Especially for t-BuOH–H_2O this seems reasonable in view of results of other studies employing $OH^\ominus$ in this binary mixture (Section 5.4).

The $OH^\ominus$ catalyzed hydrolysis of ethyl acetate has also been studied in DMSO–H_2O.[342] Again, transition state enthalpies of transfer from H_2O to DMSO–H_2O were derived from activation enthalpies and enthalpies of transfer of the reactants. Interestingly, below $n_{H_2O} = 0.85$ the enthalpy of desolvation of the transition state exceeds that for desolvation of $OH^\ominus$ and the increasing reaction rates with decreasing n_{H_2O} originate from an entropy, rather than from an enthalpy effect. In a series of other aqueous binaries, transition-state solvation effects also dominate $\delta_m \Delta G^\ddagger$.[118]

A final example of a base-catalyzed process involves the deprotonation of the internally hydrogen-bonded weak acid tropaeolin O (2,4-dihydroxy-4'-sulfonatoazobenzene 23) by $OH^\ominus$ ions.[694,695] The kinetics have been

23

investigated by the temperature jump method. They fit a scheme in which the reaction takes place by two parallel paths involving the internally hydrogen bonded form ($HA_{HB}^{2\ominus}$) and a more reactive species present in low concentrations which is presumably the non-hydrogen-bonded acid ($HA_{NHB}^{2\ominus}$). Most likely reactions (i) $\rightleftarrows$ (ii) are fast in comparison with the

other two reactions. Quite surprisingly, the reaction (iii) → (ii) showed large activation parameters ($\Delta H^\ddagger = 26.8 \pm 1.6$ kcal mol^{-1}, $\Delta S^\ddagger = 46.5 \pm 5.5$ e.u.), which are difficult to reconcile with the breaking of just one intramolecular hydrogen bond. The authors rationalize this result by invoking the different effects of $HA_{HB}^{2\ominus}$ and $HA_{NHB}^{2\ominus}$ on the water structure. Assuming that $HA_{HB}^{2\ominus}$ enhances the water structure, and that this propensity for the formation of a hydrophobic hydration sphere decreases if this species is transferred into the open form with a free OH moiety, it is proposed that the process (iii) → (ii) is accompanied by the rupture of, say, five to six hydrogen bonds instead of one. The large and positive $\Delta S^{\ddagger}$ is, of course, in agreement with this theory. It would be of great interest to investigate the tropaeolin O reaction in TA solvent mixtures as a function of n_{H_2O}.

Kinetic solvent effects on the specific acid-catalyzed hydrolysis of some aliphatic acetals (A1 mechanism) have been studied in DMSO–H_2O and dioxane–H_2O.[613] Both organic cosolvents strongly reduce the rate, a rate minimum being observed in dioxane–H_2O around $n_{H_2O} = 0.6$. Measurements of medium activity coefficients for the acetals in the solvent systems permitted the calculation of transition state activity coefficients. The behavior of the latter as a function of n_{H_2O} does not deviate significantly from that of some structurally related stable ions, thereby lending support to the view that transition-state theory provides a valid description of reactions between solvated reactants in aqueous media (compare Section 4.1). Furthermore the results confirmed earlier work in which it was demonstrated that specific solvation effects are large for positive ions in mixtures of water and aprotic solvents.[209]

As pointed out in Section 2.1, glycerol–H_2O constitutes a relatively ideal TNA mixture. This solvent mixture has been employed by Schaleger and Richards[790] as a medium for the specific acid-catalyzed hydrolysis of the cyclic acetals 24–26 (A1 mechanism). As expected for this particular binary, $\Delta H^{\ddagger}$ and $\Delta S^{\ddagger}$ for hydrolysis of 24–26, show only relatively

24, $R_1 = H$, $R_2 = R_3 = Me$
25, $R_1 = H$, $R_2 = Me$, $R_3 = i\text{-Pr}$
26, $R_1 = CH_2OH$, $R_2 = R_3 = Me$

small and compensatory changes upon changing n_{H_2O} resulting in minor changes in $\Delta G^{\ddagger}$. There are no distinct maxima in $\Delta H^{\ddagger}$ and $\Delta S^{\ddagger}$ (Table X). Via measurements of the thermodynamic quantities of transfer from water to glycerol–H_2O of the reactants, the medium effects on $\Delta H^{\ddagger}$ and $\Delta S^{\ddagger}$ were resolved into their initial-state and transition-state components. From the results it was concluded that glycerol destabilizes both the hydronium ion and the cationic transition state of the slow step. However, dioxolane 26 shows dissimilar behavior in that $\Delta H^{\ddagger}$ and $\Delta S^{\ddagger}$ are hardly medium dependent (Table X, 0–40 wt.% glycerol; $n_{H_2O} = 1.00$–0.88) owing to a fortuitous cancellation of initial- and transition-state contributions. Substituent-dependent specific solute–solvent interactions apparently turn the negative change in the partial molar enthalpy of the transition state for 24 ($\Delta H^{\ominus}_{tr,\ddagger} = -1.0 \pm 0.4$ kcal mol^{-1} between $n_{H_2O} = 1.00$ and 0.88) into a positive one for 26 ($\Delta H^{\ominus}_{tr,\ddagger} = 1.9 \pm 0.5$ kcal mol^{-1}).

TABLE X. Activation Parameters for the Hydronium-Ion-Catalyzed Hydrolysis of the Dioxolanes 24–26 in Glycerol–H_2O Mixtures at 25°C

Wt. % glycerol	$\Delta H^{\ddagger}$ (kcal mol^{-1})			$\Delta S^{\ddagger}$ (e.u.)		
	24	25	26	24	25	26
0	20.7 ± 0.2	20.0 ± 0.4	19.3 ± 0.4	7.1 ± 0.7	4.8 ± 1.3	0.1 ± 1.3
10	19.5 ± 0.3	18.9 ± 0.1	19.3 ± 0.8	3.5 ± 1.0	1.4 ± 0.5	−0.1 ± 2.7
20	19.7 ± 0.3	19.7 ± 0.4	19.5 ± 0.6	4.3 ± 1.0	2.2 ± 0.5	0.8 ± 2.0
30	17.9 ± 0.3	18.3 ± 0.1	18.9 ± 0.3	−1.6 ± 1.0	−0.3 ± 0.7	−1.1 ± 1.0
40	17.6 ± 0.3	17.6 ± 0.6	19.0 ± 0.1	−2.4 ± 1.0	−2.5 ± 2.0	−0.4 ± 0.5

The rate constants for the hydronium-ion-catalyzed hydrolysis of benzyl acetate ($A_{AC}2$ mechanism) in acetone–water pass through two extrema: one in the water-rich region (~ 12 wt. % acetone; $\Delta H^{\ddagger}$ is at a minimum at that solvent composition) and one at high acetone content (~ 80 wt. % acetone).[481] Sadek[481] and Tommila[865] explain the minimum in $\Delta G^{\ddagger}$ by invoking structure breaking by the cosolvent, which would produce an increase in the number of "free" water molecules available for the reaction. However, this explanation seems tentative at the moment and more insight into the problem may be gained through a separation of initial-state and transition-state solvation effects. Quite generally, $\Delta G_{tr,H^{\oplus}}$ is positive upon transfer from water to an aqueous binary. Below 88 wt. % water, the decrease in rate would be governed by the decrease in the stoichiometric concentration of water until a maximum in $\Delta G^{\ddagger}$ is reached at ~ 28 wt. % H_2O. Then the rate goes up again apparently owing to a sharp increase in acidity (catalysis by $Me_2COH^{\oplus}$ instead of $H_3O^{\oplus}$) of the medium as evidenced by acidity function behavior.

A number of other acid-catalyzed ester hydrolyses show extrema in activation parameters in aqueous binaries, e.g., extrema in $\Delta G^{\ddagger}$ in alcohol–water, dioxane–water, acetone–water, and DMSO–water[383,533,777,861,863,864] and extrema in $\Delta V^{\ddagger}$ in acetone–water.[48] However, these phenomena are by no means a general rule for $A_{AC}2$ processes.[862]

The perchloric acid catalyzed hydrolysis of p-substituted benzoic anhydrides in dioxane–H_2O is an example of a process for which the reaction mechanism depends on the magnitude of n_{H_2O}. As n_{H_2O} decreases, the mechanism changes from A2 ($A_{AC}2$) to A1 ($A_{AC}1$). For the HCl-catalyzed reaction in 60:40 (v/v) dioxane–H_2O a similar mechanistic change occurs when the acidity and the temperature are increased.[164]

Solvent selectivity and stereoselectivity constitutes an interesting but difficult problem for reactions in mixed aqueous solvents when both solvent components take part in the solvolysis of a substrate. In this context, the H_2SO_4-catalyzed solvolysis of 1-phenylcyclohexene-oxide (**27**) in EtOH–H_2O has recently been studied.[61] Over the whole solvent composition range, the solvolysis is completely regiospecific and involves exclusive attack on the tertiary benzylic carbon atom by water to give **28** and **29** or by ethanol to yield **30** and **31**. The products are stable under the reaction conditions

and only minor amounts of side products are produced. The molar ratios of the products **28–31** are listed as a function of solvent composition in Table XI. If the solvent selectivity (S_{sel}) is defined by

$$S_{sel} = \log \frac{k_{EtOH}}{k_{H_2O}} = \log \frac{C_{ROEt}C_{H_2O}}{C_{ROH}C_{EtOH}}$$

where the ratio of the rate constants for ethanolysis (k_{EtOH}) and for hydrolysis (k_{H_2O}) are calculated from the molar ratio of hydroxyethers (**30, 31**) to diols (**28, 29**), the overall selectivity can be plotted as a function of n_{H_2O} (Fig. 31). The plot shows that S_{sel} increases up to $n_{H_2O} = 0.83$ in the sense that the richer the solution is in water, the higher is the preference for attack by EtOH. This solvent sorting effect (compare Section 2.1) suggests that the composition of the solvation shell of the transition state leading to incorporation of the nucleophile is different from that of bulk solvent. The observation that $S_{sel} = 0$ at $n_{H_2O} = 0.5$ is probably a coincidence. Assuming that ring opening occurs after initial reversible protonation at the oxirane oxygen atom, the behavior of $\delta_m S_{sel}$ below $n_{H_2O} = 0.83$ is consistent with the view that the more polar solvent molecules (H_2O) are able to solvate the polar transition state more favorably in a medium of low polarity leading to preferential formation of the diol. If the water

TABLE XI. Molar Ratios of Products as a Function of Solvent Composition for the Acid-Catalyzed Solvolysis of 27 in EtOH–H_2O at 25°C[61]

n_{H_2O}	H_2O:EtOH (v/v)	28	29	30	31
1.000	100:0	37.4	62.6	0.0	0.0
0.984	95:5	37.6	60.8	0.5	1.1
0.967	90:10	36.1	60.4	1.0	2.5
0.929	80:20	38.8	53.3	2.2	5.7
0.883	70:30	35.0	50.4	4.6	10.0
0.830	60:40	32.9	44.8	7.5	14.8
0.765	50:50	30.9	39.8	11.1	18.2
0.684	40:60	30.2	35.1	14.1	20.6
0.500	23.5:76.5	25.0	24.2	25.0	25.8
0.263	10:90	16.4	13.5	40.4	29.7
0.146	5:95	9.8	8.1	50.5	31.6
0.000	0:100	0.0	0.0	67.5	32.5

content is increased above $n_{H_2O} = 0.83$, S_{sel} decreases again, most likely because the ethanol molecules are now heavily solvated by water leading to a reduced nucleophilicity. Figure 32 portrays the solvent selectivity for syn [$\log(k_{EtOH}^{syn}/k_{H_2O}^{syn})$] and anti [$\log(k_{EtOH}^{anti}/k_{H_2O}^{anti})$] opening of the oxirane ring. For a discussion of these data in terms of a tentative but attractive mechanistic scheme, the reader is referred to the original literature.[61] As noted already in Section 5.3, the water content of a mixed aqueous solvent may also affect the stereospecificity of an organic reaction. A dramatic example is provided by the D_2SO_4-catalyzed homoketonization of 1-acetoxy-

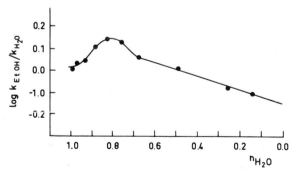

Fig. 31. Plot of overall solvent selectivity [$\log k_{EtOH}/k_{H_2O}$] vs. n_{H_2O} for the H_2SO_4-catalyzed solvolysis of 27 in EtOH–H_2O at 25°C.[61]

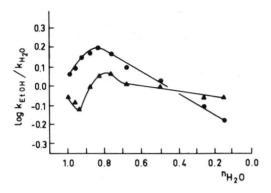

Fig. 32. Plot of selectivity for syn (●) and anti (▲) adducts vs. n_{H_2O} for the H_2SO_4-catalyzed solvolysis of **27** in EtOH–H_2O at 25°C.[61]

nortricyclane (**32**) in CH_3CO_2D–D_2O.[656] Whereas in 1:1 (v/v) CD_3CO_2D–D_2O the reaction proceeds with high retention (93% formation of **34**), 94% inversion (formation of **33**) is found in CD_3CO_2D as the solvent.

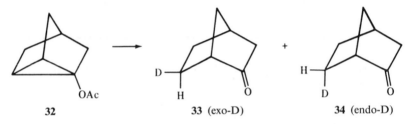

Thus, any desired product ratio of **33** and **34** could be obtained by selecting a solvent composition between the two extremes. Although it has been suggested that homoketonization in CD_3CO_2D involves three-membered ring opening in the homoenolacetate and that in aqueous systems ring cleavage of the homoenol (formed via initial ester hydrolysis) plays a role, no mechanistic explanation has as yet been offered for the remarkable solvent effect on stereospecificity.

5.4. Nucleophilic Displacement Reactions

Nucleophilic displacement reactions at saturated carbon were among the first organic reactions for which the mechanisms were investigated in great detail. Brilliant early work established the main characteristics of S_N1 and S_N2 processes.[426,436] It was particularly Winstein and his school who set the stage for more advanced studies aimed at a quantitative understanding of nucleophilic substitution processes as a function of the nature of the reagent and of the reaction medium.[156,369,914] In later years, and especially since the work of Sneen,[816] it became clear that some of the proposed mechanisms are in fact more complicated than originally assumed and might involve ion-pair intermediates. As a result, much current research is still devoted to several types of nucleophilic substitution involving, for example, reactions of secondary substrates which had previously been denoted as "borderline" between S_N1 and S_N2 (see also Section 2.2.6).[869] Some of the recent mechanistic refinements are, at least partially, based on results from studies of kinetic solvent effects.[787] In this section, we restrict ourselves to mixed aqueous solvents. The main emphasis of the discussion will be on the understanding of solvation effects in these media rather than on the utilization of kinetic solvent effects as a means to elucidate mechanistic details.

As noted previously (Section 2), the early theories[424] for solvent effects on kinetic parameters usually hinged heavily on attempts to correlate solvent effects with solvent polarity. Thus, in the Hughes–Ingold theory,[436] solvent effects were treated mainly in electrostatic terms, utilizing a continuum property of the medium, namely, the dielectric constant, as a measure for solvent polarity. Starting from a neutral substrate, it was assumed that solvation of the polar transition state for an S_N1 solvolysis was the dominating factor in determining kinetic solvent effects. This approach had only limited success and little predictive power in the quantitative sense. A significant advance was made when Grunwald and Winstein proposed the use of empirical scales of solvent polarity (Section 2.2.6).[369,815] This concept has been further developed ever since and has proved to be a versatile one in mechanistic studies,[787] although in certain mixed aqueous solvents like TFE–H_2O, Y values should be used with reservation.[736] However, the Winstein–Grunwald equation has contributed but little to our insight into the nature of solvation effects in aqueous binaries despite the fact that recently a fair correlation between the empirical m values and corresponding $\Delta C_p{}^{\ddagger}$ values could be established for hydrolysis in water.[628] Clearly a thermodynamic approach was required and it was also Winstein who performed pioneering and influential work in this area.[292,914] The fundamental observation was made that even for an S_N1 solvolysis (t-BuCl in EtOH–H_2O), $\delta_m \Delta G^{\ddagger}$ could be dominated by changes in the free energy of transfer of the neutral organic halide. This was, of course, not reconcilable with the Hughes–Ingold theory. A third major contribution was made by Winstein and his co-workers when they dissected $\delta_m \Delta G^{\ddagger}$ into contributions from $\delta_m \Delta H^{\ddagger}$ and $T\delta_m \Delta S^{\ddagger}$. It was found that smooth variation of $\Delta G^{\ddagger}$ with n_{H_2O} for simple nucleophilic displacement processes in highly aqueous binaries may obscure much larger and much more complex variations of $\Delta H^{\ddagger}$ and $\Delta S^{\ddagger}$. The same conclusion was also reached by Tommila[866] and by Hyne and Robertson,[434] and the intriguing $\Delta H^{\ddagger}/\Delta S^{\ddagger}$ patterns now appear in standard textbooks on physical organic chemistry.[28] The unique solvent properties of water were recognized and the weakness of a simple electrostatic treatment of kinetic solvent effects was demonstrated. This line of approach was further developed by Arnett and his students. Several papers from the period 1963–1965 constitute a landmark in the history of the study of solvent effects in aqueous media.[35,36,38] Now the function $\delta_m \Delta H^{\ddagger}$ was also separated into initial state $(\delta_m H_i{}^{\ominus})$ and transition state $(\delta_m H_{\ddagger}{}^{\ominus})$ contributions (Section 4.3) in an attempt to understand the origin of the minimum in $\Delta H^{\ddagger}$ around $n_{H_2O} = 0.85$ for solvolysis of t-BuCl in EtOH–H$_2$O (Fig. 33). The curious result was that the heat of solution of the transition state for t-BuCl solvolysis scarcely varied in the range $n_{H_2O} = 0.80$–1.00. Therefore the minimum in $\Delta H^{\ddagger}$ originated primarily from an endothermic change in $\Delta H_{tr,i}^{\ominus}$ in this solvent composition range. Since $\delta_m H_i{}^{\ominus}$ and $\delta_m H_{\ddagger}{}^{\ominus}$ reflect the combined effects of solute–solvent, solvent–solvent, and solute–solute interactions, partial molar heats of solution at infinite dilution were measured for a series of electrolytes (both inorganic salts and salts containing hydrophobic cations or anions) and nonelectrolytes in EtOH–H$_2$O between $n_{H_2O} \sim 0.5$ and 1.0. This was done in order to be able to compare the behavior of these functions with that of the transfer enthalpies of t-BuCl and the corresponding transition state taking into account the respective volumes of these species. Interestingly, the small variation of $\delta_m H_{\ddagger}{}^{\ominus}$ is seen to resemble the minor change of $\delta_m H^{\ominus}$

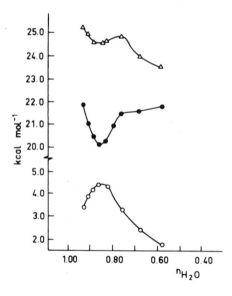

Fig. 33. Enthalpy parameters for the solvolysis of t-BuCl in EtOH–H$_2$O at 25°C; $\Delta H_{s,i}^{\ominus}$ (○), $\Delta H^{\ddagger}$ (●), and $\Delta H^{\ddagger} + \Delta H_{s,i}^{\ominus}$ (△).[35]

for $Me_4\overset{\oplus}{N}\overset{\ominus}{C}l$ between $n_{H_2O} = 0.60$ and 0.95. Recently, Abraham[1] also arrived at the conclusion that the solvation of tetraalkylammonium halide ion pairs reasonably mimics the solvation behavior of transition states for nucleophilic substitution reactions of neutral substrates (Section 4.3).

The trends in Arnett's data for the heats of solution are consistent with more recent measurements of similar thermodynamic quantities for TA mixtures, which have been qualitatively accounted for in terms of cavity formation and polar and nonpolar interactions (Sections 3.5 and 4.3). For t-BuCl solvolysis in EtOH–H$_2$O, the different behavior of $\Delta H_{tr,i}^{\ominus}$ and $\Delta H_{tr,\ddagger}^{\ominus}$ reflects a process in which a relatively nonpolar initial state is transferred into a polar or even pseudoionic transition state with partial loss of the hydrophobic hydration envelope around the substrate molecule. This also applies to solvolysis of i-PrBr in EtOH–H$_2$O, which shows a minimum in $\Delta H^{\ddagger}$ at $n_{H_2O} \sim 0.88$ (50°C).[261] Now the transition state is stabilized through interaction with the solvent as a nucleophile and, presumably, a less extended degree of solvent reorganization is involved as compared with t-BuCl hydrolysis.

In terms of Ben-Naim's definition of water structure (Section 2.2.7), the degree of water ordering will diminish at elevated temperatures. Thus, extremum behavior of $\Delta H^{\ddagger}$ and $\Delta S^{\ddagger}$ is expected to become more pronounced at lower temperatures. This turned out to be the case for solvolysis of t-BuCl in CH$_3$OH–H$_2$O at 5, 7.5, and 10°C.[428]

For solvolysis of more polar substrates like methyl benzenesulfonate[434] and dimethyl-t-butylsulfonium iodide,[431] $\delta_m\Delta H^{\ddagger}$ in EtOH–H$_2$O is determined by initial-state as well as transition-state solvation effects. For the latter substrate, $\Delta H^{\ddagger}$ now passes through a *maximum* rather than a minimum, apparently since the charge in the cationic initial state becomes dispersed in the transition state, leading to a reversal in the behavior of $\Delta H_{tr,i}^{\ominus}$ and $\Delta H_{tr,\ddagger}^{\ominus}$. A similar situation pertains to $\Delta V^{\ddagger}$. For solvolysis of

$$Me_3\overset{\oplus}{C}SMe_2 \xrightarrow{\text{slow}} Me_3\overset{\delta\oplus}{C}\cdots\overset{\delta\oplus}{S}Me_2 \longrightarrow Me_3C^{\oplus} + Me_2S$$

$$\text{fast} \downarrow H_2O$$

$$Me_3COH + H^{\oplus}$$

t-BuCl, benzyl chloride, and p-chlorobenzyl chloride in EtOH–H$_2$O, $\Delta V^{\ddagger}$ exhibits a *minimum* in the range $n_{H_2O} = 0.7$–0.8,[586] whereas $\Delta V^{\ddagger}$ passes through a *maximum* at $n_{H_2O} \sim 0.8$ for the sulfonium salt.[227] Hyne[227] argues that, in contrast to t-BuCl solvolysis, increased electrostriction on going to the transition state does not outweigh the increase in volume due

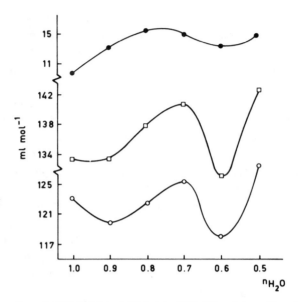

Fig. 34. Dissection of $\Delta V^{\ddagger}$ (●) into initial-state ($V_i^{\ominus}$, ○) and transition-state ($V_{\ddagger}^{\ominus}$, □) components for the solvolysis of Me_2-t-$BuS^{\oplus}$ in EtOH–H_2O at 50.25°C.[227]

to bond stretching in the sulfonium ion solvolysis. For the latter solvolysis, both the initial and transition state are essentially ionic in nature, leading to similar behavior of the partial molal volumes, $V_i^{\ominus}$ and $V_{\ddagger}^{\ominus}$, of both states (compare Section 4.3) on changing n_{H_2O} from 0.5–1.0 (Fig. 34). Most likely, charge dispersal between the sulfur and carbon atoms occurs in the activation process and weakens the electrostrictive interaction with the solvent. Subsequent studies[541] revealed closely similar solvent dependences of the partial molal volume of the ionic substrate and of simple tetraalkylammonium halides in EtOH–H_2O. Minima in $\Delta V^{\ddagger}$ have also been found for solvolysis of benzyl chloride in a series of TA alcohol–water mixtures.[353,354,433,583,586] The magnitude of the minimum becomes more marked and occurs at higher n_{H_2O} in the series MeOH < EtOH < i-PrOH < t-BuOH. For t-BuOH–H_2O, the minimum in $\Delta V^{\ddagger}$ predominantly originates from a maximum in $\delta_m V_i^{\ominus}$. For neutral substrates both $\delta_m V_i^{\ominus}$ and $\delta_m V_{\ddagger}^{\ominus}$ usually exhibit maxima in the water-rich region of TA mixtures (not necessarily at the same n_{H_2O}) and their magnitude depends on the structure of the substrate and the mechanism of the nucleophilic displacement reaction. Excess partial molal volumes of TA mixtures also show extrema at comparable solvent composition.[496] Curiously, $\delta_m V_i^{\ominus}$ and $\delta_m V_{\ddagger}^{\ominus}$ for several nucleophilic substitution reactions in aqueous binaries show two extrema and this exotic behavior has been appropriately described as that of a "roller coaster" type.[432] Reasonable transition-state models, such as anilinium chloride for the solvolysis of benzyl chloride, do not share this roller coaster behavior, possibly because of the different solvent exposure of the charges in the species compared.[432,541]

Volumes of activation as well as their temperature dependence for solvolysis of benzyl chloride and i-propyl bromide have been compared for H_2O and D_2O.[228] The more negative $dV^{\ddagger}/dT$ for D_2O than for H_2O was proposed to be consistent with D_2O being more structured, implying that the hydration sphere in D_2O has a volume that is more sensitive to thermal disruption.

It should be noted that for a variety of nucleophilic displacement reactions the trends in the functions $\delta_m \Delta H^{\ddagger}$ and $\delta_m \Delta V^{\ddagger}$ are strikingly similar although the extrema do not occur at the same n_{H_2O}. This suggests a similar cause for the extremum behavior. Since the magnitudes and the positions of the extrema in $\Delta H^{\ddagger}$ follow the same order as the excess heats of mixing of the aqueous binaries, solvent structural effects have generally been invoked[432] to explain the extrema in $\Delta H^{\ddagger}$ (compare Section 5.2). A simple, qualitative, but fairly general rationalization is the following. Large extrema in $\Delta H^{\ddagger}$ apparently reflect greatly different solvation characteristics of initial and transition state as a result of the operation of different mechanisms for interaction with the solvent. Certainly, the effect may be modulated for different substrates by differences in solvent accessibility to the reaction site. If the initial state is nonpolar, there are only minor possibilities for specific, short-range interactions. Introduction of such a species in a water-rich solvent will result in the formation of a hydrophobic hydration sphere around the molecule. If this solvated species is transferred into a polar or pseudo-ionic transition state, the hydrophobic hydration envelope is at least partly destroyed and replaced by a hydration sphere involving water molecules oriented by polar field effects of the first kind (in terms of Friedman's and Krishnan's terminology, Section 3). This solvation change will be accompanied by a favorable enthalpy change due to polar transition-state–solvent interactions despite the fact that ΔH^{cav} may be less favorable because of the greater size of the transition state. Consequently, $\Delta H_{\text{tr},i}^{\ominus}$ and $\Delta H_{\text{tr},\ddagger}^{\ominus}$ will behave differently in the solvent composition range which contains enough water to allow hydrophobic hydration. This rationalization is further supported by $\Delta C_P^{\ddagger}$ values for several nucleophilic displacement reactions in both TA and TNA mixtures. As noted earlier (Section 4.3), partial specific heats measure primarily the temperature dependence of solute–solvent interactions and contain in the water-rich region of an aqueous binary significant contributions from solute–solvent structural effects.[78,243a] The relevance of molal heat capacity data to the study of organic solute–water interaction has been amply demonstrated.[161] Recently, the usefulnes of $\Delta C_P^{\ddagger}$ as a mechanistic probe for ionogenic nucleophilic displacement reactions has been critically evaluated[758] taking into account the possibility of ion-pair intermediates.[797] It was proposed that $\Delta C_P^{\ddagger}$ values significantly more negative than -30 cal deg^{-1} mol^{-1} reflect the effect of temperature on that component of $\Delta H^{\ddagger}$ which is determined by solvent reorganization during the activation process and which is influenced by the degree of water structure. For substrates like 1- and 2-adamantyl nitrate[511] and cis,cis-2,3-dimethylcyclopropyl bromide[758] strongly negative $\Delta C_P^{\ddagger}$ values (~ -135 to -154 cal deg^{-1} mol^{-1}) were found. The conclusion was drawn that limiting $S_N 1$ mechanisms for displacement of nitrate or halide anions in water are characterized by $\Delta C_P^{\ddagger}$ values of this magnitude.* At one time, it was proposed[19] that ion-pair return provided an anomalous negative contribution to $\Delta C_P^{\ddagger}$. However, the mixed aqueous solvent effects on $\Delta C_P^{\ddagger}$, to be discussed below, point to an explanation in terms of a temperature-dependent structural feature of water related to charge development on the anionic leaving group. Thus, $\Delta C_P^{\ddagger}$ values between -30 and -150 cal deg^{-1} mol^{-1} are open to detailed mechanistic interpretation[503,759,760] in terms of a refined $S_N 1$–$S_N 2$ dichotomy.[511–513,813] Since factors other than solvent reorganization may also contribute to $\Delta C_P^{\ddagger}$,[150,756] the quantitative interpretation of $\Delta C_P^{\ddagger}$ values is fairly complex[755,758] and its physical significance is still under active investigation.

Solvolysis of t-BuCl in TA mixtures like t-BuOH–H$_2$O, i-PrOH–H$_2$O, EtOH–H$_2$O, and THF–H$_2$O is typified by minima in $\Delta C_P^{\ddagger}$ (Fig. 35) in the region $n_{H_2O} = 0.90$–0.95.[759,760] Since $\delta_m \Delta C_P^{\ddagger}$ will be governed largely by initial-state solvation, the minimum in $\Delta C_P^{\ddagger}$ suggests that addition of a TA organic solvent component makes the solvation shell around a nonpolar substrate more thermally labile. In this context, it is relevant to note that Ahluwalia et al.[629] have shown that the limiting excess partial molal heat capacities ($\Delta \bar{C}_P^{0}$) of some electrolytes where the solvation behavior is dominated by a large hydrophobic cation (e.g., n-Bu$_4$NBr and n-Am$_4$NBr) exhibit large maxima in t-BuOH–H$_2$O at $n_{H_2O} = 0.96$ (n-Bu$_4$NBr) and at $n_{H_2O} = 0.97$ (n-Am$_4$NBr). This suggests that at these solvent compositions,

---

* t-BuCl does not give a limiting value of $\Delta C_P^{\ddagger}$ for displacement of Cl$^{\ominus}$, indicating some type of interaction with the nucleophile prior to full charge development on the leaving group.[758]

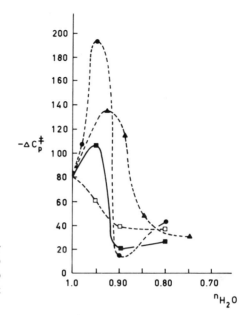

Fig. 35. Plot of $-\Delta C_P{}^{\ddagger}$ vs. n_{H_2O} for solvolysis of t-BuCl in t-BuOH–H$_2$O (●), EtOH–H$_2$O (▲), THF–H$_2$O (■), and MeCN–H$_2$O (□) at 25°C.[760]

the interactions of the hydrophobic components in solution (i.e., the $R_4N^{\oplus}$ cation and t-BuOH) lead to the formation of solvation spheres that possess a greater propensity for thermal breakdown than the corresponding hydration shells in water. More recently, Desnoyers et al.[243a] have commented extensively on heat capacity changes caused by solvation shell interaction between two hydrophobic solutes in water. In the t-BuOH–H$_2$O solvent system, positive deviation from ideality is observed, consistent with the above hypothesis.

As portrayed in Fig. 35, $\Delta C_P{}^{\ddagger}$ for t-BuCl solvolysis increases again below a critical value of n_{H_2O} when the quasi-aqueous structure is broken upon further addition of the organic cosolvent. Below $n_{H_2O} \sim 0.85$, $\Delta C_P{}^{\ddagger}$ remains relatively insensitive to solvent composition ($\Delta C_P{}^{\ddagger} \sim -30$ to -40 cal deg^{-1} mol^{-1}). In terms of the above analysis, one would expect that $\Delta C_P{}^{\ddagger}$ would increase upon initial addition of the organic cosolvent in a TNA system. This is exactly what is found for MeCN–H$_2$O (Fig. 35). Now MeCN disrupts the three-dimensional hydrogen-bond network in water and the initial solvation sphere around the substrate is rendered less sensitive towards changes in temperature. The opposite behavior of $\delta_m\Delta C_P{}^{\ddagger}$ in TA and TNA solvents is usually not paralleled by $\delta_m\Delta H^{\ddagger}$ although the minima in $\delta_m\Delta H^{\ddagger}$ become generally less marked in TNA solvent systems like MeCN–H$_2$O or DMSO–H$_2$O.[760]

At this point, it must be emphasized that the function $\delta_m \Delta H^{\ddagger}$ for any nucleophilic substitution reaction in an aqueous binary does not tell anything about the sign and magnitude of $\delta_m \Delta G^{\ddagger}$. As noted previously, reaction rates in mixed aqueous solvents may be determined by entropy rather than enthalpy effects over considerable ranges of n_{H_2O}. The situation is frequently encountered that $\delta_m \Delta G^{\ddagger}$ varies monotonically with n_{H_2O} whereas $\Delta H^{\ddagger}$ and $T\Delta S^{\ddagger}$ show much larger but largely compensatory changes (compare Section 5.2). For the majority of $S_N 1$ processes, $\Delta G^{\ddagger}$ increases as n_{H_2O} decreases in both TA and TNA solvent systems. Since $\delta_m \Delta H^{\ddagger}$ frequently passes through a minimum in the water-rich region (*vide supra*), it is evident that $\delta_m \Delta S^{\ddagger}$ will pass through a minimum albeit not necessary at precisely the same n_{H_2O}. However, there are exceptions to this rule and also $\delta_m \Delta G^{\ddagger}$ may decrease upon initial addition of the cosolvent and subsequently attain an extreme,[119,521,859] as a result of a different balance in $\Delta H^{\ddagger}/\Delta S^{\ddagger}$ compensatory behavior.

Mixed aqueous solvent effects on nucleophilic substitution reactions involving transition states containing nucleophiles other than the solvent (usually anions) have also been investigated extensively. Quantitative analysis requires more extensive thermodynamic data as compared to solvolysis processes since the kinetics now respond to solvation changes of two solvated substrates. Very much attention has been given to the TNA solvent system DMSO–H$_2$O. In early work, Tommila and his associates[867] found that the reaction of *p*-substituted benzyl chlorides with OH$^{\ominus}$ is accelerated markedly with increasing DMSO content in the range $n_{H_2O} = 0.85$–0.30. Contrastingly, acetone was found to reduce the rate ($n_{H_2O} = 0.85$–0.72). In both the TNA and the TA solvent systems, $\Delta H^{\ddagger}$ and $\Delta S^{\ddagger}$ show compensatory changes, the reduced $\Delta G^{\ddagger}$ in DMSO–H$_2$O being dominated by $\Delta H^{\ddagger}$ effects and the enhanced $\Delta G^{\ddagger}$ in acetone–water by entropy effects. The reaction of benzyl chlorides with thiosulfate as the nucleophile exhibits a similar behavior.[460] The rate increase upon addition of DMSO, which is also observed in other systems,[638] is primarily attributed to the reduced hydrogen bond stabilization of OH$^{\ominus}$ upon replacement of water by an aprotic solvent as evidenced by thermodynamic transfer parameters.[223] The reduction of $\Delta G^{\ddagger}$ by DMSO contrasts with the decrease in rate with decreasing n_{H_2O} for *solvolysis* of benzyl chlorides in DMSO–H$_2$O (vide supra).[392] This opposite behavior apparently reflects a reversal in charge-type behavior upon transferring the substrate(s) into the transition state.

The reaction of *p*-tosylmethyl benzenesulfonate (35) with OH$^{\ominus}$ in *t*-BuOH–H$_2$O ($n_{H_2O} = 0.9$–1.0) has recently been investigated by Holterman

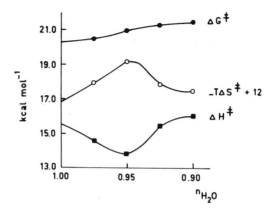

Fig. 36. Plots of $\Delta G^{\ddagger}$, $\Delta H^{\ddagger}$, and $-T\Delta S^{\ddagger}$ vs. n_{H_2O} for the reaction of **35** with $OH^{\ominus}$ ions in t-BuOH–H_2O at 25°C.[416]

and Engberts.[416] This reaction involves nucleophilic attack of $OH^{\ominus}$ on the electrophilic sulfonate sulfur atom, either via direct S_N2 substitution of sulfonate anion or via an addition–elimination-type mechanism.[415] Plots of $\Delta G^{\ddagger}$, $\Delta H^{\ddagger}$, and $-T\Delta S^{\ddagger}$ vs. n_{H_2O} are displayed in Fig. 36. Whereas $\Delta G^{\ddagger}$ increases steadily upon decreasing n_{H_2O}, $\Delta H^{\ddagger}$ and $\Delta S^{\ddagger}$ pass through extrema

$$p\text{-}CH_3C_6H_4SO_2CH_2OSO_2C_6H_5 + OH^{\ominus} \longrightarrow p\text{-}CH_3C_6H_4SO_2^{\ominus} + CH_2O + C_6H_5SO_3^{\ominus} + H^{\oplus}$$

35

located near $n_{H_2O} = 0.95$. Using the appropriate transfer functions for **35** and $OH^{\ominus}$, $\delta_m\Delta H^{\ddagger}$ could be dissected into initial-state and transition-state contributions (Fig. 37). In accordance with earlier rationalizations in terms of $\delta\Delta H^{cav}$, $\delta\Delta H^{nonpol}$, and $\delta\Delta H^{pol}$ (Section 5.2), the sum of $\Delta H_{tr,(35)}^{\ominus}$ and $\Delta H_{tr,(OH^{\ominus})}^{\ominus}$ rises more sharply in the region $n_{H_2O} = 1.0$–0.95 than that of the transition state (in which the negative charge is dispersed). In the region $n_{H_2O} = 0.95$–0.90, it is $\Delta H_{tr,\ddagger}^{\ominus}$ that rises most sharply, thereby producing a minimum in $\Delta H^{\ddagger}$ at $n_{H_2O} = 0.95$. The monotonic increase of $\Delta G^{\ddagger}$ of the reaction in t-BuOH–H_2O contrasts with the minimum in $\Delta G^{\ddagger}$ for the reaction of EtBr with $OH^{\ominus}$ in acetone–H_2O.[860] For this process $\Delta H^{\ddagger}$ exhibits a maximum in the water-rich region, which may be compared with the minimum in $\Delta H^{\ddagger}$ for *solvolysis* of EtBr in the same solvent system. This may again be reconciled with the reversal in charge type behavior upon activation. Minima in $\Delta G^{\ddagger}$ are also commonplace for the neutral solvolysis of aliphatic sulfonyl chlorides in acetone–H_2O and dioxane–H_2O and for the reaction of benzenesulfonyl chloride with some amines in both TA (EtOH–H_2O) and TNA mixtures (MeCN–H_2O) at low temperatures (0–15°C).[822,868] For nucleophilic substitution at sulfonyl sulfur in sulfonyl

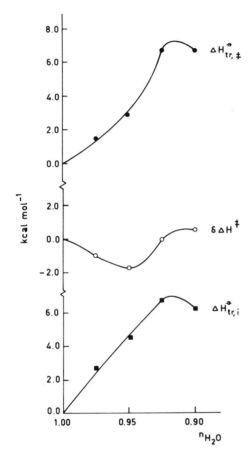

Fig. 37. Enthalpy parameters for the reaction of **35** with $OH^{\ominus}$ ions in t-BuOH–H_2O at 25°C.[416]

halides these extrema in $\Delta G^{\ddagger}$ are by no means confined to aqueous mixed solvents.[33]

Aromatic nucleophilic displacement reactions have been studied less systematically in mixed aqueous solutions[639] although protic–dipolar-aprotic solvent effects have been examined in detail.[210]

5.5. Miscellaneous Reactions

Mixed aqueous solvent effects have been investigated for some bimolecular S_E2 processes of organometallic compounds.[1] One of the reactions analyzed in considerable detail is that of tin tetraethyl with mercuric chloride in $CH_3OH–H_2O$. As argued by Abraham,[1] the data are best reconcilable with an S_E2 mechanism involving an open transition state.

TABLE XII. Thermodynamic Activation Parameters for the Reaction of Et$_4$Sn with HgCl$_2$ in CH$_3$OH–H$_2$O at 25°C

n_{H_2O}	$\Delta G^{\ddagger}$ (kcal mol^{-1})	$\Delta H^{\ddagger}$ (kcal mol^{-1})	$\Delta S^{\ddagger}$ (e.u.)
1.000	16.68[a]	12.6[b]	−13.7[b]
0.900	17.13[a]		
0.800	17.73[a]		
0.700	18.00[a]		
0.600	18.41[a]		
0.490	18.86	12.2	−22.2
0.360	19.37	12.3	−23.7
0.284	19.65	12.4	−24.1
0.200	19.99	12.3	−25.6
0.126	20.29	12.7	−25.4
0.086	20.45	13.1	−24.5
0.044	20.65	13.9	−22.4
0.001	20.83	14.3	−21.8

[a] Extrapolated value; see Refs. 3, 5.
[b] Estimated value; see Refs. 3, 5.

Thermodynamic activation parameters are listed in Table XII; unfortunately the estimated data for the water-rich region are somewhat uncertain. The rates increase as the solvent becomes more aqueous and are quite sensitive to changes in n_{H_2O}. The kinetic solvent effects were further un-

$$Et_4Sn + HgCl_2 \longrightarrow Et\overset{\overset{\delta\oplus}{SnEt_3}}{\underset{\underset{\delta\ominus}{HgCl_2}}{}} \longrightarrow EtHgCl + Et_3SnCl$$

raveled employing the method described in Section 4.3. To this end free energies of transfer for Et$_4$Sn were obtained through determinations of vapor–liquid equilibria and those for HgCl$_2$ from solubility measurements.[5,6] Transfer enthalpies were obtained calorimetrically.[3] Combination of $\Delta G^{\ddagger}$, $\Delta H^{\ddagger}$, and $\Delta S^{\ddagger}$ values with $\Delta G_{tr}^{\ominus}$, $\Delta H_{tr}^{\ominus}$, and $\Delta S_{tr}^{\ominus}$, respectively, then provides the transfer parameters for the dipolar transition state (Table XIII). It is evident that the rate acceleration with increasing n_{H_2O} is due to destabilization of the reactants. The function $\Delta G_{tr}^{\ominus}$ is particularly dominating for the hydrophobic Et$_4$Sn and $\Delta G_{tr}^{\ominus}$ varies even more strongly for n-Pr$_4$Sn and n-Bu$_4$Sn[5] (compare Section 5.2). The transition state is also destabilized with increasing n_{H_2O}, but the effect is smaller, as expected for a dipolar

TABLE XIII. Thermodynamic Quantities of Transfer[a] from CH_3OH to $CH_3OH–H_2O$ (at 25°C) for Reactants and Transition State of the Reaction of Et_4Sn with $HgCl_2$

n_{H_2O}	$\Delta G_{tr}^{\ominus}$ (kcal mol^{-1})			$\Delta H_{tr}^{\ominus}$ (kcal mol^{-1})			$\Delta S_{tr}^{\ominus}$ (e.u.)		
	Et_4Sn	$HgCl_2$	‡[b]	Et_4Sn	$HgCl_2$	‡[b]	Et_4Sn	$HgCl_2$	‡[b]
1.000	7.80	1.62	5.27	2.5[c]	4.2	5.0[c]	−17.8[d]	8.8	−0.9[d]
0.900	6.60	1.53	4.43						
0.800	5.40	1.36	3.66						
0.700	4.25	1.11	2.53						
0.600	3.35	0.86	1.79						
0.490	2.45	0.62	1.10	2.20	3.44	3.54	−0.8	9.5	8.3
0.360	1.65	0.41	0.60	1.77	2.81	2.53	0.4	8.0	6.5
0.284	1.22	0.29	0.33	1.41	2.43	1.94	0.6	7.2	5.5
0.200	0.81	0.19	0.15	1.01	1.77	0.78	0.7	5.3	2.2
0.126	0.48	0.11	0.04	0.63	1.17	0.15	0.5	3.6	0.5
0.086	0.31	0.07	0.01	0.35	0.85	0.00	0.1	2.6	0.0
0.044	0.15	0.03	0.00	0.19	0.41	0.20	0.1	1.3	0.8
0.001	0.00	0.00	0.00	0.00	0.00	0.00	0.0	0.0	0.0

[a] On the molar scale.
[b] ‡ denotes the transition state.
[c] Estimated value, see Ref. 3.
[d] Estimated value, see Ref. 3.

species. In order to arrive at some insight into the magnitude of charge separation ($\delta\pm$) in the transition state, $\Delta G_{tr,\ddagger}^{\ominus}$ values have been compared[5] with those of nonelectrolytes and ion pairs of about the same size. This suggests that $\delta\pm$ is relatively high and is perhaps only surpassed by transition states of S_N1 reactions of t-butyl halides. Abraham notes, however, that the data do not reveal a simple correlation between the electrostatic part of $\Delta G_{tr}^{\ominus}$ and the dipole moment of the transition state.[4] This seems to indicate that the free energy of solvation of a dipolar species in a mixed aqueous solvent depends on the magnitude of the positive and negative charges rather than on the distance between the charges.[5]

The decrease of $\Delta H^{\ddagger}$ upon going from CH_3OH to H_2O is also an initial-state effect which is again larger than the increase in enthalpy of the transition state. A further conclusion from the data in Table XIII is that the free energies of transfer of Et_4Sn in the range $n_{H_2O} = 0.00–0.49$ are governed by the $\Delta H_{tr}^{\ominus}$ values but that this situation is apparently reversed in the more highly aqueous solutions. This type of behavior is consistent with the formation of hydrophobic hydration spheres around Et_4Sn at high n_{H_2O} and contrasts with the solvation behavior of $HgCl_2$ and the transition state. The entropies of transfer of both $HgCl_2$ and the transition state vary markedly with n_{H_2O} in the range $n_{H_2O} = 0$–0.49, but for pure water ΔS_{tr} is much larger for Et_4Sn. This also points to the hydrophobic nature of Et_4Sn.

The reduction of trifluoroacetophenone (**36**) by 1-*n*-propyldihydro-nicotinamide (**37**)[877] is of biological relevance since it may mimic aspects of NADH reduction reactions *in vivo*. The reaction follows second-order

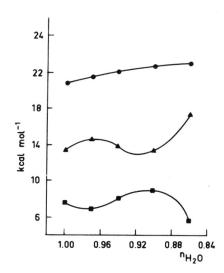

36   **37**

kinetics, is rather insensitive to ionic strength, and, moreover, is pH independent in the range pH = 7.2–9.9. Changing the solvent from DMSO to water increases k_2 by a factor of ~ 5000 (at $50.0 \pm 0.1°C$). Most noticeable is the observation that plots of $\log k_2$ vs. n_{H_2O} are experimentally indistinguishable for DMSO–H_2O and *i*-PrOH–H_2O despite the fact that, for instance, plots of ε vs. n_{H_2O} differ markedly for both solvent systems. The authors conclude that stabilization of the developing negative charge on the carbonyl oxygen in the transition state by hydrogen bonding[681]

Fig. 38. Plot of $\Delta G^{\ddagger}$ (●), $\Delta H^{\ddagger}$ (▲), and $-T\Delta S^{\ddagger}$ (■) vs. n_{H_2O} for the reaction of **36** with **37** in DMSO–H_2O containing $10^{-2}\ M\ Na_2CO_3$–$NaHCO_3$ (pH 9.9) buffer at 50°C.[877]

to water is the main driving force for the rate enhancements in highly aqueous environments. Figure 38 portrays the smooth change in $\Delta G^{\ddagger}$ as a function n_{H_2O} and which is the result of irregular changes of $\Delta H^{\ddagger}$ and $T\Delta S^{\ddagger}$. Thermochemical changes for the substrates with solvent composition have not been determined so far.

Kinetic solvent effects on organic reactions accelerated through neighboring group participation have recently been reviewed.[168] For anchimerically assisted solvolytic processes, the Grunwald–Winstein relationship (Section 2.2.6) has been a popular vehicle for analyzing medium effects. As expected, neighboring-group-assisted ionization reactions resemble limiting k_c solvolyses and are less dependent on solvent nucleophilicity than the solvolyses of k_s substrates.[168] Here k_c is the rate constant for a reaction involving neighboring-group participation after rate-determining ionization and k_s is the rate constant for a reaction involving nucleophilic assistance by the solvent.

Rate constants and activation parameters for the intramolecular acid-catalyzed hydrolysis of (Z)-2-carboxy-N-methyl-N-phenylethenesulfonamide (38) and N-methyl-N-phenylmaleamic acid (39) have recently been determined[359] in t-BuOH–H$_2$O in the range $n_{H_2O} = 1.0$–0.8. Hydrolysis of 38

38: X = SO$_2$
39: X = CO

most likely proceeds via reversible protonation of the nitrogen atom of the sulfonamide moiety, followed by nucleophilic attack of the carboxylate anion on the sulfonyl sulfur atom to yield an intermediate which contains a pentacoordinated sulfur. Largely based on an analysis of substituent effects,[887] it is highly likely that the rate-determining step constitutes the expulsion of the amine leaving group from this intermediate, to yield a cyclic mixed anhydride. The formation of a reactive intermediate cyclic anhydride is also a feature of the mechanism of solvolysis of 39, but this

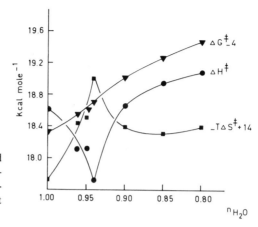

Fig. 39. Plot of $\Delta G^{\ddagger}$, $\Delta H^{\ddagger}$, and $-T\Delta S^{\ddagger}$ vs. n_{H_2O} for the intra-molecular carboxyl-catalyzed hydrolysis of **38** in t-BuOH–H$_2$O at 25°C.[359]

species is now produced via slow breakdown of a tetrahedral intermediate.[485] Despite the different mechanistic pathways, plots of $\Delta G^{\ddagger}$, $\Delta H^{\ddagger}$, and $\Delta S^{\ddagger}$ for both reactions are closely similar (Figs. 39 and 40). The monotonic increase of $\Delta G^{\ddagger}$ with decreasing n_{H_2O} is not linear with solvent polarity parameters like $(\varepsilon - 1)/(2\varepsilon + 1)$, Z, or E_T (Section 2). As often observed, $\Delta H^{\ddagger}$ and $\Delta S^{\ddagger}$ show mirror image behavior and exhibit extrema at $n_{H_2O} = 0.95$, the solvent composition for which one can expect maximal formation of hydrophobic hydration spheres.

The diffusional separation of radical pairs created in a solvent cage has earlier been cited (Section 2.2.4) as a process depending on solvent viscosity

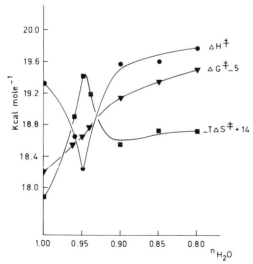

Fig. 40. Plot of $\Delta G^{\ddagger}$, $\Delta H^{\ddagger}$, and $-T\Delta S^{\ddagger}$ vs. n_{H_2O} for the intra-molecular carboxyl-catalyzed hydrolysis of **39** in t-BuOH–H$_2$O at pH 2.85 (25°C).[359]

(η). A detailed study has now been made of the photolysis of azomethane in $t\text{-BuOH-H}_2\text{O}$ as a function of $n_{\text{H}_2\text{O}}$[658] and the results have been compared with those for a variety of other solvents.[504]

The mechanistic pathway may be depicted as

$$CH_3\text{---}N\text{==}N\text{---}CH_3 \xrightarrow{h\nu} [\cdot CH_3N_2CH_3\cdot] \underset{k_d}{\overset{k_c}{\Big\langle}} \begin{array}{l} CH_3CH_3 \\[1ex] 2CH_3\cdot \xrightarrow{\text{solvent}} 2CH_4 \end{array}$$

in which k_c is the rate constant for radical–radical reaction within the solvent cage and k_d the effective rate constant for diffusion from the solvent cage. The ratio $CH_4/2C_2H_6$, which represents k_d/k_c, may be taken as a measure for the rate of diffusion of methyl radicals from the solvent cage relative to dimer formation within the cage, if it is assumed that the latter process is solvent independent. In a series of organic solvents, the magnitude of the cage effect increases with increasing η, and k_d/k_c is fairly linear with the fluidity $(1/\eta)$ of the medium (correlation coefficient $r = 0.937$). Nodelman and Martin[658] have shown that a considerably better correlation $(r = 0.982)$ is obtained if k_d/k_c is plotted vs. $c(P_{\text{eff}}V^{2/3}\mu^{1/2})^{-1}$, in which c is the lattice factor $(c = 2$ for cubic closest packing), P_{eff} is an "effective" pressure calculated from the cohensive energy density (Section 2.2.4), and μ is the reduced mass appropriate for relative diffusion of solvent and solute molecules. For our discussion, the most significant and perhaps unexpected result is the minimum of k_d/k_c at $n_{\text{H}_2\text{O}} = 0.90$, a solvent composition markedly different from that of maximum viscosity $(n_{\text{H}_2\text{O}} \sim 0.60$, Fig. 41). This implies that in the range $n_{\text{H}_2\text{O}} = 0.90$–0.60, the rate of diffusion from the solvent cage is increasing while the viscosity is increasing. However, Nodelman and Martin's semiempirical diffusion equation is able

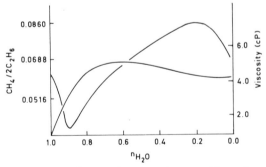

Fig. 41. Plots of $k_d/k_c = CH_4/2C_2H_6$ for photodecomposition of azomethane (curve with minimum at $n_{\text{H}_2\text{O}} = 0.90$) and viscosity vs. $n_{\text{H}_2\text{O}}$ in $t\text{-BuOH-H}_2\text{O}$ at 25°C.[658]

to correlate k_d/k_c successfully over the entire range of n_{H_2O} in t-BuOH–H_2O. Practical application of the equation hinges on the availability of several experimentally accessible solvent parameters: the internal pressure, the cohesive energy density, the activation energy for viscous flow, and the activation energy for viscous flow at constant volume. In the framework upon which the diffusion theory has been developed, the minimum in k_d/k_c at $n_{H_2O} = 0.90$ in t-BuOH–H_2O may be ascribed to the stiffening of the hydrogen-bond network induced by t-BuOH. This structure making is accompanied by a greater fractional contribution of the cohesive energy of hydrogen bonds leading to an enhanced P_{eff}, a smaller effective volume, and slower diffusion. The work by Nodelman and Martin represents one of the most successful treatments of solvent effects in the nonideal TA solvent t-BuOH–H_2O based upon experimentally obtainable solvent parameters.

ACKNOWLEDGMENTS

I am grateful to G. Berti, B. G. Cox, B. Perlmutter-Hayman, R. E. Robertson, and M. C. R. Symons for sending unpublished information and to M. H. Abraham for correspondence. I am also much indebted to my co-workers, who shared, in good friendship and cooperation, my interest in organic reactions in aqueous solvent systems.

CHAPTER 5

Solvent Structure and Hydrophobic Solutions

D. Y. C. Chan, D. J. Mitchell, B. W. Ninham, and B. A. Pailthorpe

Department of Applied Mathematics, Research School of Physical Sciences
Institute of Advanced Studies, The Australian National University
Canberra, A.C.T. 2600, Australia

1. INTRODUCTION

In a paper on aqueous solutions of nonpolar solutes published in 1973 F. H. Stillinger[827] acknowledged his debt to H. S. Frank in these words:

The author has enjoyed a protracted and constructive dialogue with Professor H. S. Frank on the nature of water and aqueous solutions. This interaction has helped the former to maintain in his work a respectable balance between the sterile intricacy of formal theory and the seductive simplicity of poetic "explanation."

Stillinger's second sentence comes as near to capturing the elusive essence and quintessential difficulties of the problem as we have seen. The editor's brief to these authors goes further. In not quite so many words: The biologist and the chemist are too busy with their own pursuits to worry about the intricacies of modern physics as applied to their own subjects. Yet they urgently need to be confronted with those conceptual frameworks, those predictive dictionaries of observation that mathematical physicists call a theory. Then—and sternly—the onus rests with the theoreticians to bridge the present communication gaps.

The chasm that separates the various perceived paths to understanding, geometric versus statistical, is very wide. On the one hand "structure" conjures up images of molecular packing and hydrogen-bond networks, and on the other it suggests statistical mechanical distribution functions and long-range forces. The jump from thermodynamics to interaction mechanism is large. Witness the continuing debate that surrounds even the definition of hydrophobic interaction.[328] In the Lewis–Randall[337] picture (constant pressure) the "interaction" may have one sign,[516] in the McMillan–Mayer (constant solvent chemical potential) the opposite sign, a situation exacerbated by confusion on standard states. Again it is confidently asserted and reasserted that water is peculiar. Undoubtedly it is. Yet to push this article of faith to the point of dogma is almost tantamount to the claim that *no* present physical theory can explain *any* solvent effects due to water.

This essay attempts to draw and weave together several threads spun out of advances in surface and liquid-state physics. For dilute solutions of aqueous nonelectrolytes an extremely subtle mosaic is beginning to emerge and some progress has definitely been made. More is in sight. This has come about because of several recent developments. First, the separate experiments of LeNeveu, Rand, and Parsegian and co-workers[545,546] on lecithin multilayers and those of Adams and Israelachvili[9,10] on mica in aqueous electrolyte solutions have provided the first *direct* evidence of structural (hydration) forces between interfaces. Second, the nature and delicacy of long-range van der Waals forces between molecules and surfaces has now been explored and understood both theoretically[587] and experimentally.[10,439] Third, liquid-state physics has now reached such a stage that one has both some confidence in the predictive capacity of present theories and a knowledge of the limitations of those theories.[53] Fourth, attempts to characterize and quantify the structural forces in liquids responsible for hydrophobic interactions in terms of a phenomenological theory have been at least partially successful.[591] This, a mean field theory, provides a conceptual framework of appealing simplicity, which is important. There is now a real expectation that the subtleties of hydrophobic forces, which play such an important part in maintaining protein conformational stability can be captured and encapsulated in simple analytic forms.

The hope, the wish, the belief that this Eldorado of chemistry can be so reached colors all perception, dangerously. A head-on assault must fail. The strategy must be to build base camps by exploring dilute solutions of lesser complexity first. Exactly soluble statistical mechanical models illumine the interplay between intermolecular potentials, solute and solvent distribution functions, and the thermodynamic classification of solutions. These throw into focus the differences between normal and hydrophobic solutions. Long-range van der Waals interactions between apolar media across water bring short-range forces into sight, at a distance. Model calculations of short-range forces bring us closer. Enough soliloquy, and to work.

The article is organized as follows: We begin with a few formal definitions of the thermodynamic quantities necessary in analyzing solution behavior. The relations between these quantities and solvent–solute distribution functions are explored to provide a link between solvent structure and solution properties. Next the phenomenological classification of solutions into various types is outlined. The way in which these types can arise is illustrated by some model systems that can be solved exactly. These models adumbrate the interplay between intermolecular potentials, temperature, and pressure in determining dilute solution properties. We then move on to discuss the relation between the continuum theory of interparticle forces and the long-range part of the interaction between hydrophobic solutes in aqueous solutions. Short-range solvent mediated solute–solute potentials of mean force obtained for some model systems will be discussed. These potentials show such universality and simplicity that one is led to believe that a description of hydrophobic interactions by the simple analytic forms given by phenomenological mean-field theory might well be adequate for many purposes. The mean-field theory is introduced and discussed. Finally we come back to Stillinger's dictum, and summarize the state of play.

2. THERMODYNAMICS: FORMALISM AND NOTATION

The formal theory of solutions is as arid and devoid of interest as it is necessary. Indeed it is probably that subject most likely to be voted most boring, by acclamation. It takes on real interest only when experimental quantities are linked via the distribution functions of statistical mechanical to molecular mechanisms. To set notation for our later study, we here define and sketch the relationship between thermodynamic quantities like excess free energies of mixing, partial molar volumes, and heats of solution to distribution functions. This is necessary because, although a number of good treatments of the subject exist,[337,488] confusion about standard states, choice of concentration units, and the meaning of molecular forces in solution can easily arise.

2.1. Definitions

The chemical potential or the partial free energy of a solute molecule (species 2) will be written as

$$\mu_2(X_2, P, T) = \mu_2{}^0(P, T) + kT \ln(fX_2) \tag{1}$$

Both $\mu_2{}^0$ and the activity coefficient f will depend on the choice of units for the concentration X_2. We take X_2 to be the solute mole fraction $N_2/(N_1 + N_2)$, where N_1 and N_2 are the number of solvent (species 1) and solute (species 2) molecules, respectively. At a prescribed temperature T and pressure P the quantity $\mu_2{}^0$ depends only on the choice of reference system, but not on the mole fraction X_2. We shall adopt as the reference state that of pure unmixed solvent and solute molecules at the given P and T.* For this choice of the initial state, the partial free energy of mixing can be written as

$$\Delta\mu_2(X_2, P, T) = \mu_2{}^0(P, T) + kT \ln(fX_2) - \mu_2(1, P, T) \qquad (2)$$

This is just the change in Gibbs free energy in transferring one solute molecule from the pure solute to a solution of mole fraction X_2. This notation will be adopted generally, e.g., ΔH_2 denotes the difference in enthalpy between the mixture and that of the pure solvent plus pure solute (with the same N_1, N_2, P, and T).

The superscript E will be used to denote excesses over the ideal solution values. Thus, for instance, the partial excess free energy of a solute molecule is

$$\Delta\mu_2{}^E(X_2, P, T) = \Delta\mu_2(X_2, P, T) - \Delta\mu_2{}^{\text{ideal}}(X_2, P, T)$$
$$= \mu_2{}^0(P, T) + kT \ln f - \mu_2(1, P, T) \qquad (3)$$

2.2. Thermodynamics and Distribution Functions

We now exhibit the link between thermodynamics and the distribution functions of statistical mechanics. This can be done, e.g., through the McMillan–Mayer equations[488]

$$kT\left(\frac{\partial n_i}{\partial \mu_j}\right)_{\mu_{k \neq j}} = n_i \delta_{ij} + n_i n_j \int h_{ij}(\mathbf{r}) \, d^3\mathbf{r} \qquad (4)$$

where n_i is the number density of species i and $g_{ij} = 1 + h_{ij}$ is the pair distribution function. [In general $g_{ij}(\mathbf{r})$ and $h_{ij}(\mathbf{r})$ may depend on the relative orientations of the molecules, in which case the integral in eqn. (4) will include an additional normalization factor for the angular integration.]

---

* Ben-Naim[86] has argued cogently that the only sensible convention for the solute reference state is that of the ideal gas. Uniform adoption of this convention would certainly remove confusion. For the purposes of this article we persist with the more natural reference state above.

In evaluating thermodynamic quantities from eqn. (4) it is convenient to take the densities n_i and temperature T as independent variables. Integration of eqn. (4) gives the chemical potentials and pressure as functions of the n_i. It is then a straightforward but tedious matter to determine the μ_i, partial molar volume $V_i = (\partial \mu_i/\partial P)_{X_j, T}$, densities, and other quantities as functions of the experimentally convenient variables P, T and mole fractions X_i.

For dilute solutions, relatively simple expressions connect solution thermodynamic quantities with the distribution functions. We introduce a new quantity $\mu_2'(X_2, P, T)$ which will be seen to be closely related to the solvent–solute distribution function. We define μ_2' by [cf. eqn. (1)]

$$\mu_2(X_2, P, T) = \mu_2^g(n_2, T) + \mu_2'(X_2, P, T) \tag{5}$$

where n_2 is the solute density corresponding to the solute mole fraction X_2 and μ_2^g is the chemical potential of an ideal gas at density n_2:

$$\mu_2^g(n_2, T) = \mu_2^*(T) + kT \ln n_2 \tag{6}$$

Here $\mu_2^*(T)$ is independent of n_2 but again depends on the choice of reference state. From its definition μ_2' has the property that

$$\lim_{P \to 0} \mu_2'(0, P, T) = \lim_{n_1 \to 0} \mu_2'(0, P, T) = \lim_{n_1, n_2 \to 0} \mu_2'(X_2, P, T) = 0 \tag{7}$$

Combining eqns. (1), (5), and (6) gives μ_2 and μ_2^0 in terms of μ_2':

$$\mu_2(X_2, P, T) = \mu_2^*(T) + kT \ln n_2 + \mu_2'(X_2, P, T) \tag{8}$$

$$\mu_2^0(P, T) = \mu_2^*(T) + kT \ln n_1^0 + \mu_2'(0, P, T) \tag{9}$$

and it follows from eqn. (3) that the partial excess free energy of a solute molecule at infinite dilution $(X_2 \to 0)$ is

$$\Delta \mu_2^E(0, P, T) = \mu_2^0(P, T) - \mu_2(1, P, T)$$
$$= kT \ln(n_1^0/n_2^0) + \mu_2'(0, P, T) - \mu_2'(1, P, T) \tag{10}$$

Here n_i^0 is the number density of pure component i. The required expression for $\mu_2'(0, P, T)$ follows from the McMillan–Mayer equation [eqn. (4)] as

$$\left(\frac{\partial \mu_2'}{\partial \mu_1}\right)_{n_2} = \left(\frac{\partial \mu_2}{\partial \mu_1}\right)_{n_2} = -\left(\frac{\partial n_2}{\partial \mu_1}\right)_{\mu_2} \Big/ \left(\frac{\partial n_2}{\partial \mu_2}\right)_{\mu_1} = -\frac{n_1 \tilde{h}_{12}}{1 + n_2 \tilde{h}_{22}} \tag{11}$$

where

$$\tilde{h}_{ij} \equiv \int h_{ij}(\mathbf{r}) \, d^3r \tag{12}$$

Whence for $X_2 \to 0$

$$\frac{\partial \mu_2'(0, P, T)}{\partial \mu_1} = n_1{}^0 \tilde{h}_{12}^0 \tag{13}$$

Using eqn. (7) this may be integrated to give

$$\mu_2'(0, P, T) = -\int_{-\infty}^{\mu_1} n_1{}^0 \tilde{h}_{12}^0 \, d\mu_1 = -\int_0^P \tilde{h}_{12}^0 \, dP \tag{14}$$

The partial molar volume of the solute at infinite dilution follows from eqns. (9) and (14) as

$$V_2{}^0 = \frac{\partial \mu_2{}^0}{\partial P} = \frac{kT}{n_1{}^0} \frac{\partial n_1{}^0}{\partial P} + \frac{\partial \mu_2'(0, P, T)}{\partial P}$$

$$= \varkappa_T{}^0 kT - \tilde{h}_{12}^0 \tag{15}$$

where $\varkappa_T{}^0$ is the isothermal compressibility of the pure solvent. A knowledge of $V_2{}^0$ will thus give some information about solvent structure near a solute molecule.

Finally the integral $\tilde{h}_{22}^0$ is directly related to the osmotic pressure. This follows from the Gibbs–Duhem equation and eqn. (4):

$$\left(\frac{\partial \pi}{\partial n_2}\right)_{\mu_1, T} = n_2 \left(\frac{\partial \mu_2}{\partial n_2}\right)_{\mu_1, T} = \frac{kT}{1 + n_2 \tilde{h}_{22}} \approx kT(1 - n_2 \tilde{h}_{22}^0 + \cdots)$$

whence

$$\pi = n_2 kT(1 - \tfrac{1}{2} n_2 \tilde{h}_{22}^0 + \cdots) \tag{16}$$

Thus the second osmotic virial coefficient $(B_2{}^* = -\tfrac{1}{2}\tilde{h}_{22}^0)$ provides the most direct information about solute–solute interactions. By contrast the activity coefficients (whether in molar, molal, or mole-fraction scales) all differ from $\tilde{h}_{22}^0$ by volumetric terms which must be known if any inference about solute–solute interactions is to be drawn from such data.

2.3. Remarks on Structure

Having outlined the connection between thermodynamic properties of dilute solutions and molecular distribution functions, it may be appropriate to comment on the concept of "structure" in liquids or solutions as used here. The notion of structure in a crystalline solid is clear. It can be characterized by the periodic lattice arrangement of the molecules of the material. The molecules in a liquid are disordered. The long-range order characteristic of a solid is absent, yet short-range density correlations do occur. These are probed directly in X-ray and neutron scattering experiments. In liquid water, the average coordination number is slightly larger than 4 and it is found that, on average, deviations from the ideal tetrahedral arrangement are quite small.[645] Thus structure in a bulk fluid is just the nonuniformity or nonrandomness in the spatial *and* orientational distribution of molecules about other neighboring molecules and is described by the pair distribution function $g_{11}(\mathbf{r})$. In bulk liquids, the density distribution as a function of distance from a typical molecule, averaged over all orientations, say, shows damped oscillations and decays within a few molecular diameters. This reflects the lack of long-range correlations in liquids where correlations vanish beyond a few molecular diameters (with the proviso that the system is not too close to the critical point).

Similarly, the solvent–solute correlation function $g_{12}(\mathbf{r})$ describes solvent structure, in other words the average distribution of solvent molecules about a solute molecule. A measure of structure induced by a solute molecule can be obtained by a comparison of the bulk solvent correlation function $g_{11}(\mathbf{r})$ with the solvent–solute correlation function $g_{12}(\mathbf{r})$. In later examples we shall see that either $g_{12}(\mathbf{r})$ can exhibit damped oscillations similar to $g_{11}(\mathbf{r})$, or there may be a *depletion* of solvent molecules near a solute particle. The latter type of behavior leads to potentials of mean force between solutes that are smooth (near exponential) rather than oscillatory functions of the solute–solute separation, even though the interaction is mediated by discrete solvent particles. The notion of solvent structure is not new. The hydration shell of ions in aqueous electrolytes, dielectric saturation near surfaces of colloidal particles, density profiles at the free liquid interface or at the boundary of two immiscible fluids are different manifestations of solvent structure. We shall return to these points again in a later section.

3. SOLVENT STRUCTURE AND SOLUTION THERMODYNAMICS: MODELS

The reader who by now is not totally convinced of the essential correctness of the opening sentence of Section 2 is beyond redemption. The thermodynamic framework is established. We now wish to contrast briefly the properties of aqueous solutions of hydrophobic solutes with those of other "normal" solutions, and then go on to study the thermodynamics of several simple model systems with two separate but related ends in mind: (a) To use distribution functions to illustrate the idea of indirect solvent mediated forces; and (b) To try to elucidate factors at the molecular level that contribute to the variety of solution behavior. The examples chosen will help in building an intuition concerning solvent structure and lay the groundwork for the more refined calculations to follow.

3.1. Characterization of Hydrophobic Solutions

The distinction between hydrophobic and normal solutions can be made by a comparison of the several partial excess quantities of the solution at infinite dilution. These are the partial excess free energy $\Delta \mu_2^E$, enthalpy ΔH_2^E, entropy ΔS_2^E, heat capacity (at constant pressure) $\Delta C_{p_2}^E$, and the partial volume change ΔV_2^E. Typical values of the various excess quantities are given in Table I. (The initial state is chosen to be the unmixed pure solute plus pure solvent.) Examples of normal solutions are: mixtures of inert gases, nonpolar hydrocarbons, and mixtures of polar solutes and polar solvents. The numbers quoted in Table I are merely indicative of the range of values that one would encounter in a broad spectrum of observed solution properties. Detailed discussions of the various properties may be found in a number of reviews.[326,327,772]

Here we note the fact that hydrophobic solutions are characterized first by the *large positive* free energy of mixing which is due to the large negative entropy term. The generally accepted interpretation of this phenomenon by Frank and Evans,[320,326,327] based upon their analysis of aqueous solutions, is that hydrophobic or lyophobic solutes can give rise to extensive "structuring" of surrounding solvent molecules, which accounts for the large entropy change upon solution. This picture is further supported by large values of the heat capacity which reflect the energy needed to "melt down" such a structured solvent. Evidently these concepts need to be

TABLE I. Partial Excess Quantities for the Process: Pure Solute + Pure Solvent → Infinitely Dilute Solution

Quantity	Hydrophobic	Normal
$\Delta \mu_2^E$ (cal mol^{-1})	≥ 5000	$\sim \pm 600$
ΔH_2^E (cal mol^{-1})	$\sim \pm 1000$	$\sim \pm 1000$
ΔS_2^E	$T\mid \Delta S_2^E \mid \gg \mid \Delta H_2 \mid$	$\mid \Delta H_2 \mid > T \mid \Delta S_2^E \mid$
$\Delta C_{p_2}^E$ (cal mol^{-1} deg^{-1})	≥ 50	$\sim \pm 3$
ΔV_2^E (cm^3 mol^{-1})	≤ -15	$\sim \pm 1$

made definite. In addition, for dilute aqueous solutions of apolar molecules, there is still considerable doubt as to the limiting value of $(\partial V_2/\partial X_2)$ at very low concentrations $[X_2 = 10^{-4}]$. It seems that hydrophobic solute–solute interactions persist to unexpectedly low concentrations. This would be an indication that solute-induced solvent structure extends over large distances and could give rise to a long-range solute–solute interaction. Also, at first sight one might expect all thermodynamic quantities associated with the process of solute–solute interaction to have the opposite sign to those associated with the process of solution.[474] This would mean that the concentration dependences of thermodynamic excess functions would have the opposite signs to those shown in Table I. This is indeed the case, except for the specific heat $\Delta C_{p_2}^E(m_2)$ (which is positive) and $\Delta V_2^E(m_2)$ (which is negative), and would suggest that solute–solute interactions are not necessarily a reversal of the solution process. Indeed, $\Delta C_{p_2}^E(m_2, T)$ exhibits an extremely complex course, with maxima and minima.

For normal solutions, on the other hand, one generally finds that both the enthalpy ΔH_2^E and the entropy ΔS_2^E are of the same sign so that the effects of these contributions to the free energy will oppose each other. Nevertheless one has $|\Delta H_2^E| > T|\Delta S_2^E|$ so that $\Delta \mu_2^E$ and ΔH_2^E have the same sign; usually $\Delta C_{p_2}^E \simeq 0$.

3.2. Effects of Molecular Properties

The formalism of Section 2.2 shows that solution properties are dictated by molecular distribution functions which themselves are determined by several factors at the molecular level. Those that spring to mind are:

(a) "chemical" association (hydrogen bonding in water),

(b) shape and size of solute molecules,

(c) energies of interaction of two molecules *in vacuo*, which can be both distance and orientation dependent (e.g., dipolar, Debye, London forces),[587]

(d) connectivity or dimensionality (e.g., difference in nature of hydrogen-bond networks, three-dimensional about small solute molecules, two-dimensional about large solutes or at an interface), and

(e) many-body forces (e.g., dipolar especially, Axilrod–Teller three-body dispersion forces).[587]

In addition external conditions (temperature, pressure) play a role. It is impossible to focus attention on all these competing factors simultaneously, and in particular to isolate which are most responsible for the special properties of aqueous solutions. Let us then proceed by a process of elimination. It will be instructive to take several of the above factors at a time, and thereby gradually build up an intuition. Major suspicion centers on factor (a) as the prime source of peculiarity. At first, we therefore remove it from consideration entirely to see how far we can proceed in generating the spectrum of solution thermodynamics with model solutions. If these models are to reveal the essential physics of the problem, a desirable if not prime requirement is analytic simplicity. Few such examples can be constructed.

3.3. Hard-Sphere Mixtures

Artificial it may be, but this is one of the few three-dimensional models of a solution for which good approximate analytic expressions exist for thermodynamic properties. It is relatively easy to examine the effect of the solute–solvent size ratio and molecular distribution functions or solvent structure on the thermodynamics of such dilute solutions.

The hard-sphere mixture thermodynamics can be found directly from good approximate equations of state which are known analytically.[539,588] Table II gives a list of thermodynamic properties of an infinitely dilute solution of hard spheres of varying solute/solvent size ratio. Clearly the dilute hard-sphere mixture is almost ideal, and not of great interest.

To see the effect of solvent structure consider a dilute solution where both solute and solvent molecules are distinguishable hard spheres of *equal* diameter R. The solution is clearly ideal, so that its second osmotic virial

TABLE II. Excess Thermodynamics of Mixing of Infinitely Dilute Solutions of Hard Spheres with Solvent Diameter R_1, Typical Liquid Density $\pi \varrho R_1^3/6 = 0.25$, and Pressure $\beta P/\varrho = 0.78$ as a Function of R_2/R_1 ($\beta = 1/kT$)

R_2/R_1	$\beta \mu_2^E$	$\beta \Delta H_2^E$	$(\pi R_1^3/6)^{-1} \Delta V_2^E$
0.5	−0.242	−0.085	−0.109
0.75	−0.058	−0.017	−0.022
1.0	0	0	0
1.25	−0.052	−0.012	−0.015
1.5	−0.198	−0.041	−0.053

coefficient is given by $B_2^* = -\frac{1}{2}V_1^0$. But from eqn. (16) we have

$$B_2^* = -\frac{1}{2} \int [g_{22}(\mathbf{r}) - 1]\, d^3\mathbf{r} \qquad (17)$$

where $g_{22}(r)$ is the solute–solute radial distribution function. If we ignore solvent effects g_{22} will be uniform outside the hard-sphere diameter. We write

$$g_{22}(r) = \begin{cases} 0, & r < R \\ 1, & r > R \end{cases}$$

and have $B_2^* = 2\pi R^3/3$. The two results for B_2^* are incompatible. For a dense fluid a characteristic value of V_1^0 (corresponding to a face-centered-cubic array) is $V_1^0 \equiv R^3/2^{1/2}$, whence $B_2^* = R^3/2^{3/2}$. The contradiction stems from the assumption that $g_{22}(r)$ depends only on the direct hard-sphere interaction. In a hard-sphere solution, $g_{22}(r)$ has damped oscillations due to structuring of the solvent molecules about a solute. Consequently the potential of mean force[612] between two solute molecules

$$W_{22}(r) = -kT \ln g_{22}(r) \qquad (18)$$

extends over several particle diameters; quite unlike the bare hard-sphere solute–solute potential (Fig. 1). The long-range tail arises from the correlated nature of the solvent. This indirect solvent-mediated or structural force takes varying forms and will recur repeatedly in subsequent discussions. Other examples of indirect particle interaction well known in the physics literature are mentioned in Ref. 591. We return to a detailed study of distribution functions in Section 6.

We emphasize that for some solutions, e.g., glycerol, sucrose, urea—those with positive B_2^*, it is always possible to fit second osmotic virial

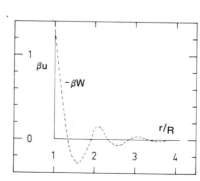

Fig. 1. Pair potential $u(r)$ and potential of mean force $W(r)$ for a fluid of hard spheres with $\varrho R^3 = 0.75$. Here r is the distance between centers and R is the hard-sphere diameter.

coefficients B_2^* to an "effective hard-sphere" model.[516,591] However, such models cannot account for negative B_2^* or the temperature dependence of virial coefficients.

3.4. One-Dimensional Solutions

The next model solutions we consider are those in one dimension. Such systems have the advantage that they are exactly soluble and are particularly suitable for investigating solution properties over all interesting regimes of the strength of molecular interaction. We choose first a mixture of particles with only nearest-neighbor interactions. Similar models have been used to mimic bulk properties of water,[67] solution properties,[572] and solvent-mediated structural forces.[175] The disadvantages of a nearest-neighbor one-dimensional model are that phase separations do not occur and size effects cannot be taken into account so that "hard" particles in one dimension are ideal. The range of behavior exhibited by this model covers normal solutions. In addition some features of hydrophobic solutions can be captured. This model neglects one very important feature of hydrophobic solutions, namely, the possibility of clathrate formation.[226] Subsequently we develop a model that admits this process, and provides further insight.

3.4.1. Nearest-Neighbor Models

Given a nearest-neighbor interaction potential between species i and j: $u_{ij}(x)$ which is a function of the separation x, the partial excess free energy of the solute at *infinite* dilution is known to be[395]

$$\Delta\mu_2^E(0, P, T) = -kT \ln(F_{12}^2/F_{11}F_{22}) \qquad (19)$$

where

$$F_{ij} \equiv \int_0^\infty \exp\{-[u_{ij}(x) + Px]/kT\} \, dx \qquad (20)$$

and P is the pressure. The corresponding enthalpic contribution to the excess free energy is

$$\Delta H_2^E = 2\eta_{12} - \eta_{11} - \eta_{12} \qquad (21)$$

with

$$\eta_{ij} = F_{ij}^{-1} \int_0^\infty [u_{ij}(x) + Px] \exp\{-[u_{ij}(x) + Px]/kT\} \, dx \qquad (22)$$

The "volume" change upon solution is

$$\Delta V_2^E = 2V_{12} - V_{11} - V_{22} \tag{23}$$

where

$$V_{ij} = F_{ij}^{-1} \int_0^\infty x \exp\{-[u_{ij}(x) + Px]/kT\} \, dx \tag{24}$$

For calculations, we choose potentials of the form

$$u_{ij}(x) = \begin{cases} \infty, & x < R \\ -A_{ij}/x^2, & R < x < 2R \\ 0, & x > R \end{cases} \tag{25}$$

Here R is the "hard core" diameter of the particle and the x^{-2} distance dependence is chosen to mimic a one-dimensional array of plates interacting under dispersion forces.[175]

The effects of varying the pressure and interaction parameters on the thermodynamic behavior of the solution are summarized in Table III.

If A_{12} is *less* than but not too close to $\frac{1}{2}(A_{11} + A_{22})$ one finds that $\Delta\mu_2^E$, ΔH_2^E, ΔS_2^E, ΔV_2^E are all positive and $\Delta C_{p_2}^E$ is negative (case 1). Conversely, when A_{12} is *greater* than but not too close to $\frac{1}{2}(A_{11} + A_{12})$

TABLE III. Typical Thermodynamic Inequalities of Nearest-Neighbor One-Dimensional Solutions at Infinite Dilution[a]

A. *Effects of molecular potential*

Case	Thermodynamic properties
(1) $A_{12} < \frac{1}{2}(A_{11} + A_{22})$	$\Delta\mu_2^E$, ΔH_2^E, ΔS_2^E, $\Delta V_2^E > 0$, $\Delta C_{p_2}^E < 0$
(2) $A_{12} > \frac{1}{2}(A_{11} + A_{22})$	$\Delta\mu_2^E$, ΔH_2^E, ΔS_2^E, $\Delta V_2^E < 0$, $\Delta C_{p_2}^E > 0$
(3) $A_{12} \approx \frac{1}{2}(A_{11} + A_{22})$	ΔS_2^E, $\Delta V_2^E < 0$, $\Delta\mu_2^E$, $\Delta C_{p_2}^E > 0$, $\Delta H_2^E \approx 0$

B. *Effect of pressure* [cases (1) and (2) only]

$$\text{High pressure } (PR/kT > 1) \begin{cases} |\Delta H_2^E| \gg T|\Delta S_2^E| \\ \Delta V_2^E, \Delta C_{p_2}^E \text{ small} \end{cases}$$

$$\text{Low pressure } (PR/kT < 1) \begin{cases} |\Delta H_2^E| \approx T|\Delta S_2^E| \\ |\Delta H_2^E| \gg |\Delta\mu_2^E| \end{cases}$$

[a] See text for discussion.

(case 2) all the above inequalities are reversed. It is clear that for both cases $|\Delta H_2^E| > |\Delta \mu_2^E|$ and $|\Delta H_2^E| > T|\Delta S_2^E|$. In other words, the change in entropy opposes that in enthalpy but the latter still dominates— similar to the behavior in normal solutions. Further, at high pressures we also have $|\Delta H_2^E| \gg T|\Delta S_2^E|$, reminiscent of regular solutions. When the interactions between all species in the mixture are similar, $A_{12} \approx \frac{1}{2}(A_{11} + A_{22})$, various possibilities arise. Of special interest is the case when the enthalpy almost vanishes and all the thermodynamic quantities obey the same inequalities as those of a hydrophobic solution (cf. Table III, case 3). It is interesting to note that the Hamaker constants for the rather small dispersion interaction due to electronic correlations in water/hydro-carbon systems also satisfies case 3 (cf. Section 4). However, unlike hydro-phobic solutions, the *magnitudes* of the nonvanishing terms in case 3 a much smaller than those in case 1 or 2. For case 3 varying the temperature or pressure changes the thermodynamic inequalities to those of case 1 or 2. We remark parenthetically that case 2 is excluded if the interactions are real dispersion forces (for dispersion forces always $A_{12} \le (A_{11} + A_{22})/2$).

In the present model the second virial coefficient for $\Delta \mu_2^E(X_2, P, T)$ (the coefficient proportional to X_2 in a Taylor series expansion) vanishes identically—another deficiency. At infinite dilution the solute–solvent distribution function g_{21}^0 and solute–solute potential of mean force W_{22}^0 are generally oscillatory functions, reminiscent of bulk fluids, and show no hint of the structural interactions to be discussed in Section 6.

3.4.2. Clathrates in One Dimension

A major deficiency of the above one-dimensional model as a source of intuition about hydrophobic solutions is that, even when the various thermodynamic quantities satisfy the required inequalities (case 3 in Table III), the entropic contribution to the excess free energy $T|\Delta S_2^E|$ is very small ($\sim 0.2\,kT$/particle) compared to "normal" one-dimensional solution ($\sim 2\,kT$/particle). This is due to the lack of a clathrate-forming mechanism in this model. In dilute aqueous solutions of hydrophobic molecules the solutes are believed to occupy interstitial sites in the aqueous environment, consistent with the experimental $\Delta V_2^E < 0$. Water molecules adjacent to the solute reorient to preserve the hydrogen bonding at the expense of a decrease in entropy. In other words, the rather open structure of water allows the solvent to maintain the *energetic* part of the solvent–solvent interaction even in the presence of a solute molecule. Structurally this is reflected in the cage formed around the solute.[326,327,827]

In the one-dimensional model that we have just investigated only nearest-neighbor interactions were considered. The introduction of a solute particle between two solvent molecules "breaks up" a solvent–solvent "bond" and replaces it by two solvent–solute "bonds." One way[175] to overcome this situation so as to preserve the solvent–solvent "bond" even in the presence of the solute is to relax the restriction upon nearest-neighbor interactions when solutes are inserted. In other words, neighboring solvent molecules still interact even when separated by a solute molecule (see Fig. 2). This is a simple device to mimic the clathrate structure in aqueous solutions. Other solvent–solvent interactions are still restricted to nearest neighbors. A similar model which allows for the formation of solvent *n*-mers has been put forward by Lovett and Ben-Naim[572] to study solute effects in a mixture model of water.

In order to exhibit the thermodynamics of this solution we choose as the nearest-neighbor solvent–solvent interaction the square well potential

$$u_{11}(x) = \begin{cases} \infty, & x < R_1 \\ -\varepsilon_{11}, & R_1 < x < 2R_1 \\ 0, & x > 2R_1 \end{cases} \tag{26}$$

For simplicity the solute–solvent interaction is treated as a hard core repulsion

$$u_{12}(x) = \begin{cases} \infty, & x < \tfrac{1}{2}(R_1 + R_2) \\ 0, & x > \tfrac{1}{2}(R_1 + R_2) \end{cases} \tag{27}$$

The size of the solute R_2 is chosen to be *less than* R_1 so as to allow the

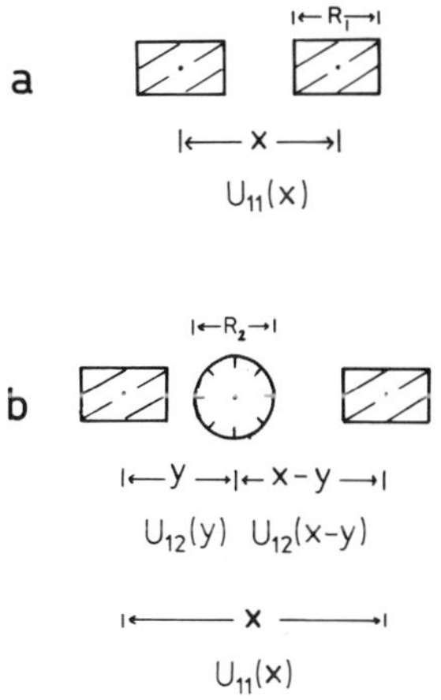

Fig. 2. Solvent–solvent interactions in the one-dimensional clathrate model: (a) with no solute and (b) with a solute 2 present. The squares represent solvent molecules of size R_1 and the circles solutes of size R_2.

solvent molecules on either side of a solute to interact with each other. The thermodynamics of this model is straightforward. Equation (19) for the partial excess free energy of the solute is generalized to

$$\Delta\mu_2^E(0, P, T) = -kT \ln(F_{121}/F_{11}F_{12}) \tag{28}$$

where F_{121} is the configuration integral corresponding to the solvent–solute–solvent complex (Fig. 2),

$$F_{121} = \int_0^\infty dx \int_0^x dy \; e^{-\beta U_{12}(y)} \, e^{-\beta U_{12}(x-y)} \, e^{-\beta U_{11}(x)} \, e^{-\beta PX} \tag{29}$$

Following the notation of Kozak et al.[516] we write

$$\beta\Delta\mu_1^E(X_2, P, T) = \beta\Delta\mu_1^E(0, P, T) + BX_2^2 + \cdots \tag{30}$$

where

$$B = \frac{1}{2}\left[\frac{F_{21}^2}{F_{121}} - 1\right] \tag{31}$$

It can be shown that B is always negative or zero unlike hydrophobic solutions and vanishes when only nearest-neighbor interactions are allowed, i.e., when $F_{121} = F_{12}^2$.

The model yields more respectable "aqueouslike" thermodynamics $(T \,|\, \Delta S_2^E \,| \gg \Delta H_2^E)$. Typical results are given in Table IV. Results from the nearest-neighbor one-dimensional solution which exhibit normal solution properties are included for comparison. We note that at higher temperatures, and when solute and solvent are interchanged, the solution

TABLE IV. Comparison of Thermodynamic Properties of Clathrate and Nearest-Neighbor One-Dimensional Models ($\beta PR_1 = 1.0$, $\beta\varepsilon_{11} = 4$, $R_2/R_1 = 0.6$. Energies are in Units of kT)

Property	Clathrate model	Nearest-neighbor model
$\Delta\mu_2^E(0, P, T)$	2.09	3.55
ΔH_2^E	0.06	4.52
$T\Delta S_2^E$	−2.03	0.97
$\Delta V_2^E/R_1$	−0.77	0.57
$\Delta C_{p_2}^E/k$	4.87	0.59

reverts to normal behavior ($|\Delta H_2^E| \gg T|\Delta S_2^E|$)—this is reminiscent of the vanishing of the hydrophobic effect at elevated temperatures and the normal behavior of dilute solutions of water in apolar solvents.

An additional remarkable feature of the model is revealed by a detailed study of the solute–solute pair distribution function.[175] It has been found that, in the regime in which the model exhibits "aqueouslike" thermodynamic properties, the solute–solute potential of mean force is attractive, and has a strong minimum when the solutes are separated by one solvent molecule rather than at contact. This is reminiscent of hydrophobic solution behavior inferred from specific heat data (cf. Section 3.1). {Note that no inference concerning the solute–solute potential of mean force can be made directly from knowledge of the coefficient B alone. B differs from B_2^* [cf. eqns. (16) and (17)] by volumetric terms.}

The one-dimensional model of this section has many defects as a model of water. These weaknesses are in fact its strength, since the only property that it shares with water is the possibility of clathrate formation. This provides a strong hint that the large entropy change and some other features characteristic of hydrophobic solutions can be traced back primarily to this distinguishing aspect of water. Obviously the model can be decorated to mimic the properties of water in a more or less realistic manner. A simple modification that allows for a density maximum to occur at high enough pressure is the following.[175]

At constant pressure P, the temperature dependence of the density is governed by the manner in which the enthalpy function $H(x) \equiv Px + u_{11}(x)$ varies with intermolecular separations. Detailed analysis of the equation of state reveals that if the intermolecular potential $u_{11}(x)$ has a hard core at $x = R$ and an energy minimum located at the same position, then, irrespective of the pressure, the system cannot have a density maximum as a function of temperature. However, if the minimum of $u_{11}(x)$ is located at a *larger* intermolecular separation than that of the hard core, the system may exhibit a density maximum at constant pressure, provided the enthalpy function $H(x)$ also has an *absolute* minimum at a position away from the hard core, in addition to the minimum at $x = R$, e.g.,

$$u_{11}(x) = \begin{cases} \infty, & x < R_1 \\ 0, & R_1 < x < 2R_1 - a \\ -\varepsilon, & 2R_1 - a < x < 2R_1 \\ 0, & x > 2R_1 \end{cases}$$

In water a situation similar to that just posed is brought about by two factors: the strong and highly directional hydrogen bonds, which operate effectively at intermolecular separations that are larger than close packed, and the relatively weak dispersion interactions, which are effective at smaller separations. In the next section we shall consider the implications of the second factor in the long-range behavior of intermolecular interactions.

4. EVIDENCE FROM VAN DER WAALS FORCES

At this point it is instructive to pause and digress. We ask if the modern theory of long-range molecular interactions can tell us anything about the hydrophobic effect. The answer is an unequivocal yes. There are already in the Lifshitz theory of condensed media strong hints that shed some light on the peculiarity of interactions between organic materials in water. At worst, whatever the details of the force law between two hydrophobic molecules in water, at large distances this force must go on to a form identical with the van der Waals forces calculated on the basis of continuum theory. The tail (or asymptotic form) of the hydrophobic potential is immediately accessible.

The nature of long-range van der Waals forces between molecules and between bodies immersed in liquids has been the subject of an extensive literature[587] (both theoretical and experimental) over the past 15 years. Especially subtle and delicate are the forces acting between bodies in water. It is not our purpose to review the literature. We shall simply quote results and point out special features that bear on our problem. In general these forces depend on the dielectric properties of the bodies and intervening medium, on size and shape, anisotropy, temperature, electrolyte concentrations, retardation, inhomogeneity, spatial dispersion, and other factors, all of which have been extensively explored.

To begin, consider the problem of two semi-infinite half-spaces 2 separated by a liquid 1 of thickness l [Fig. 3(a)]. This problem will be seen later to bear upon the interaction of 2 molecules in solution. The distance l requires definition. If the media 2 constitute a perfect solid made up of polarizable molecules, if medium 1 is a vacuum, and if the only forces acting are London dispersion forces, then l is well defined at least up to a molecular diameter. On the other hand if the media 2 are liquids, the most appropriate choice for this interface is that given by the Gibbs dividing surface [Fig. 3(b)], and the media can be taken as of uniform density (and presumably dielectric properties) up to this interface provided l is itself very much greater than the width of the density profile. Normally, far from

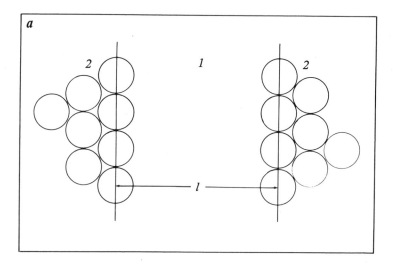

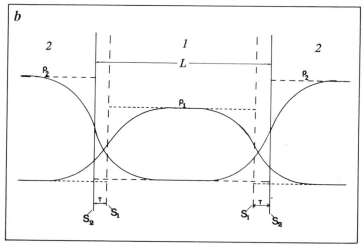

Fig. 3. Interaction of two half-spaces 2 across medium 1 when (a) the 2–1 interface is sharp and (b) when medium 2 is a fluid. There are density profiles at each interface for case (b) and we adopt a triple film representation. $S_{1,2}$ locate the Gibbs dividing surfaces for each species.

any critical points, we can expect this profile width to be of the order of a molecular diameter.

Subject to this proviso, and ignoring retardation (important beyond about 50 Å) the free energy of interaction per unit area measured with respect to a zero at infinite separation can be written approximately as

(c indicates classical, q quantum mechanical)

$$G \simeq G^c + G^q$$

$$= -\frac{kT}{16\pi l^2} \left[\frac{\varepsilon_1(0) - \varepsilon_2(0)}{\varepsilon_1(0) + \varepsilon_2(0)} \right]^2 - \frac{\hbar}{16\pi^2 l^2} \int_0^\infty d\xi \left[\frac{\varepsilon_1(i\xi) - \varepsilon_2(i\xi)}{\varepsilon_1(i\xi) + \varepsilon_2(i\xi)} \right]^2 \quad (32)$$

The formulas rely on *measured* dielectric permittivity ε as a function of frequency. [The conversion of measured response functions $\varepsilon(\omega)$ to response functions $\varepsilon(i\xi)$ on the imaginary frequency axis in the complex ω plane is simply a mathematical artifice that facilitates computation.] We remark first that the zero-frequency term G^c is explicitly temperature dependent; the second term G^q has a weak implicit temperature dependence via the permittivities ε. G^q takes account of ultraviolet visible and infrared frequency contributions. For interactions across a vacuum, or for most inorganic materials in water, the interaction is dominated by G^q (in vacuum $G^q \sim 50G^c$, while in water $G^q \sim 10G^c$). But for some organic materials in water the situation is rather different. Here, as shown by extensive investigation and experiment, the relative similarity of dielectric properties of, e.g., hydrocarbons, and water in the ultraviolet drastically *reduce* G^q to the point that G^c can give up to three-quarters of the total interaction free energy. The van der Waals interaction is very much weaker than is usual (cf. the nearest-neighbor model of Section 3.4.1).

We now consider evidence from the continuum theory concerning the influence of many-body forces, temperature, and solvent effects on molecular interactions in solution.

4.1. Many-Body Forces

The connection of eqn. (32) with London dispersion forces can be made explicit as follows. Consider G^q. If characteristic absorption occurs only in the ultraviolet, e.g., for hydrocarbons, we have

$$\varepsilon_j(\omega) = 1 + \frac{n_j^2 - 1}{1 - \omega^2/\omega_j^2} \simeq 1 + 4\pi\varrho_j\alpha_j(\omega), \qquad j = 1, 2 \quad (33)$$

where n_j is the refractive index, and ω_j the measured absorption frequency which can be taken as the first ionization potential, ϱ_j is the density, and $\alpha_j(\omega)$ the dynamic polarizability. Then putting $\omega = i\xi$ and substituting into eqn. (32) we have

$$G^q = -\frac{\hbar}{l^2} \int_0^\infty \frac{(\varrho_1\alpha_1 - \varrho_2\alpha_2)^2}{[2 + 4\pi(\varrho_1\alpha_1 + \varrho_2\alpha_2)]^2} \, d\xi \quad (34)$$

If we drop terms in α_1, α_2 in the denominator, this expression reduces further to the sum of two-body London dispersion forces. The terms in $(\varrho_1\alpha_1 + \varrho_2\alpha_2)$ measure 3, 4, and higher many-body forces. Even for the simplest dielectric media these contributions are *not* small. Indeed, neglect of many-body forces (built naturally into Lifshitz theory) results in errors of a factor of at least 2. This is a feature of interactions in solution (vastly magnified for temperature-dependent forces) and underlines an important point: Even if solute and solvent molecules interact through a two-body potential that provides a good description of bulk thermodynamic properties of each medium separately, the solute–solute interaction *cannot* be modeled quantitatively through a statistical mechanics based on a two-body Hamiltonian.

For organic materials interacting across water, we remark that detailed analysis shows that about 20%–30% of the contribution to G^q comes from infrared fluctuations, and the remainder from the visible and ultraviolet.[349,587,657]

4.2. Temperature-Dependent Forces and Salt Effects

The nature of the temperature-driven free energy G^c due to classical dipole–dipole interactions is much more subtle. Consider[684] two *low-density* media 2 made up of water molecules interacting across a vacuum 1. At low densities the dipoles μ have an effective static polarizability $\mu^2/3kT$, hence

$$\varepsilon_2(0) = 1 + 4\pi\varrho(\mu^2/3kT), \qquad \varepsilon_1(0) = 1 \tag{35}$$

where μ is the dipole moment, whence to leading order in the density

$$G^c \simeq - \frac{\pi}{36l^2} \frac{\mu^4\varrho^2}{kT} \tag{36}$$

and this is precisely what we would have obtained by summing pairwise the Keesom intermolecular interaction energies $V(r) = -2\mu^4/3kTr^6$.

The entropy and enthalpy associated with the free energy G^c are

$$S = -\left(\frac{\partial G^c}{\partial T}\right) = \frac{G^c}{T}, \qquad H = 2G^c \tag{37}$$

The temperature dependence of the interaction as predicted by pairwise summation is proportional to $1/T$. However, for water at liquid densities this is quite wrong. To see this we decompose the Lifshitz expression for G^c into entropic and enthalpic contributions; we have

$$H = -T^2 \frac{\partial(G^c/T)}{\partial T} \tag{38}$$

If medium 1 is water, 2 hydrocarbon, $\varepsilon_1(0) \simeq 80$, $\varepsilon_2(0) \simeq 2$, $\Delta_0 = [\varepsilon_1(0) - \varepsilon_2(0)]/[\varepsilon_1(0) + \varepsilon_2(0)] \sim 0.95$ at 20°C. Further $\partial\varepsilon_2(0)/\partial T \simeq -0.0016$; $\partial\varepsilon_1(0)/\partial T \simeq -0.37$, whence $\partial\Delta_0/\partial T \simeq -1.8 \times 10^{-4}$, $H \simeq 0.1G^c$, $-TS \simeq 0.9G^c$. The relative contributions of enthalpy and entropy are quite unlike those obtained from summing two-body Keesom potentials. The entropy of interaction increases on approach of two hydrocarbon bodies, and the free energy change is almost entirely due to the entropy change, even when contributions from G^q are included.[742] The enthalpy change has the same sign as the free energy change. This situation is characteristic of the hydrophobic interaction, and shows that the long-range part of the interaction is behaving as it should. Such behavior is *not* peculiar to water, but holds for *any* medium of high dielectric constant and large $\partial\varepsilon/\partial T$, whether an associated liquid or not. (*On the other hand the anomalous behavior of ΔC_p as two hydrocarbon bodies come together is not predicted from Lifshitz theory.*) Further, although the interaction increases with temperature as it should, it does not exhibit the observed decrease at elevated temperatures, at which water more closely resembles a "normal" polar liquid. These two facts taken together imply some structural change in the distribution of water molecules in the immediate neighborhood of hydrocarbons or surfaces as they come together. This is already clear from a consideration of the liquid/vapor surface free energy. If temperature-dependent forces are absent, we can calculate the surface energy or surface tension as follows: Consider two half-spaces of nonpolar material, e.g., hydrocarbons, at "contact," and separate the two half-spaces to an infinite distance apart to create two surfaces. The work done against the interaction energy at the "contact" distance l_c becomes the surface energy γ, of the two half-spaces at infinite separations; that is

$$2\gamma = \frac{\hbar}{16\pi^2 l_c^2} \int_0^\infty d\xi \left[\frac{\varepsilon(i\xi) - 1}{\varepsilon(i\xi) + 1} \right]^2 \tag{39}$$

For a whole series of aliphatic hydrocarbons for which temperature-dependent interactions (across vacuum) are unimportant, this result gives good predictions of the surface energy.[438] The contact distance l_c which is expected to be the separation between polarizable units —CH_2 groups, is taken to be ≈ 2 Å throughout. Now when this approach is used to calculate the surface free energy of water, it fails miserably even when the temperature-dependent term involving the high static dielectric constant of water is included. This observation merely lends weight to the earlier remark about the structural rearrangement of water in the vicinity of interfaces and hydrophobic molecules.

We turn briefly to the effect of inorganic salts on aqueous solutions. It is well known that the solubility of organic materials in water is very ion specific, that is it can increase or decrease depending on the type of ions added—the salting-in and salting-out effect.[327] There is no theory of specific ion effects on the interaction between organic materials across water. What is known, however, is the effect of added salt on the long-range interaction between macroscopic bodies.[587] The temperature-dependent term G^c in eqn. (32) changes from a l^{-2} form at small separations to an exponentially screened $e^{-2\varkappa l}/l$ distance dependence when $l > 1/\varkappa$. Here $\varkappa$ is the inverse Debye screening length. In general, the addition of indifferent electrolyte to the intervening aqueous medium lowers the magnitude of and decreases the range of the van der Waals interaction. The long-range entropically driven force becomes negligible with increasing salt concentration.

5. EFFECT OF SOLVENT STRUCTURE ON LONG-RANGE FORCES

We have already alluded to the limitations in the assumption of sharp boundaries. If one or more of the media is a liquid, there will be a density profile at the phase boundaries. The width of the profile is of the order of a molecular diameter. Provided the separation l between the interfaces is sufficiently large, the effects of density profiles can be ignored. This is not the case when we consider interactions at smaller separations. Here we wish to consider a couple of examples to illustrate the importance of these density profiles (solute-induced solvent structure) on the van der Waals interaction. This will give us some feeling for the range of structural forces.

5.1. Solute Interactions via Lifshitz Theory

First, we study the form of the long-range interaction between two solutes in a dilute solution. For simplicity we consider the nonretarded limit and focus attention on temperature-independent van der Waals forces. The interaction energy per unit area of two half-spaces of material

3 interacting across material 1 is [cf. eqn. (32)]

$$G^q = -\frac{\hbar}{16\pi^2 l^2} \int_0^\infty d\xi \left(\frac{\varepsilon_3 - \varepsilon_1}{\varepsilon_3 + \varepsilon_1}\right)^2 \tag{40}$$

We further assume that medium 3 is a dilute solution of molecules of type 2 in medium 1 and that medium 1 is also sufficiently dilute so that we can write

$$\varepsilon_1 = 1 + 4\pi\varrho_1\alpha_1, \qquad \varepsilon_3 = 1 + 4\pi\varrho_1\alpha_1 + 4\pi\varrho_2\alpha_2 \tag{41}$$

It is readily shown that to leading order in the solute density, the interaction energy G^q may be represented by a pairwise summation of an effective two-body solute-solute potential of the form[624]

$$W_{22}(r) = -\frac{3\hbar}{\pi r^6} \int_0^\infty d\xi \left(\alpha_2 + \varrho_1\alpha_1 \int h_{12}(\mathbf{r})\, d^3\mathbf{r}\right)^2 \tag{42}$$

This can be identified as formally equivalent to the expression for the vacuum London interaction between two molecules having effective polarizabilities

$$\alpha_2{}^{\text{eff}} = \alpha_2 + \varrho_1\alpha_1 \int h_{12}(\mathbf{r})\, d^3\mathbf{r} \tag{43}$$

To see the physical significance of this expression let us take the form for $h_{12}(\mathbf{r})$ for a structureless solvent, i.e.,

$$h_{12}(\mathbf{r}) = g_{12}(\mathbf{r}) - 1 = \begin{cases} -1, & r < R_{12} \\ 0, & r > R_{12} \end{cases} \tag{44}$$

where $R_{12} = (R_1 + R_2)/2$ is the average of the solvent and solute molecular diameters. With this simple expression for $\alpha_2{}^{\text{eff}}$ it is clear that the term $\varrho_1\alpha_1 \int h_{12}(\mathbf{r})\, d^3\mathbf{r}$ accounts for the displacement of the solvent by the solute particle. In general h_{12} will have a complicated form due to the structuring of a solvent molecule about the solute. The implications of this result are clear. Solvent structure affects the long-range solute–solute potential to the extent that the solutes behave like dressed particles with an effective polarizability that depends on the form of the solute–solvent distribution function $h_{12}(\mathbf{r})$. Equation (42) holds only when the density profiles (or corresponding orientations) of solvent molecules around each solute molecule do not overlap. At small distances there will be strong overlap, and the interaction must take a different form. It is interesting to note that the Lifshitz expression for the van der Waals interaction includes implicitly many-body effects as well as structural effects (for point molecules only) through the use of measured dielectric data.

5.2. Planar Interactions via Statistical Mechanics

Next we consider the effect of solvent structure upon the long-range form of the interaction potential between half-spaces (walls) or large particles across a liquid. We seek the first-order correction term beyond $1/l^2$ due to the effects of structuring in the intervening liquid. The geometry and notation are defined in Fig. 4. Only the large distance limiting behavior of the vacuum potentials are required. We assume these to be dominated by pairwise dispersion interactions. The walls (2) are treated as a rigid solid although we do recognize its molecular nature [see Fig. 3(a)]. The final result for the interaction energy per unit area valid in the limit $l \to \infty$ is[624]

$$E(l) = -\frac{1}{12\pi}\left[\frac{A_{22}}{(l+2d-R_1)^2} - \frac{2A_{12}}{(l+d-R_1)^2} + \frac{A_{11}}{(l-R_1)^2}\right.$$
$$\left. + \frac{(A_{21}-A_{11})}{l^3}\int_0^\infty h_{21}'(x)\,dx + \cdots\right] \tag{45}$$

where the A_{ij} are Hamaker constants. The first three terms (in l^{-2}) are reminiscent of the Hamaker result for a triple film. The last term in l^{-3} arises from the nonuniform density profile of solvent molecules near each wall described by the distribution function $1 + h_{21}'(x)$.

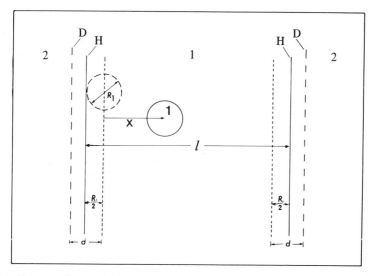

Fig. 4. Geometry for two half-spaces 2 interacting across a liquid 1. The solvent molecules are of diameter R_1. H defines the plane of hard contact of solvent molecules with the wall. Located a distance d behind H is the dielectric interface D which defines the half-space 2.

An intuitive discussion of eqn. (44) using a triple film picture[685] has already been presented.[624,625] Consider two half-spaces of liquid 2 separated by another liquid 1, for instance, a water–oil–water system. The two liquids are immiscible and there will be a density profile at each interface as illustrated schematically in Fig. 3(b). For convenience the 2–1 interfaces may be chosen as the Gibbs dividing surfaces for species 2 [S_2 in Fig. 3(b)] whose separation is L. To a good approximation, when L is large the actual density profile for each species can be replaced by a step function uniform up to the Gibbs dividing surface for that species. These will be at different positions for each species and their separation will be denoted by τ. Thus the system of Fig. 4 can be regarded as a triple film in which the region of thickness τ is a vacuum. The interaction energy per unit area $E(L)$ is then

$$E(L) = -\frac{1}{12\pi}\left[\frac{A_{22}}{L^2} - \frac{2A_{21}}{(L-\tau)^2} + \frac{A_{11}}{(L-2\tau)^2}\right] \tag{46}$$

and, expanding to order $1/L^3$, this becomes

$$E(L) = -\frac{1}{12\pi}\left\{\frac{(A_{22} - 2A_{21} + A_{11})}{L^2} - 4\tau\frac{(A_{21} - A_{11})}{L^3} \cdots\right\} \tag{47}$$

In order to relate this to eqn. (45) we note that, for the system defined by Fig. 4, $L = l + 2d - R_1$. Comparison of eqns. (45) and (47) suggests the identification:

$$\tau = d - \int_0^\infty h'_{21}(x)\,dx \tag{48}$$

Previous studies using Lifshitz theory provide estimates of the Hamaker constants for hydrocarbon interacting across water[685]: $A_{11} = 5.8 \times 10^{-13}$ erg, $A_{22} = 7.5 \times 10^{-13}$ erg, and we take $A_{12} \simeq (A_{11}A_{22})^{1/2}$, which for the present argument is an acceptable approximation. A plausible estimate for τ is a molecular diameter R_1,[623,625] whence eqn. (47) yields

$$-12\pi L^2 E(L) \underset{L\to\infty}{\sim} \left(1.1 - \frac{30R_1}{L} + \cdots\right) (\times 10^{-14}\,\text{erg}) \tag{49}$$

The asymptotic formula eqn. (49) breaks down for distances $L \leq 20$ diameters. For systems with dissimilar Hamaker constants, this breakdown of continuum theory is not nearly so dramatic. Oil (hydrocarbons) in water are peculiar. The form of the short-range structural interaction which takes over at small separations is the subject of the next section.

6. SHORT-RANGE SOLUTE–SOLUTE INTERACTION

Unlike the long-range interaction between continuous media, which is well understood both theoretically and experimentally, the nature of the interaction between solutes or surfaces at separations below about 20 solvent diameters is still very much of a puzzle. Expertise developed in the field of liquid-state physics should be able to make significant contributions in this area. In addition, direct experimental investigations of interactions at short range are making healthy progress.[10,546] There is urgent need to develop a complementary set of theoretical concepts to interpret these experimental observations.

In this section, we look back on some of the results of theoretical investigation of solvent-mediated interactions. One line of investigation into properties of fluids near surfaces or of solutions uses direct computer "experiments"—Monte Carlo or molecular dynamics—a technique already familiar in the study of bulk liquids.[53] However, the majority of this work concentrates more on the density profile of the fluid rather than addressing the interaction problem directly. A first attempt to simulate the hydrophobic interaction has been made by Dashevsky and Sarkisov,[225] who obtained the potential of mean force between two methanelike solute molecules in a model water solvent, which force was a smooth, monotonic function, largely entropic, and significantly larger than the bare solute–solute interaction. It is possible that owing to the small sample size this result may be an artifact reflecting the effect of the walls. Related work in this area is the Monte Carlo calculation of the ion–ion potential of mean force in polar fluids.[609,689]

Another class of theories which concentrates on the interaction aspect uses the integral equation approach.[53.896] The molecular distribution functions are obtained from the Ornstein–Zernike (OZ) equations which have to be supplemented by approximate closure relations, e.g., the Percus–Yevick (PY) or hypernetted chain (HNC). For nonpolar fluids, at least, this approach yields quite acceptable results for bulk properties of liquids. The PY approximation has the additional appeal of relative analytical simplicity as well as being exactly soluble for hard-sphere systems. The natural development of the integral equation approach to study mixtures and interfaces has been hampered by the problem that early attempts using the PY approximation gave nonsensical results (negative pair distribution functions) at liquid densities when the solute to solvent size ratio is too high. However, a recent observation allows the PY result to be given a physical interpretation in terms of the HNC approximation.[623] (The exact nature of this reinterpretation—a mere technicality—is of no interest here.) We summarize here what has been gleaned concerning distribution functions from this integral equation approach. We remark first that short-range structural forces between surfaces in water are exponential,[10,546] which is at first sight confusing, since experience with bulk fluids would lead us to expect oscillatory distribution functions. Let us therefore consider two surfaces separated by various model solvents to see under what conditions such behavior occurs theoretically. These surfaces can be considered to be the limit of very large solute molecules.

6.1. Hard-Sphere Solvent

Our first model comprises a hard-sphere fluid between two hard walls. The potential of mean force per unit area is presented in Fig. 6 for a typical liquid density. Note the damped oscillations, with period of the order of the solvent molecular diameter. Similar behavior is observed for h_{12} (Fig. 5).

6.2. Lennard-Jones Solvent

We now move on to investigate the inclusion of attractive interactions between species,[624] and consider a model in which the solvent–solvent

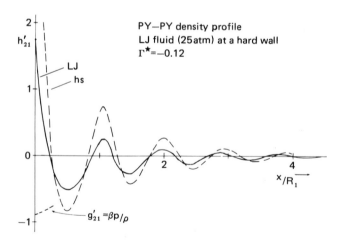

Fig. 5. Density profile h'_{21} for a fluid at an interface calculated from the PY equation. The broken curve is for a hard sphere fluid ($\varrho^* = 0.8$) at a hard wall. The solid curve is for a L.J. fluid ($T^* = 0.73$, $\varrho^* = 0.84$) at a hard wall ($\beta M'_{21} = 0$ and $d = \sigma$). Here X is the normal distance from the center of a solvent molecule to the wall and Γ^* is the adsorption excess. The exact contact density is indicated.

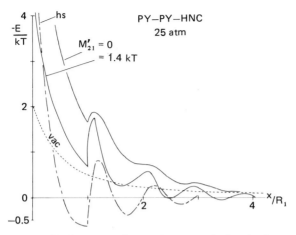

Fig. 6. Structural contributions to $-\beta E$. Broken curve is for the interaction of two hard walls across a hard-sphere fluid. Solid curves are for two attractive walls separated by a L.J. fluid ($T^* = 0.73$, $\varrho^* = 0.84$): the curves are labeled with the values of βM_{21}. The vacuum interaction is also shown (dotted curve) for $\beta A_{22}/12\pi = 2$. Here X is the separation of the walls. The full interaction is the sum of the vacuum and solid curves.

interaction is described by the Lennard-Jones (LJ) potential

$$u_{11}(r) = 4\varepsilon_1\left[\left(\frac{\sigma}{r}\right)^{12} - \left(\frac{\sigma}{r}\right)^6\right] \tag{50}$$

The solvent-molecule–wall potential assumes the form[624]

$$u_{12} = \begin{cases} -\dfrac{M_{12}}{(x/\sigma + 1)^3}, & x > 0 \\ \infty, & x < 0 \end{cases} \tag{51}$$

where x measures the distance from the molecules to the surface (see Fig. 4; d is taken to be σ). The choice of LJ fluid parameters is $T^* = kT/\varepsilon_1 = 0.73$, $\varrho^* = \varrho\sigma^3 = 0.84$. This choice represents a liquid state of the LJ fluid midway between the triple and critical points. The relative strength of the solvent–wall attraction parameter M_{12} is chosen to mimic the dispersion or temperature-independent part of the van der Waals interaction in oil–water systems. Since $T^* = 0.73$ corresponds to a Hamaker constant $A_{11}/12\pi kT = 1.84$ at $T = 86°C$, by assuming that the geometrical mean rule $A_{12} = (A_{11}A_{22})^{1/2}$ holds, we get $M_{12}/kT = 4.8$; we choose 1.4 as dictated by the stability of the PY equation.[624] In Fig. 5 we show a typical density profile h'_{12} of a LJ fluid at a hard wall. The peak near the surface (incorrectly) indicates strong adsorption at the wall, but the surface excess Γ^* is negative indicating surface depletion. The result is quite different from the hard-sphere–hard-wall system. Indeed the adsorption excess of LJ particles at these attractive walls is in fact negative—indicating a preference of the solvent to be away from the surface. In Fig. 6 we show the interaction free energy per unit area between two flat surfaces. For comparison we also plot in the energy per unit area for the van der Waals interaction between two hydrocarbon media across vacuum. (The corresponding van der Waals interaction across the solvent would be very much weaker still.) For small separations ($\leq 10\sigma$) the structural or solvent-mediated interaction dominates the vacuum dispersion interaction. The qualitative features of the potential of mean force are quite different from those of the hard-sphere–hard-wall system. For our choice of parameters the structural interaction mediated by the LJ fluid is a relatively smooth function of the distance between the surfaces whereas that for the hard-sphere fluid is highly oscillatory. Finally we note that the calculated contact pressure [initial slope of $E(x)$] overestimates the known exact limit by a factor of about 5. It is not possible to assess the contact energy.

What are the limitations of this calculation?

(a)   The PY approximation for h'_{21} is badly in error at contact.[624] If adjustment is made for this by an appropriate interpolation the results of Fig. 6 are essentially unchanged beyond 1–2 molecular diameters. The calculations are subject to a "weak overlap" approximation, a severe restriction.

(b)   In Section 4, we have seen that many-body forces are important in interactions involving oil–water systems, hence the use of a two-body potential in this calculation suggests that the results will only serve as qualitative guides.

(c)   The fact that interactions mediated by discrete particles can be fairly smooth is more puzzling. However, as one is able to recover the Deryaguin limit (anticipated from physical considerations) for the potential of mean force between large spherical solutes, this is an indication that this approach is self-consistent.[623,624] This smooth potential of mean force warrants further attention.

6.3. Sticky Hard Spheres

The numerical computation that went into the calculation for these LJ systems is quite substantial. Therefore this approach is not readily applicable to study a wider range of solvent parameters and to explore the circumstances for which the oscillatory or smooth form of the potential of mean force hold. A model fluid suited to this purpose is that of a mixture of hard spheres with surface adhesion[62,176,216] ("sticky" spheres). Here the potential is defined through the equation

$$e^{-u_{ij}/kT} = \begin{cases} \dfrac{R_{ij}}{12\tau_{ij}}\,\delta(r - R_{ij}), & r \leq R_{ij} \\ 1, & r > R_{ij} \end{cases} \qquad (52)$$

where $R_{ij} = \frac{1}{2}(R_i + R_j)$ with R_i and R_j solvent and solute particle diameters and τ_{ij} a measure of the strength of the adhesion. The limit of $\tau_{ij} \to \infty$ corresponds to hard spheres while $\tau_{ij} \to 0$ corresponds to strong adhesion. This model exhibits a bulk phase transition similar to that of a real fluid and has sufficient physical content to serve as a model system. The interaction between planes can be obtained by taking the limit $R_2 \to \infty$, $\varrho_2 \to 0$. In Fig. 7 we plot the potential of mean force between two "sticky" walls interacting across a fluid of "sticky" spheres. The parameters are chosen[176] to correspond to a LJ fluid between two x^{-3} attractive walls. From these

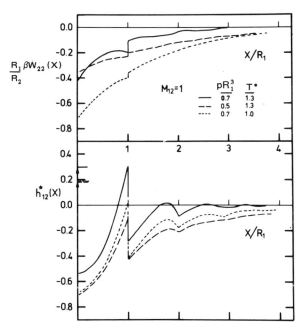

Fig. 7. The solvent-wall correlation function h'_{12} and potential of mean force W_{22} for two large spheres in a "sticky" sphere solvent with weak solvent-wall attraction ($\beta M_{12} = 1$). The δ-function contributions to h'_{12} are indicated by arrows at $x = 0$ and their coefficients by horizontal lines.

results and further calculations given in Ref. 176 we conclude that weak solvent–solute attraction and strong solvent–solvent attraction (i.e., low pressure) favors a near monotonic solute–solute potential of mean force. The converse, strong solvent–solute attraction, weak solvent–solute attraction, weak solvent–solvent attraction (i.e., high pressure) favors oscillatory behavior.[175] The smooth behavior has a fairly simple physical explanation. When solvent–solvent attraction is strong, it becomes energetically unfavorable for solvent particles to remain between the surfaces. They can interact more strongly with other solvent particles by moving into the bulk. This causes a density depletion between the surfaces in spite of the loss in entropy. The resultant imbalance between the pressure between the surfaces and that in the bulk solvent is the source of the attractive potential of mean force. It is interesting to note that in a dilute solution of nonpolar solutes in water, the solvent–solvent interaction is relatively strong because of hydrogen bonding. Thus one may speculate that the effective solute–solute potential in this case may also turn out to be a smooth function of distance.

6.4. Finite-Size Solutes

Similar results have been obtained for finite-size solutes. For example Deutch and co-workers[765] have investigated, in HNC approximation, the potential of mean force between hard-sphere solutes in a solvent of hard spheres with an attractive Yukawa potential. At liquid densities and with reasonable solvent–solvent attraction, they found significant depletion in the density of solvent species in the neighborhood of a solute particle. In the same regime the solvent-mediated solute–solute interaction is attractive and decreases monotonically.

For the sticky-sphere model with a hard-sphere solute we find similar trends (cf. Fig. 8). It would be interesting to study a sticky-sphere model with angular-dependent surface adhesion, and/or a decoration which exhibits a density maximum (cf. Section 3.4.2).

In recapitulation, we see that the short-range behavior of solvent-mediated solute–solute interaction is theoretically accessible. However, with the present state of the art, anything more complex than Lennard-Jones fluids becomes rapidly intractable. Nonetheless we have already learned a great deal through studying such simple systems, though the availability of a theory that does not require a large amount of numerical computation will be a significant advance. A first attempt at such a theory will be described in the next section.

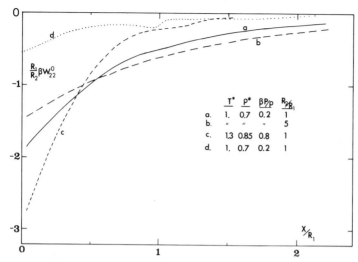

Fig. 8. Potential of mean force between solutes in a sticky-sphere solvent showing the effects of solute size (a, b), pressure (a, c), and solute–solvent attraction (a, d); for curve d the solute–solvent attraction is about one-quarter the solvent–solvent attraction. Nb. $x = r - R_2$.

7. MEAN-FIELD THEORY OF SOLVENT STRUCTURE

Since extension of the numerical techniques like those of the previous section to complex systems like aqueous solutions is a formidable task, attempts have been made to develop alternative approaches capable of describing effects of solvent structure in terms of simple analytical expressions. Chronologically, attempts to characterize structural forces in fact predate some of the detailed calculations given in the previous section. One such attempt is implicit in scaled particle theory. A critical review of this theory lies outside the theme of this article, and we refer the reader to the literature.[88,327,707,827] A different kind of approach is a mean-field theory first introduced by Marčelja in connection with his study of lipid-mediated protein interactions in membranes.[589] This relies on a number of ideas which we wish to discuss first before proceeding on to examining the utility of this theory.

From examples already discussed we see that the introduction of a solute into a solution results in a significant perturbation of the solvent structure near the solute. Although in many cases (the hard-sphere mixture is a good example) the direct solute–solvent interaction is short ranged, the perturbation of the solvent structure will spread beyond the first neighbors and be propagated some distance away from the solute. The spatial scale of this perturbation is referred to as the *correlation length* ξ. In the Lennard-Jones as well as some "sticky"-sphere systems we saw that this quantity enters as a characteristic decay length in the distribution functions: $\exp(-r/\xi)$.

Interaction between two solute molecules takes place when the distance between these molecules becomes comparable to the correlation length ξ. It is possible to express the free energy of the system as a function of the total perturbation of solvent structure, which in turn depends on the solute–solute separation. Although in the examples we have seen so far, the solvent perturbation takes the form of density variations, it should be stressed that the foregoing remarks will apply equally to perturbations in relative orientations or any other relevant quantities.

The mean-field theory of Marčelja *et al.*[590–592] exploits the above ideas and introduces the concept of a spatially dependent *order parameter* η. The choice of the relevant order parameter for a given system depends on the physical situation and on the nature of the perturbation. It may be a scalar such as density or a vector such as molecular orientation, or dielectric polarization if electric fields are involved. A structural perturbation is described as a deviation of the order parameter from its bulk value. The free energy of the system is expressed as a function of this parameter. In the complicated situations represented by aqueous solutions it may be necessary to use a number of interdependent order parameters to describe perturbations in the solvent density and orientation as well as in the dielectric polarization. The problem of solvent structure and solute–solute interaction is reduced to finding variations in the order parameter due to the presence of the solute.

When the deviations of the order parameter from its bulk value are small and slowly varying the change in the free energy density* due to the presence of a solute may be written as

$$\Delta f = \frac{1}{2} \frac{\partial^2 f_{\text{bulk}}}{\partial^2 \eta} \left[(\delta \eta)^2 + \xi^2 (\nabla \delta \eta)^2 + \cdots \right] \tag{53}$$

Minimization of this free energy with respect to $\delta \eta$ gives

$$\xi^2 \nabla^2 (\delta \eta) - (\delta \eta) = 0 \tag{54}$$

---

* The concept of a free energy density and the ideas used here are familiar in the physics literature, e.g., in superconductivity, nucleation theory, liquid crystals, gas–liquid interfaces, and other areas.

For nonpolar solutions, the appropriate order parameter is the density and we take $\delta\eta = \varrho h_{21}$.

For this case the correlation length is given by

$$\xi^2 = \frac{1}{6\varkappa kT} \int r^2 h_{11}(\mathbf{r}) \, d^3\mathbf{r}, \qquad \frac{1}{2} \frac{\partial^2 f_{\text{bulk}}}{\partial^2\eta} = \frac{1}{\varkappa\varrho^2} \tag{55}$$

where $\varkappa$ is the isothermal compressibility of the solvent. The correlation length is only a function of solvent properties. These results may be justified using statistical mechanics. The potential of mean force between two solute molecules is obtained as follows: Equation (54) together with appropriate boundary conditions can be solved for $\delta\eta$, whence integration of eqn. (53) over all space yields the free energy as a function of separation; similarly for solution properties. Unlike order at two surfaces yields repulsive forces, like order leads to attractive forces. The appropriate boundary condition for the solvent density profile near a hard wall is $\delta\eta/\varrho = (P/\varrho kT - 1)$ at the surface of the solute molecule. In general the boundary condition is not known exactly.

Owing to the analytical structure of the differential equation for $\delta\eta$, this mean-field theory is only capable of predicting smooth, exponential density profiles and potentials of mean force. From the examples already studied it appears that this situation obtains for low pressures and strong solvent–solvent attractions. Because of the deceptive simplicity of the theory, one would like to test out its predictive capacity on some simple systems. To this end we compare the correlation length ξ as predicted by eqn. (55) and the exponential decay length in the potential of mean force (computed numerically, see Section 6) between two "hard" planes separated by (a) a Lennard-Jones fluid and (b) a sticky-sphere fluid. These systems are chosen to be in the regime where we know already that exponential forces exist—i.e., weak solute–solvent attraction, strong solvent–solvent attraction, and low pressure. The results are tabulated in Table V. Here we see that predictions of the correlation length ξ are good, even though ξ is only of the order of a molecular diameter. This is remarkable. Strictly speaking mean-field theory should only be valid when the correlation length is of the order of many molecular diameters.[591]

A different but related application is the prediction of the observed repulsive pressure between lecithin bilayers in water.[591–592] In that experiment the repulsive pressure has an exponential distance dependence with a decay length of 1.9 Å. Whatever the nature of the solvent [water, LJ, sticky fluid], it does appear that at least one order parameter is characterized by a correlation length of the order of a molecular diameter.[678]

TABLE V. Comparison of Correlation Lengths Predicted by Eqn. (54) and Determined by the Decay Length of the Exponential of Mean Force between Two Hard Planes across a Lennard-Jones and a Sticky-Sphere Fluid[a]

$\varrho\sigma^3$	T^*	Numerical	ξ/σ Mean-field theory
		(a) Lennard-Jones fluid	
0.84	0.73	~ 1	1.7
		(b) Sticky-sphere fluid	
0.5	1.30	1.81	1.79
0.7	1.30	0.81	0.77
0.7	1.0	1.43	1.40
0.75	1.05	0.95	0.92
0.85	1.30	0.54	0.44

[a] Results for small hard-sphere solutes are more complicated, but are equally good.

(Another, in water, is of the order of 10 Å, which we have not considered.)[10,678] This observation lends weight to the view that it may be possible to describe hydrophobic solution properties in terms of mean-field theory.

Application to Water

An attempt to apply these ideas to aqueous solutions has been made in Ref. 591. The theory was motivated by a generalization of the Pople model,[713] which generalization takes some account of the cooperativity of the hydrogen-bond network in water. It gives practically identical thermodynamics to that of the original Pople model, and a reasonable description of the surface free energy of water. The correlation length was taken to be 2 Å as in the lecithin experiment. Neglecting (small) volumetric terms, the free energy of transfer of hydrocarbon molecules from pure hydrocarbon to water were calculated from eqns. (53) and (54). For a spherical molecule (e.g., methane) of radius r the free energy of transfer is

$$F(\text{sphere}) = 4\pi r^2 c\left(1 + \frac{\xi}{r}\right) \qquad (56)$$

TABLE VI. Hydrophobic Free Energies of Transfer of Simple Hydrocarbons (cal mol^{-1})

Hydrocarbon	Experiment	Mean-field theory
CH_4	2510–3150	3161
C_2H_6	3320–3860	3956
C_3H_8	4900	4900
Long chains (per CH_2 group)	825–884	832

where c is a constant related to bulk properties of water and the boundary condition. For a cylindrical molecule the corresponding expression is (per unit length)

$$F(\text{cylinder}) = 2\pi rc \, \frac{K_1(r/\xi)}{K_0(r/\xi)} \tag{57}$$

where K_1 and K_0 are modified Bessel functions. Results are listed in Table VI (for details see Ref. 591). For ethane and propane approximate solutions were obtained numerically by superposing single sphere solutions to resemble the shape of these molecules. The parameter c was determined by fitting the experimental value for the solvation energy of propane, and bare molecular radii were used.

The only advance over earlier simple-minded treatments* is that mean-field theory does attempt to take account of *correlations* in water structure. In addition, the hydrophobic free energy is not related just to the surface area of the solute molecule; and contributions from separate CH_2 groups are not additive for short chains—two hydrophobic groups have a smaller free energy if they are in close juxtaposition. Given one free parameter c, the agreement with such an insensitive function as free energy—rather than enthalpy and entropy—is hardly a big surprise.* One might then question whether anything is really achieved, because the whole crux of the problem

---

* One could simply multiply the "area" of a molecule by the interfacial tension per unit area and correct for finite size by a thermodynamic argument. This can be made to give good answers for free energies. However the argument is erroneous. Since the entropic contribution to interfacial tension is positive, this approach predicts the wrong *sign* for the entropy of solution of hydrophobic molecules. Nothing could be more disastrous!

is really hidden in the unknown boundary conditions and correlation length. Such a criticism would hold, however, for all theories where appeal must be made to related properties of the solute and solvent.

The sole advantage of the mean-field theory is that it does allow a wider conceptualization of the problem of aqueous solutions. If the solubility data are interpreted to give the "correct" boundary conditions one can then determine the interaction free energies and at the same time give good reasons for the observed sign of the temperature dependence of virial coefficients.[591] The search for simple formulas to describe such interactions is of course of paramount importance in biological problems.

8. CONCLUSION

In this review we have tried to highlight some recent ideas long nurtured by the good Editor, which bear on the theory of solutions and solvent structure. There is of course no one path to understanding, particularly in dealing with such a frustratingly elusive problem as hydrophobic solutions. Several other approaches deserve comment, and we refer the reader to the literature cited below for detailed accounts.

Scaled particle theory[707,827] has enjoyed some popularity over the past five years and has appeared to exhibit quite surprising success in accounting for some properties of aqueous solutions. The theory, originally devised to deal with hard-sphere liquids, is not a molecular theory, and draws on known (measured) thermodynamic quantities for the solvent to account for solution properties in an indirect manner. The theory is exact for an infinitely large solute molecule (a planar surface) where the concept of surface tension is a well-defined macroscopic concept. In the case of infinitely small solute molecules the solvent molecules can "see" around the solute, and as we have seen, this is a precondition for hydrophobic solution thermodynamics. For this case too scaled particle theory is exact, since only bulk solvent properties enter into the final calculations. It is therefore not surprising that this theory gives a reasonable account of the thermodynamics of hydrophobic solutions in which the solute molecules are small. However, it does not give much insight into molecular mechanisms in general, and fails for intermediate solute molecule sizes, where solute–solute, solute–solvent, and solvent–solvent distribution functions all rear their ugly heads.

The pioneering work of Ben-Naim[82,85–86] in a whole series of papers over the years deserves special comment. Some of the ideas developed here are already implicit in his work, and much credit is due to him for breaching the walls of prejudice which formerly separated statistical mechanics from chemistry. The one-dimensional clathrate model, or rather one very similar, had already been introduced by Lovett and Ben-Naim.[572] Most of his work, however, takes a quite different emphasis to that developed here, and focuses attention on so-called mixture models of water, whose status is still unclear despite a long and honorable history.

Finally the work of Pratt and Chandler[720] approaches the problem from yet a different point of view. Their work builds on the earlier highly successful theory of Weeks, Chandler, and Andersen[902] for simple fluids. The work of Pratt and Chandler has the merit that it is the first to attempt to use *measured* distribution functions for bulk water and gives excellent *numerical* results for solubility properties of methane and higher hydrocarbons using only one sensitive parameter. This (numerical) agreement is somewhat puzzling, since the theory uses an angle-averaged interaction between solute and solvent, apparently in direct contradiction to the conventional wisdom. The reason for the success of their theory may well be profound, but could be accidental. We cannot be quite sure which.

Our own approach attempts to steer a middle road between the two extremes represented by the Ben-Naim and Pratt and Chandler approaches. We have tried to capture the essence of the hydrophobic interaction by exploring and contrasting how what is known about molecular interactions conspires through distribution functions to give rise to normal and hydrophobic solution behavior. If the conclusion of Sections 3 and 4 is correct—that a simple one-dimensional clathrate structure taken together with the known nature of van der Waals forces in oil–water systems accounts for a significant portion of the peculiar properties of aqueous solutions—then the essence of the problem does stand revealed. The sign of thermodynamic quantities is largely dictated by the nature of long-range van der Waals forces. These reflect very complicated many-body effects characteristic of any high-dielectric-constant solvent, which are not accessible quantitatively through a two-body statistical mechanics, without parameters which have to be adjusted for each and every situation. The magnitude of the thermodynamic quantities and fine details like the specific heat do reflect the special properties of water as a solvent, in its ability to form clathrate cages about small molecules. But of course this was all known.

Therefore, in closing, we quote A. Comte who in 1830 (*Philosophie Positive*), wrote:

Every attempt to employ mathematical methods in the study of chemical questions must be considered profoundly irrational and contrary to the spirit of chemistry. If mathematical analysis should ever hold a prominent place in chemistry—an aberration which is happily almost impossible—it would occasion a rapid and widespread degeneration of that science.

ACKNOWLEDGMENTS

Come back to the second sentence of this essay and substitute Felix Franks for H. S. Frank. We can think of no higher compliment. We are much indebted to our friend and colleague Stjepan Marčelja with whom much of this work was initiated. We are grateful to Tom Healy for drawing our attention to the views of A. Comte.

NOTE ADDED IN PROOF

The recent molecular dynamics simulation studies of A. Geiger, A. Rahman, and F. H. Stillinger [*J. Chem. Phys.* **70**, 263 (1979)] represent a major advance in our understanding of hydrophobic solutions. Strong angular correlations stand revealed and are explicit and finally confirm the earlier intuition.

Computer Simulation of Water and Aqueous Solutions

D. W. Wood

Mathematics Department
University of Nottingham
Nottingham, NG7 2RD

1. INTRODUCTION

During the last five years or so a forceful attempt (in fact a brute force attempt) has been made to obtain a structural view of liquid water and aqueous solutions by the application of a "first principles" approach starting from an initial statement of the total intermolecular potential energy of an assembly of N water molecules, or N water molecules and n solute molecules. The calculations that have been performed have been truly massive computer simulations of either (a) the time evolution of an isolated assembly of molecules or (b) the direct numerical evaluation of the equilibrium thermodynamic functions and spatial correlations using a Monte Carlo sampling of points in classical configuration space. Neither (a) or (b) constitutes a true molecular theory because in both methods the linkage between the microscopic description and the derived equilibrium and non-equilibrium properties vanishes in the black box complexity of the computation. The ambition of any true theory of a many-body system must be to relate the microscopic details of atomic and molecular interactions to the behavior of the observed thermodynamic functions. Statistical mechanics is the means by which this objective can in principle be realized; however, such a relation is seldom, if ever, accomplished to our satisfaction. Even in cases where an excessive modeling of the particle interactions yields an exact mathematical treatment, the key to such relationships is frequently obscured owing to the removal of all but the very coarse grained features of the Hamiltonian function in the modeling process.

Constructing models of many-body systems and using a mathematical analysis to evaluate the statistical mechanics is sometimes referred to as "mathematical statistical mechanics." The computations of types (a) and (b) above are not calculations of this type, but it seems proper to regard them as numerical approximation methods in statistical mechanics. The method (a) is known as the method of molecular dynamics, the ensemble is the microcanonical ensemble, and expectation values of observables are defined as averages over the time evolution of a single phase point moving on a constant energy surface in phase space. The Monte Carlo method is essentially a numerical scheme to evaluate the ensemble averages in the canonical ensemble, in fact a sophisticated scheme of numerical integration for integrals defined over the configuration subspace of phase space. The two approaches referred to above, mathematical and numerical, are in fact both simulation methods. The mathematical models simulate as much reality in the total N-body Hamiltonian function as the mathematical analysis will allow. Efforts to construct such models of liquid water and aqueous solutions have been made in recent years by Bell[69-71] and by Ben-Naim[79,80,82] (see also Fleming and Gibbs,[311] and Herrick and Stillinger[401]), but this article will review only the numerical simulation methods (a) and (b). Such methods are not restricted by the complexities of Hamiltonian functions, which can be made as realistic as any *ab initio* method will permit, but they are restricted to treatments of a finite, and usually very small, number of particles.

The author has included in this review a detailed account of the molecular dynamics technique for applications to molecular assemblies; the molecular dynamics method is the more powerful of the two schemes since time-dependent properties of the assembly can also be revealed. Technical details of the Monte Carlo method are relegated to the references.

2. SOME ASPECTS OF STATISTICAL MECHANICS FOR MOLECULAR ASSEMBLIES

We consider an assembly of N identical rigid molecules; the Hamiltonian of the assembly has the form

$$\mathscr{H}_N(\mathbf{x}_1, \ldots, \mathbf{x}_N, \mathbf{p}_1, \ldots, \mathbf{p}_N) = \frac{1}{2m} \sum_{i=1}^{N} \mathbf{p}_i \cdot \mathbf{p}_i + \frac{1}{2} \sum_{i=1}^{N} \mathbf{w}_i \cdot \mathbf{I}_i \cdot \mathbf{w}_i$$
$$+ U_N(\mathbf{x}_1, \ldots, \mathbf{x}_N) \quad (1)$$

where x_i defines the set of configurational coordinates of the ith molecule, m is the mass of each molecule, I_i is the inertial moment tensor, and $U_N(x_1, \ldots, x_N)$ is the total potential energy of the N-body assembly. All the statistical mechanical calculations reported in this article are performed within the framework of classical mechanics [the potentials U_N in some cases will have been obtained from *ab initio* quantum mechanical calculations, but either molecules (molecular dynamics) or phase points (Monte Carlo) move in *classical* phase space]; in this framework the classical canonical partition function of the assembly is given by

$$Z_N = \frac{1}{N! \, h^{Nf}} \int \cdots \int \exp\{-\beta \mathscr{H}_N\} \, dp \, dx \qquad (2)$$

$$= \frac{1}{N! \, h^{Nf}} Z_{N,\text{trans}} Q_N \qquad (3)$$

where $Z_{N,\text{trans}}$ is trivially obtained by integrating over the fN momentum components, and Q_N is the configurational partition function

$$Q_N = \int \cdots \int \exp[-\beta U_N(x_1, \ldots, x_N)] \, dx_1 \cdots dx_N \qquad (4)$$

The difficulties of all statistical mechanical theories originate in attempts to evaluate the configurational partition function (4); this is a formidable mathematical problem even for atomic assemblies where

$$\int dx_i = \int_v dr_i \qquad (5)$$

but even more so for a molecular assembly where the configuration space is $6N$-dimensional and

$$\int dx_i = \int_v dr_i \int_0^{2\pi} d\varphi_i \int_0^{\pi} \sin\theta_i \, d\theta_i \int_0^{2\pi} d\psi_i \qquad (6)$$

where the three additional degrees of freedom, the Euler angles φ_i, θ_i, and ψ_i, specify the orientation of the rigid molecule with respect to a fixed Cartesian frame. The doubling of the number of degrees of freedom in changing from an atomic to a molecular assembly is the first and obvious technical problem encountered in seeking a statistical mechanical theory of a molecular liquid. This difficulty turns out to be crucial and inhibits the development of well-known approximate methods for molecular assemblies [see eqns. (25) and (26) below].

In contrast to the sheer technical difficulties imposed by (6), a fundamental obstacle appears when we seek to determine the total intermolecular potential energy function $U_N(x_1, \ldots, x_N)$. If we ignore the boundary interactions with the containing vessel, and if there are no external fields present, then U_N can be resolved into a sum of many body potentials in the form

$$U_N = \sum_{i<j=1}^{N} V_2(x_i, x_j) + \sum_{i<j<k=1}^{N} V_3(x_i, x_j, x_k)$$
$$+ \sum_{i<j<k<m=1}^{N} V_4(x_i, x_j, x_k, x_m) + \cdots \tag{7}$$

The energy U_N can be regarded as the stabilization energy of the assembly; the absolute total potential energy would include the single-body energies $V_1(x_i)$. The n-body potential terms in (7) are defined by successive applications of (7) to $1, 2, \ldots$, and n-body assemblies, thus

$$U_2(x_i, x_j) = V_2(x_i, x_j) \tag{8a}$$

$$U_3(x_i, x_j, x_k) = V_2(x_i, x_j) + V_2(x_j, x_k) + V_2(x_k, x_i)$$
$$+ V_3(x_i, x_j, x_k) \tag{8b}$$

$$U_n(x_{i_1}, x_{i_2}, \ldots, x_{i_n}) = \sum_{j=2}^{n-1} \sum_{i_1 < \cdots < i_n}^{n} V_j(x_{i_1}, x_{i_2}, \ldots, x_{i_j})$$
$$+ V_n(x_{i_1}, \ldots, x_{i_n}) \tag{8c}$$

Almost all applications of statistical mechanics use only the first term in the expansion (7); in this approximation the total force acting on any particle is the sum of the forces acting between "the particle" and the remaining $N - 1$ other particles. This is the assumption of "pairwise" additive forces which is almost always invoked in theories of the liquid state. Such theories (for reviews see Temperley et al.[856] and Hansen and MacDonald[381]), which have made good progress in recent years, are confined to atomic liquids; even for nonpolar atomic liquids it seems likely that an accurate comparison between theory and experiment would require three-body forces to be included.[268] In the case of water the evidence for the importance of at least the V_3 and V_4 terms in (7) is overwhelming, and even if V_2 was known for a pair of water molecules it is likely that by itself this knowledge would be of little value in forming a molecular theory of liquid water. Thus a pair potential V_2 for point dipole–dipole interactions shows a low-temperature ordered state in the hexagonal close packed crystal,[673] whereas the potential U_N for water must account for the prefer-

ence of water molecules to order themselves by hydrogen-bond formation
into a structure which, on average, is locally tetrahedronal, like the ice
lattice. The large difference between the dipole moment of the free water
molecule, and the average dipole moment per molecule in ice (1.84 D and
2.60 D, respectively) is also evidence for the nonadditivity of U_N. (For
reference to this phenomenon see Coulson and Eisenberg,[204] Weissman
and Cohan,[903] and Crowe and Santry.[214]) SCF quantum mechanical
calculations of the ground-state energies of small clusters of water molecules
over many relative geometries also show directly that a significant contribution to U_N comes from beyond the pair interactions. Hankins *et al.*[379]
carried out accurate SCF calculations on pairs and triplets of water molecules and found that the three molecule nonadditivities are large in magnitude. Kistenmacher *et al.*[490] have performed extensive direct Hartree–
Fock computations on clusters of three and four water molecules to examine
the nonadditive effect. These authors concluded that the three- and four-
body contributions to U_4 (for example) are quite small in the geometries
considered, and deduce that the pair additive assumption used in conjunction with the computed Hartree–Fock pair energies yields a good approximation to the total stabilization energy. (These pair potentials are discussed
in Section 4.3.1.) Kistenmacher *et al.* find a three-body contribution to
U_N of about 10%, which to the present author seems large enough to be
of thermodynamic importance.

The total intermolecular potential energy of N water molecules is an
impossibly complicated function containing short-range repulsive forces,
intermediate-range dispersion and dipole–dipole forces, hydrogen-bond
formation at close range, and important many-body potential terms. To
most people this would appear to be a good point at which to abandon any
attempt to form a molecular theory of liquid water. Clearly some form of
functional modeling of U_N will be required which replaces U_N by what
is known as an *effective* potential, expressed as a sum of *effective* pair
potentials $V_e(\mathbf{x}_i, \mathbf{x}_j)$:

$$U_N(\mathbf{x}_1, \ldots, \mathbf{x}_N) = \sum_{i<j=1}^{N} V_e(\mathbf{x}_i, \mathbf{x}_j) \qquad (9)$$

$V_e(\mathbf{x}_i, \mathbf{x}_j)$ is *not* the true pair potential between two water molecules, rather,
it is hoped that two variable functions V_e can be constructed between objects
that are waterlike rather than "exact" water molecules. It is to be hoped
that the statistical mechanics of these waterlike "pair additive" objects
may be evaluated in some way and compared with the behavior of liquid
water. We must be sufficiently optimistic that the functions V_e exist, and

sufficiently careful not to put into $V_e(\mathbf{x}_i, \mathbf{x}_j)$ those features that we *wish* to see in the structure of liquid water. It must be realized that an effective pair potential will be both temperature and density dependent; this dependence will usually appear in the values chosen for parameters in V_e, and not in the form of an explicit dependence. The determination of effective pair potentials for strongly nonadditive interactions has been discussed in general terms by Stillinger,[826] who also considers an application to liquid water. Stillinger proposes an extremum condition for an optimal choice of $V_e(\mathbf{x}_i, \mathbf{x}_j)$. Under this minimization condition the partition function and Helmholtz free energy are invariant under the two potentials, $\frac{1}{2} U_N + \frac{1}{2} \sum_{ij} V_e(i, j)$, and $\sum_{ij} V_e(\mathbf{x}_i, \mathbf{x}_j)$; and the Helmholtz free energy using only the pair potential $V_e(\mathbf{x}_i, \mathbf{x}_j)$ is bounded below by the true free energy obtained using the potential U_N. Stillinger considers a perturbative scheme to determine the optimal function $V_e(\mathbf{x}_i, \mathbf{x}_j)$ iteratively. It is unlikely that Stillinger's scheme could be fully implemented in a full-scale calculation, but the scheme can be used to examine special features of V_e.

The great advantage in being able to use the pair additive form (9) is that all of the thermodynamic observables of the assembly can be expressed in terms of the generic pair distribution function

$$\varrho^{(2)}(\mathbf{x}_1, \mathbf{x}_2) = \frac{N(N-1) \int d\mathbf{x}_3 \cdots \int d\mathbf{x}_N \exp[-\beta \sum_{ij} V_e(\mathbf{x}_i, \mathbf{x}_j)]}{\int d\mathbf{x}_1 \cdots \int d\mathbf{x}_N \exp[-\beta \sum_{ij} V_e(\mathbf{x}_i, \mathbf{x}_j)]} \tag{10}$$

which is simply the probability that volume elements $d\mathbf{x}_1$ and $d\mathbf{x}_2$ are simultaneously occupied by one molecule. This is no easier to determine than the partition function, but is employed also as a means of describing the equilibrium (averaged) molecular structure existing in the liquid state in the form of the "pair correlation" function

$$g(\mathbf{x}_1, \mathbf{x}_2) = \left(\frac{8\pi^2}{\varrho}\right)^2 \varrho^{(2)}(\mathbf{x}_1, \mathbf{x}_2) \tag{11}$$

where ϱ is the density; and where the conditional probability of observing a molecule in a configuration coordinate between $\mathbf{x}_2$ and $\mathbf{x}_2 + d\mathbf{x}_2$, given the existence of a molecule at $\mathbf{x}_1$, is

$$\varrho^{(1)}(\mathbf{x}_2) g(\mathbf{x}_1, \mathbf{x}_2) \, d\mathbf{x}_2 \tag{12}$$

Here $\varrho^{(1)}(\mathbf{x}_2)$ is the conditional local density at $\mathbf{x}_2$ with a particle fixed at $\mathbf{x}_1$. (For references to molecular and atomic correlation functions see Refs. 82, 381, 856.) The pair correlation function (11) reflects the averaged and inhomogeneous structure of $N - 1$ molecules in the external field of a fixed molecule. To discuss the equilibrium structure of a molecular assembly it is often more useful to define a set of internuclear pair correlation functions $g_{\alpha\beta}(r)$, where r is the internuclear separation between nuclear species α on one molecule and β on another molecule. The probabilistic definition of $g_{\alpha\beta}(r)$ is such that

$$\varrho_\alpha\varrho_\beta g_{\alpha\beta}(r)\, d\mathbf{r}_\alpha\, d\mathbf{r}_\beta \tag{13}$$

is the probability that different volume elements $d\mathbf{r}_\alpha$ and $d\mathbf{r}_\beta$ separated by a distance r simultaneously and respectively contain nuclear species for distinct molecules, where ϱ_α and ϱ_β are the densities of the respective species in the whole assembly. This definition is a simple extension of (11), and $g_{\alpha\beta}(r)$ approaches unity at large r; also the average number of $\alpha\beta$ pairs up to a distance r from the α nucleus is

$$n_{\alpha\beta}(r) = 4\pi\varrho_\beta \int_0^r s^2 g_{\alpha\beta}(s)\, ds \tag{14}$$

In the case of water we can define the three pair correlations $g_{OO}(r)$, $g_{OH}(r)$, and $g_{HH}(r)$. It should be emphasized that these are theoretical definitions which are not necessarily accessible to experimental determination; thus X-ray scattering experiments on liquid water will yield results for center-of-mass scattering, and hence structure factors closely related to $g_{OO}(r)$, whereas the elucidation of $g_{OH}(r)$ and $g_{HH}(r)$ requires neutron scattering data on water and heavy water.

The thermodynamic functions can be expressed in terms of $g(\mathbf{x}_1, \mathbf{x}_2)$; the formulas are straightforward extensions of the well-known expressions employed for atomic assemblies with spherically symmetric potentials,[194, 381,856] some examples are as follows:

(a) the total internal energy $\langle \mathscr{H}_N \rangle$

$$\langle \mathscr{H}_N \rangle = \frac{3NkT}{2} + \frac{1}{2} \int\int V_e(\mathbf{x}_1, \mathbf{x}_2)\varrho^{(2)}(\mathbf{x}_1, \mathbf{x}_2)\, d\mathbf{x}_1\, d\mathbf{x}_2 \tag{15}$$

(b) the pressure

$$\frac{pV}{NkT} = 1 - \frac{1}{6NkT} \int\int [\mathbf{r}_{12} \cdot \mathbf{\nabla}_{r_{12}} V_e(\mathbf{x}_1, \mathbf{x}_2)\varrho^{(2)}(\mathbf{x}_1, \mathbf{x}_2)]\, d\mathbf{x}_1\, d\mathbf{x}_2 \tag{16}$$

where $\mathbf{r}_{12}$ is the center-of-mass separation;

(c) the isothermal compressibility (this expression is independent of the pairwise additive assumption)

$$\varkappa_T = \left(\frac{V}{N}\right) + \int [g(r_{12}) - 1]\, d\mathbf{r}_{12} \qquad (17)$$

where $g(r_{12})$ is the center-of-mass pair correlation function; and

(d) the static dielectric constant ε_0[487]

$$\frac{(\varepsilon_0 - 1)(2\varepsilon_0 + 1)}{3\varepsilon_0} = \frac{4\pi N}{V}\left(\alpha + \frac{\mu_d^2 g_K}{3kT}\right) \qquad (18)$$

where α is the molecular polarizability, μ_d is the permanent dipole moment, and

$$g_K = 1 + \frac{N}{8\pi^2 V}\int g(\mathbf{x}_1, \mathbf{x}_2)\mathbf{b}_1 \cdot \mathbf{b}_2\, d\mathbf{x}_2 \qquad (19)$$

In (19), $\mathbf{b}_1$ and $\mathbf{b}_2$ are unit vectors in the directions of the permanent dipole moments (see page 370).

In addition to the usual thermodynamic properties the pair correlation function $\varrho^{(2)}(\mathbf{x}_1, \mathbf{x}_2)$ [or $g(\mathbf{x}_1, \mathbf{x}_2)$] can be used to investigate any type of fine detail regarding the average relative positioning of pairs of molecules or molecular fragments. If $B(\mathbf{x}_i, \mathbf{x}_j)$ is a characteristic function such that

$$B(\mathbf{x}_i, \mathbf{x}_j) = 1 \qquad (20)$$

when i and j satisfy a specified property, and

$$B(\mathbf{x}_i, \mathbf{x}_j) = 0 \qquad (21)$$

otherwise, then the average number of pairs of molecules exhibiting this property is

$$n = \frac{1}{2} \iint B(\mathbf{x}_1, \mathbf{x}_2)\varrho^{(2)}(\mathbf{x}_1, \mathbf{x}_2)\, d\mathbf{x}_1\, d\mathbf{x}_2 \qquad (22)$$

The expression (22) is simply a formal expression for *counting* the average number of pairs exhibiting the desired property in a given equilibrium state. An example of (22) is used in Section 6 to discuss the hydrogen-bonding networks in liquid water. Later on we shall be interested in learning how the equilibrium potential energy is distributed among molecular pairs. Defining $n(V_1, V_2)$ as the average number of molecules neighboring a given molecule having pair potential energies (with the given molecule) in the range V_1 to V_2, we can write

$$n(V_1, V_2) = \int_{V_1}^{V_2} p(V)\, dV \tag{23}$$

where $p(V)$ is the counting function

$$p(V) = \frac{\varrho}{8\pi^2} \int \delta[V - V_e(\mathbf{x}_1, \mathbf{x}_2)] g(\mathbf{x}_1, \mathbf{x}_2)\, d\mathbf{x}_2 \tag{24}$$

This resolution of potentials acting between molecular pairs was recently introduced by Rahman and Stillinger in molecular dynamics calculations (for references see Section 3).

Assuming that a sufficient degree of optimism is on hand to accept the construction of a model effective pair potential, the question arises of how to implement a specific calculation of $g(\mathbf{x}_1, \mathbf{x}_2)$ or the internuclear correlations (13). The time-honored ambition of theoreticians to accomplish their enquiries using a mathematical analysis is immediately thwarted by the scale of the numerical computation involved in an evaluation of the formalism. To see this we can briefly consider what methods are available in the liquid density range. There is only one possible approach, and this is to adapt the closure approximation integral equations which have been successful in applications to atomic liquids with spherically symmetric potentials.[194,381,744,856] The implications of such a scheme extended to a molecular liquid are considered by Ben-Naim and Stillinger.[421] The same fate befalls all the various integral equations; two of the earliest equations are the Percus–Yevick equation[692]

$$e^{\beta V_e(\mathbf{x}_1,\mathbf{x}_2)} g(\mathbf{x}_1, \mathbf{x}_2)$$
$$= 1 + \frac{N}{8\pi^2 V} \int d\mathbf{x}_3 [g(\mathbf{x}_1, \mathbf{x}_3) - 1](1 - e^{\beta V_e(\mathbf{x}_3,\mathbf{x}_2)}) g(\mathbf{x}_3, \mathbf{x}_2) \tag{25}$$

and the hypernetted chain approximation

$$\ln g(\mathbf{x}_1, \mathbf{x}_2) + \beta V_e(\mathbf{x}_1, \mathbf{x}_2)$$
$$= \frac{N}{8\pi^2 V} \int d\mathbf{x}_3 [g(\mathbf{x}_1, \mathbf{x}_3) - \ln g(\mathbf{x}_1, \mathbf{x}_3) - 1 - \beta V_e(\mathbf{x}_1, \mathbf{x}_3)]$$
$$\times [g(\mathbf{x}_2, \mathbf{x}_3) - 1] \tag{26}$$

A numerical solution of (25) or (26) proceeds by iteration from an initial trial function for $g(\mathbf{x}_1, \mathbf{x}_2)$. In the case of an atomic liquid the pair correlation function is the center-of-mass correlation and one-dimensional $[g(r)]$. Here the integrals in (25) and (26) are one-dimensional, the single variable being the interatomic separation r. These integrals can be rapidly evaluated by a computer, but even here convergence at high liquid densities is often very slow; it can in fact be so slow that convergence appears to have been achieved when it has not. These equations are also highly unstable. For a molecular liquid the integration in (25) and (26) is the sixfold integral denoted by (6). Thus independently of any difficulties with convergence and stability the sheer size of the computation defeats any attempts at a numerical solution (see page 331). Ben-Naim and Stillinger[421] have estimated that about 4 minutes is required on present high-speed computers to obtain $g(\mathbf{x}_1, \mathbf{x}_2)$ using (25) at *one* pair of points $\mathbf{x}_1$ and $\mathbf{x}_2$. A million such calculations only yields a resolution of 10 points on each axis (each of the six variables), thus a *single* iteration of (25) would require well over 1000 hours of computer time! Clearly an approach using an angular-dependent potential is not feasible and is likely to remain so for the foreseeable future. It may seem strange that a brute force simulation of the dynamics of several hundred molecules is computationally quite possible when a mathematical approximation scheme is similarly impossible, and yet this is the outcome of any serious attempt to apply statistical mechanics to dense molecular liquids.

3. THE MOLECULAR DYNAMICS METHOD

The molecular dynamics method was first implemented by Alder and Wainwright.[24,25,888] The technique is to obtain a computer solution of the time evolution of a finite number of atoms or molecules fully isolated from their surroundings. The corresponding ensemble is the microcanonical ensemble (constant energy and volume); thus the computer solution traces the trajectory of a single representative point constrained to move on a constant energy surface $[\mathscr{H}_N(\mathbf{p}, \mathbf{q}) = E]$ in classical phase space. The method of obtaining equilibrium thermodynamic expectation values is therefore the same as that originally envisaged by Boltzmann, where, if $F(\mathbf{p}, \mathbf{q})$ is any property defined on phase space, the ensemble average $\langle F \rangle$ is the infinite time average

$$\langle F \rangle = \lim_{c \to \infty} \frac{1}{c} \int_0^c F\big(\mathbf{p}(t), \mathbf{q}(t)\big) \, dt \tag{27}$$

which is equal to the microcanonical ensemble average

$$\langle F \rangle = \frac{\int \cdots \int \delta\big(\mathscr{H}_N(\mathbf{p}, \mathbf{q}) - E\big) F(\mathbf{p}, \mathbf{q}) \, d\mathbf{p} \, d\mathbf{q}}{\int \cdots \int \delta\big(\mathscr{H}_N(\mathbf{p}, \mathbf{q}) - E\big) \, d\mathbf{p} \, d\mathbf{q}} \tag{28}$$

if the system is ergodic.[295,484] Of course the limit in (27) cannot be taken, and in fact it is replaced by very small values of c; these will be obtained either as times that appear sufficient, or more likely as the times that can reasonably be achieved. It is possible that the representative point becomes trapped into a small domain of the whole surface $\mathscr{H}_N(\mathbf{p}, \mathbf{q}) = E$.

In a long series of papers Alder and co-workers[22,23,26,27,273,417] have applied the molecular dynamics method to simple spherically symmetric potentials for structureless particles with only impulsive forces acting. The square well potential,

$$V(r) = \begin{cases} \infty, & r \leq \sigma_1 & \text{(29)} \\ -\varphi, & \sigma_1 < r \leq \sigma_2 & \text{(30)} \\ 0, & r > \sigma_2 & \text{(31)} \end{cases}$$

is such a case, where $\sigma_1 = \sigma_2$ is the special case of hard spheres. The advantage of using such model potentials is that the equations of motion of the N particles can be solved *exactly*. This is because the particles move with constant velocity in the time intervals between collisions, a collision occurring if two atoms reach a separation of σ_1 (a repulsive collision) or a separation of σ_2 (an attractive collision). Molecular dynamics calculations are only possible on a small number of particles—an upper limit of about 3500 degrees of freedom seems to operate—although recently Hockney has introduced a new technique applicable to Coulombic potentials between classical point charges[411,412] but also adaptable to arbitrary potentials,[409] which can handle as many as 10,000 particles. So far this new method (the PPPM method) does not appear to have been applied to a molecular assembly, and applications have been restricted to two-dimensional assemblies.[410] The interaction between the particles and the boundary of the "box" are removed by making the assembly infinitely periodic; these are the so-called periodic boundary conditions, which are illustrated in Fig. 1. The molecular dynamics cell is the center cell of volume l^3, and in three dimensions it is surrounded by 26 nearest-neighbor image cells, each one a periodic repeat of the molecular dynamics cell. Image cells of all orders may need to be included in some calculations. Particles in the molecular dynamics cell interact both with themselves and their periodic images in the image cells. By means of this trick a few hundred particles are converted to an infinite, albeit periodic, system, which seems to have little effect when equilibrium properties of the bulk phase are considered. It is generally accepted that errors arising from the imposition of periodic boundary conditions are of order $1/N$ and also that averages obtained using

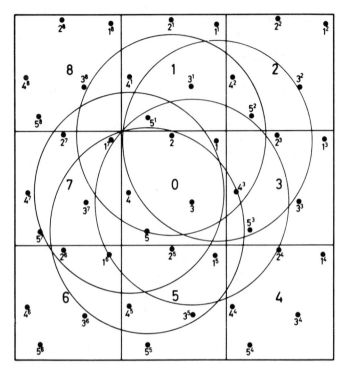

Fig. 1. The periodic boundary conditions employed in molecular dynamics and Monte Carlo simulations, illustrated here for a five-particle assembly. The central box is the molecular dynamics cell containing the five particles $j = 1, \ldots, 5$; the image cells are numbers $i = 1, \ldots, 8$ (a two-dimensional version) and contain the particle images j^i. The potential cutoff sphere of radius r_c is centered on each particle j, and only interactions within this sphere are included.

(27) are only weakly dependent upon N beyond about $N = 100$. This excludes the critical point region; the density fluctuations are constrained by maintaining a constant number density in each cell. Any spatial fluctuations with wavelengths of the order of l cannot be examined by this method (or any other!).

With only impulsive forces acting, the simulation is just a matter of forming an ordered list of the collision times between particle pairs. Initially positions $\mathbf{r}_i(0)$ and velocities $\mathbf{v}_i(0)$ $(i = 1, 2, \ldots, N)$ are assigned to each particle. The $\frac{1}{2}N(N - 1)$ interparticle separations then evolve according to

$$\mathbf{r}_{ij} = \mathbf{r}_{ij}(0) + t\mathbf{v}_{ij} \qquad (\mathbf{r}_{ij} = \mathbf{r}_i - \mathbf{r}_j, \quad \mathbf{v}_{ij} = \mathbf{v}_i - \mathbf{v}_j) \qquad (32)$$

If $t_{ij}^{(\alpha)}$ is the time required for an attractive $(\alpha = 2)$ or a repulsive $(\alpha = 1)$

collision to occur between the i, j pair, then

$$v_{ij}^2 t_{ij}^{(\alpha)} = -b_{ij} \pm (b_{ij}^2 - v_{ij}^2 c_{ij}^{(\alpha)})^{1/2} \quad (b_{ij} = \mathbf{r}_{ij} \cdot \mathbf{v}_{ij}, \; c_{ij}^{(\alpha)} = r_{ij}^2 - \sigma_\alpha^2) \quad (33)$$

Each pair of particles in the molecular dynamics cell represents an infinite number of pairs separated by distances

$$\mathbf{r}_{ij}(r, s, t) = \mathbf{r}_{ij} + l(r\mathbf{i} + s\mathbf{j} + t\mathbf{k}) \quad (r, s, t = \pm 1, \pm 2, \ldots) \quad (34)$$

Initially $t_{ij}^{(\alpha)}$ for a given i and j is found by selecting the minimum distance from the set (34); the other pairs in (34) all have separations $> \frac{1}{2}l$. A time τ is selected which is too short for any separation $\mathbf{r}_{ij}$ to change by $\frac{1}{2}l - \sigma_2$, and a list is compiled of all the $t_{ij}^{(\alpha)} < \tau$. The time is now advanced by

$$\min_{i,j,\alpha} t_{ij}^{\alpha} = t_{IJ}^{(\alpha)} \quad (35)$$

at this point the velocities of I and J must be altered in accordance with the type of collision that has taken place (see Fig. 2). All the $t_{ij}^{(\alpha)}$ are now contracted to $t_{ij}^{(\alpha)} - t_{IJ}^{(\alpha)}$ and the $2N - 3$ pairs $t_{Ij}^{(\alpha)}$ and $t_{iJ}^{(\alpha)}$ are recalculated. This procedure is repeated until the elapsed time reaches τ, at which point the whole list $t_{ij}^{(\alpha)}$ is recalculated and the whole cycle is recommenced. The logical sequence of comparisons and calculations is shown in Fig. 2.

The exact algorithmic solution above is only possible for purely impulsive forces; clearly for a simulation of "real" atoms or molecules a continuous potential function must be employed, which, hopefully, will allow comparisons between theory and experiment. The molecular dynamics method was first adapted to cope with a "realistic" potential by Rahman[729] in 1964, and applied to an assembly of N structureless particles with the central potential $V(r)$ chosen to simulate liquid argon. The potential employed here is the Lennard-Jones pair potential

$$V(r) = 4\varepsilon\left[\left(\frac{\sigma}{r}\right)^{12} - \left(\frac{\sigma}{r}\right)^{6}\right] \quad (36)$$

which is qualitatively similar to interatomic interactions in most gases. The two terms in (36) represent hard-core repulsive forces and longer-range dispersive forces; the potential has a minimum of $-\varepsilon$ at $r = 2^{1/6}\sigma$, and σ is a measure of the "size" of the particle. For argon Rahman uses the values $\varepsilon/k = 120°K$, and $\sigma = 3.4$ Å. This Lennard-Jones assembly has been extensively studied using molecular dynamics[548,730,879-880] (and by Monte Carlo methods[919]). Some reliable method of approximating the continuous dynamics of the N particles by discrete jumps in position vectors $\Delta\mathbf{r}_i$ and

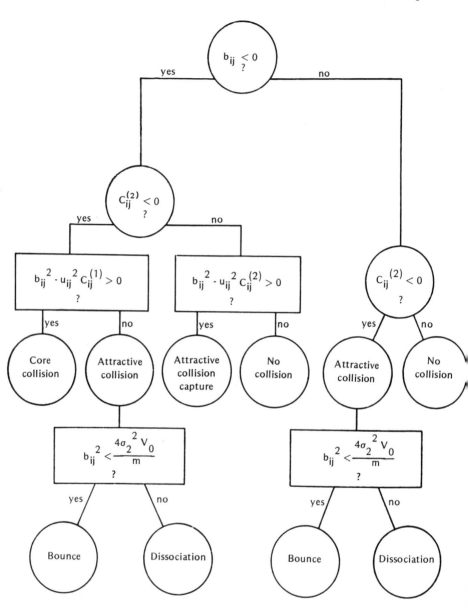

Fig. 2. The logical sequence of decisions to determine the collision times $t_{ij}^{(\alpha)}$ in (33), from which $t_{IJ}^{(\alpha)}$ in (35) is determined. The time is advanced by this amount and the velocities altered in accordance with the type of collision occurring between I and J. This algorithm represents an *exact* solution to the time evolution of the assembly. [Reproduced with permission from *J. Chem. Phys.* **31**, 459 (1959).]

velocities Δv_i over small time intervals Δt must be employed to obtain the phase point trajectory. Also because of the periodic image system special methods of computing the total force acting on a given particle may be needed for long-range potentials; for example, if Coulombic potentials are present, the image particles of all orders may need to be included. Special methods that exploit the periodicity of the system are available for this.[133,608,920] However, for intermediate and short-range potentials like (36) a device commonly used is to cut off the interparticle potential $V(r)$ at some "cutoff radius" r_c (see Fig. 1): thus for argon the potential actually employed is

$$V(r) = \begin{cases} 4\varepsilon\left[\left(\dfrac{\sigma}{r}\right)^{12} - \left(\dfrac{\sigma}{r}\right)^{6}\right], & r < r_c \qquad (37) \\ 0, & r \geq r_c \qquad (38) \end{cases}$$

at r_c V is no more than a few percent of ε. The cutoff introduces a discontinuity in $V(r)$, but the total energy is still conserved. The equations to be numerically integrated are Newton's equations

$$m\ddot{r}_i = \mathbf{F}_i = -\nabla_{r_{ij}} \sum_{j \neq i} V(r_{ij}) \qquad (39)$$

or equivalently

$$m\ddot{x}_i = 24\varepsilon \sum_{j \neq i} \frac{(x_i - x_j)}{r_{ij}^2}\left[2\left(\frac{\sigma}{r_{ij}}\right)^{12} - \left(\frac{\sigma}{r_{ij}}\right)^{6}\right] \qquad (40)$$

with similar equations for y_i and z_i. The equations (40) are reduced to $6N$ simultaneous first-order equations by writing

$$m\dot{v}_{\alpha i} = F_{\alpha i} \qquad (41)$$
$$v_{\alpha i} = \dot{a}_i \qquad (\alpha = x, y, z, \quad i = 1, 2, \ldots, N) \qquad (42)$$

A control over the reliability of a numerical scheme for solving two or three thousand simultaneous equations of the type (41) and (42) is to monitor constants of the motion; these are the total energy and the total linear momentum. Another device is to reverse the time direction after a time interval ξ has elapsed and compare the initial points $\mathbf{r}_i(0)$, $\mathbf{v}_i(0)$ with $\mathbf{r}_i(\xi + (-\xi))$ and $\mathbf{v}_i(\xi + (-\xi))$. It turns out that very simple algorithms suffice to satisfy these checks with properly chosen time intervals Δt. For structureless particles the algorithm introduced by Verlet[879] is probably the most frequently used, and employs the Taylor expansion

$$\mathbf{r}_i(t \pm \Delta t) = \mathbf{r}_i(t) \pm \mathbf{v}_i(t)\,\Delta t + \frac{\Delta t^2}{2m}\mathbf{F}_i + O(\Delta t^3) \qquad (43)$$

where $\mathbf{F}_i$ is the total force acting on the ith particle (39). By adding and subtracting those two equations we obtain

$$\mathbf{r}_i(t + \Delta t) = -\mathbf{r}_i(t - \Delta t) + 2\mathbf{r}_i(t) + \frac{\Delta t^2}{m} \mathbf{F}_i \qquad (44)$$

$$\dot{\mathbf{r}}_i(t) = \frac{1}{2\Delta t} [\mathbf{r}_i(t + \Delta t) - \mathbf{r}_i(t - \Delta t)] \qquad (45)$$

where the errors are, respectively, $O(\Delta t^4)$ and $O(\Delta t^3)$. Some method of starting the motion is required; beyond this the phase point trajectory is found recursively.

For practical computations it is convenient to use dimensionless variables; on choosing σ as the unit of length, and using $r^* = r/\sigma$, (43) suggests that the dimensionless variable t/τ, $\tau = \sigma m^{1/2}/\varepsilon$ should be used. Computer trials on two particles for the argon assembly show that a time interval of $\Delta\tau = 0.3$–0.4 can be used corresponding to a real molecular time of $t \sim 10^{-14}$ sec. To avoid excessive calculations on all relevant pairs of $V(r_{ij})$ after each time step is completed, linking algorithms can be employed (and checked against energy conservation). Verlet[879] introduces a sphere of radius $R > r_c$ ($r_c \sim 2.25\sigma$ for argon calculations), and only those atoms j for which $r_{ij} < R$ are used to compute $\mathbf{F}_i$ for an elapsed time interval, which is chosen to be less than the time required for atoms to traverse the skin depth $R - r_c$. Other listing algorithms are described by Hockney et al.[412]

When the assembly has "relaxed" into an equilibrium state (see below), the calculation proper begins by accumulating as many phase points $\{\mathbf{r}_i(t), \mathbf{v}_i(t)\}$ as is practical (probably not more than 5000); this will extend over a real time interval of 10^{-11}–10^{-10} sec. We can illustrate the use of these data in relation to (27); consider the center-of-mass pair correlation function $g(r)$ for atomic systems; if $n_i(r)$ is the number of neighboring atoms to i in a spherical shell of radius δr then the correlation function is given by

$$g(r) = \frac{\Delta t}{t} \sum_k g_k(r) \qquad (46)$$

where

$$g_k(r) = \frac{1}{N} \sum_{i=1}^{N} \frac{n_i(r)}{4\pi r^2 \, \delta r} \frac{V}{N} \qquad (47)$$

and is the average correlation function per particle recorded at the snapshot of the assembly as seen at a time $k\Delta t$. A typical dynamic property is the velocity autocorrelation function, which tells us features of the average atomic motion occurring from a given position, and is defined by

$$\psi(t) = \langle \mathbf{v}(0) \cdot \mathbf{v}(t) \rangle / \langle v(0)^2 \rangle \tag{48}$$

and calculated using the scheme

$$\psi(t) = \frac{1}{200N} \sum_{j=1}^{200} \sum_{k=1}^{N} \frac{\mathbf{v}_k(t_j) \cdot \mathbf{v}_k(t_j + t)}{v_k(t_j)^2} \tag{49}$$

where we have supposed that 200 phase points are chosen as initial points
[$t = 0$ in (48)], $t_j = j \, \Delta t$, and we are averaging over $200N$ trajectories
in the phase space of a *single* atom. Both $\psi(t)$ and $g(r)$ obtained by Rahman[729] for liquid argon are shown in Figs. 3 and 4.

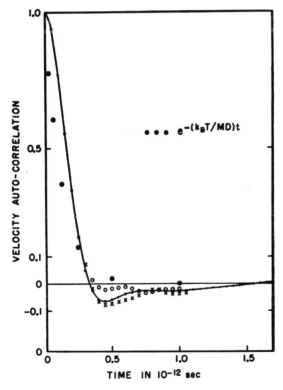

Fig. 3. The velocity autocorrelation function (48) obtained by Rahman[729] for liquid
argon at 94.4°K and a density of 1.374 g/cm³. The continuous curve is the mean of 64
curves; the two members of this set that have maximum departures from the mean are
shown as circles and crosses. The Langevin-type exponential function is also shown.
[Reproduced with permission from *Phys. Rev.* **136**, A405 (1964).]

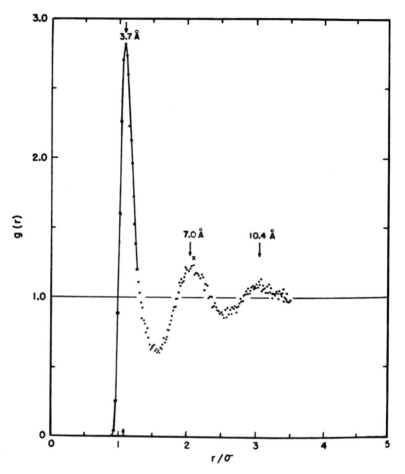

Fig. 4. The radial pair correlation function (46) obtained by Rahman[729] for liquid argon at 94.4°K and 1.374 g/cm³. [Reproduced with permission from *Phys. Rev.* **136**, A405 (1964).]

The next level of complexity attempted using molecular dynamics was the simulation of diatomic liquids. Extensive calculations on liquid carbon monoxide and liquid nitrogen were initially performed by Harp and Bern,[100,384,385] and recently a detailed study of liquid nitrogen has been reported by Cheung and Powles.[177] With an assembly of N diatomic molecules we have $5N$ degrees of freedom; these are the coordinates of the center of mass r_i, and θ_i, φ_i, the polar angles of the molecular axis with respect to a Cartesian frame at the center of mass (see Fig. 5). This places a practical upper limit of about 500 such molecules in a molecular dynamics calculation. Vibrational motion of the atoms *along* the nuclear axis can also be included without increasing the number of degrees of freedom. Berne and Harp[100] included harmonic forces between the two nuclei, but found that the internuclear separation remained to within 10^{-4} Å of the equilibrium separation, and their calculations are essentially those for an assembly of N rigid rotators. The potentials employed by Berne and Harp[100] always included the Lennard-Jones potential (37) and (38) (r_c typically in the range 2.25σ–2.5σ) between the centers of mass of the molecules $\{\varepsilon/k = 87.5°K, \sigma = 3.702$ Å for nitrogen, and $\varepsilon/k = 109.9°K, \sigma = 3.585$ Å for carbon monoxide$\}$. The other pair potentials that were considered are the following:

(a) the dipole–dipole interaction (see Fig. 5)

$$V_{DD}(i, j) = \frac{-\mu^3}{r_{ij}^3} [2 \cos \theta_i \cos \theta_j - \sin \theta_i \sin \theta_j \cos(\varphi_i - \varphi_j)] \qquad (50)$$

($\mu = 0.1172$D for CO, and $\mu = 0$ for N_2);

(b) the quadrupole–dipole interaction

$$V_{QD}(i, j) = \frac{3\mu Q}{4r_{ij}^4} [\cos \theta_i(3 \cos^2 \theta_j - 1) + \cos \theta_j(3 \cos^2 \theta_i - 1)$$
$$- 2 \sin \theta_i \sin \theta_j \cos(\varphi_i - \varphi_j) (\cos \theta_i + \cos \theta_j)] \qquad (51)$$

($Q = 2.43 \times 10^{-26}$ esu for CO, and $Q = 2.05 \times 10^{-26}$ esu for N_2); and finally

(c) the quadrupole–quadrupole interaction

$$V_{QQ}(i, j) = \frac{3Q^2}{4r_{ij}^5} [1 - 5 \cos^2 \theta_i - 5 \cos^2 \theta_j - 15 \cos^2 \theta_i \cos^2 \theta_j$$
$$+ 2(\sin \theta_i \sin \theta_j \cos(\varphi_i - \varphi_j) - 4 \cos \theta_i \cos \theta_j)^2] \qquad (52)$$

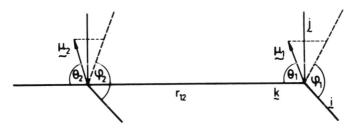

Fig. 5. The polar coordinates of a diatomic molecule employed in the potential functions (50), (51), and (52).

For CO all four branches of the above potential were employed, but for N_2 only the Lennard-Jones and quadrupole–quadrupole interaction are relevant. The dynamical equations of motion to be solved numerically are the $10N$ simultaneous first-order canonical equations of motion, using the total Hamiltonian function

$$\mathscr{H}_N = \sum_{i=1}^{N} \frac{1}{2m} (p_{x_i}^2 + p_{y_i}^2 + p_{z_i}^2) + \sum_{i=1}^{N} \frac{1}{2I} \left(p_{\theta_i}^2 + \frac{p_{\varphi i}^2}{\sin^2 \theta} \right)$$
$$+ \sum_{i<j=1}^{N} V(\mathbf{x}_i, \mathbf{x}_j) \tag{53}$$

where $\mathbf{x}_i = (x_i, y_i, z_i, \theta_i, \varphi_i)$, and I is the moment of inertia about the center of mass and perpendicular to the nuclear axis. The canonical equations of motion are

$$\dot{\alpha}_i = p_{\alpha_i}/m, \quad \dot{\theta}_i = p_{\theta_i}/I, \quad \dot{\varphi}_i = p_{\varphi_i}/I \sin \theta \quad (\alpha_i = x_i, y_i, \text{ and } z_i) \tag{54}$$

and

$$-\dot{p}_{x_i} = \frac{\partial V}{\partial \alpha_i}, \quad -\dot{p}_{\theta_i} = \frac{\partial V}{\partial \theta_i}, \quad -\dot{p}_{\varphi i} = \frac{\partial V}{\partial \varphi_i} \tag{55}$$

In all essential respects the technique is the same as for spherically symmetric particles, the dimension of the molecular dynamics cell l is chosen in accordance with the required bulk density $[l = (mN/\varrho)^{1/3}]$ and is about 30 Å in the Berne and Harp calculations,[100] and the potential $V(\mathbf{x}_i, \mathbf{x}_j)$ is cut off at a center-of-mass cutoff range r_c. Careful experiments with two molecules are required in deciding on an appropriate numerical algorithm for integrating (54) and (55); Berne and Harp have given a very useful summary of their experience with a number of algorithms. The rapid rotational motion of diatomic molecules (particularly for light nuclei) reduces the time interval of integration Δt; for CO and N_2 the real time interval is of order 10^{-15} sec.

Recent molecular dynamics calculations on diatomic molecules have employed a so-called atom–atom potential, which is simply the Lennard-Jones potential (36) acting between the four nuclear pairs associated with the atoms in the two molecules; Cheung and Powles[177] and Barojas[59] have employed this model for liquid nitrogen, thus neglecting the electric quadrupole–quadrupole interaction (52) that is present for nitrogen. Very recently Singer et al.[811] have studied the behavior of the diatomic system in general terms using this atom–atom potential, and in particular have reported on how the properties of the model depend upon the reduced bond length; these authors select a range of bond lengths for their calculations to be compared with F_2, Cl_2, and Br_2. Cheung and Powles[177] present a very thorough comparison of their calculations with experimental results on thermodynamic functions and the structure factor measured in X-ray and neutron scattering. Where comparisons can be made, the overall agreement is good. The only clear discrepancy between experiment and the model calculations is in the correlation times related to reorientational motion [for example, the angular velocity autocorrelation functions—see (48)]. For a diatomic assembly the averaged equilibrium spatial structure can be described using the correlation functions between nuclear centers (13); an example of this resolution is shown in Figs. 6 and 7 on results obtained by Harp and Berne[384,385] for carbon monoxide.

Using molecular dynamics to simulate a diatomic liquid is a significantly more ambitious program than the corresponding point mass calculations; there is approximately a tenfold increase in the number of integration steps Δt needed to span a given elapsed time, and also an increase in the time required to compute the forces acting on each molecule by using a more complicated pair potential function $V_e(\mathbf{x}_i, \mathbf{x}_j)$. A further extension of molecular dynamics to cope with a polyatomic molecular assembly is yet more ambitious, and the recent simulations of triatomic liquid water initiated by Rahman and Stillinger[732] (for further references see Section 6) should perhaps be viewed as extending the method to its present limits. Assuming some form of effective pair potential $V_e(x_i, x_j)$ (see Section 4) between two water molecules, the added strain on the method is again the diminished time step Δt to span a given elapsed time, and the time spent in the force loop. Each water molecule requires the six coordinates in (6) to specify its position and orientation; the molecular frame is chosen so that the inertial moment tensor I is diagonal (I_1, I_2, I_3) as illustrated in Fig. 8.

The equations of motion now yield 12 first-order equations per molecule, and a workable number for N (for most people!) is now not much greater than 250. The equations of motion are the Newton–Euler coupled equations;[355]

$$m\dot{\mathbf{v}}_i = \mathbf{F}_i = -\boldsymbol{\nabla}_{r_i} \sum_{k \neq i} V_e(\mathbf{x}_i, \mathbf{x}_j) \quad \left(\boldsymbol{\nabla}_{r_i} = \mathbf{i}\frac{\partial}{\partial x_i} + \mathbf{j}\frac{\partial}{\partial y_i} + \mathbf{k}\frac{\partial}{\partial z_i}\right) \quad (56)$$

$$\frac{d\mathbf{r}_i}{dt} = \mathbf{v}_i \quad (57)$$

are the equations for center-of-mass motion. The rotational equations require the computation of the total moment vector $\mathbf{M}_i$ in the molecular

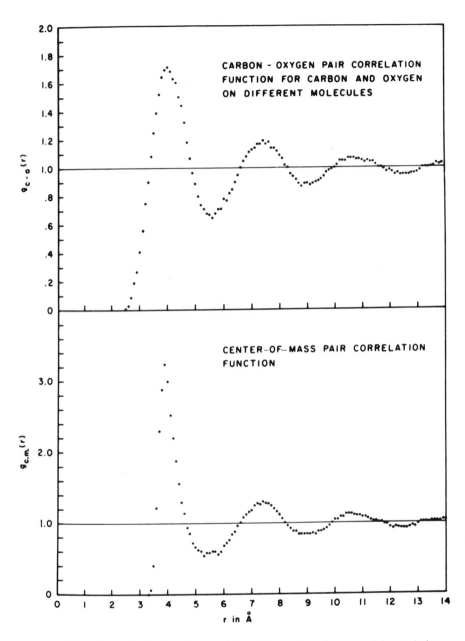

Fig. 6. The pair correlation function $g_{CO}(r)$, and the center-of-mass radial correlation function obtained by Harp and Berne[385] for liquid carbon monoxide in a molecular dynamic simulation using 512 molecules. [Reproduced with permission from *Phys. Rev. A* **2**, 975 (1970).]

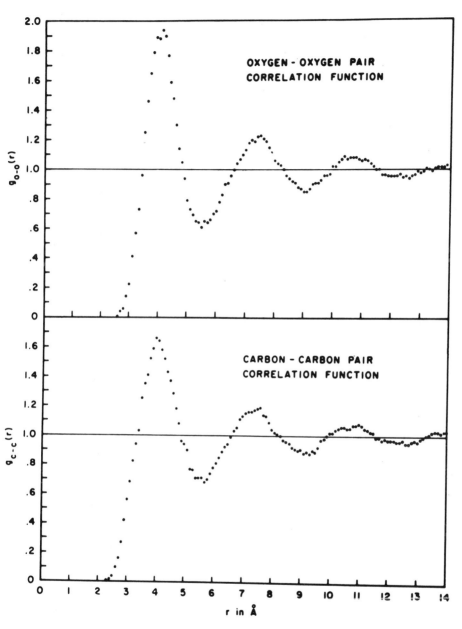

Fig. 7. The pair correlation functions $g_{OO}(r)$ and $g_{CC}(r)$ corresponding to Fig. 7. [Reproduced with permission from *Phys. Rev. A* **2**, 975 (1970).]

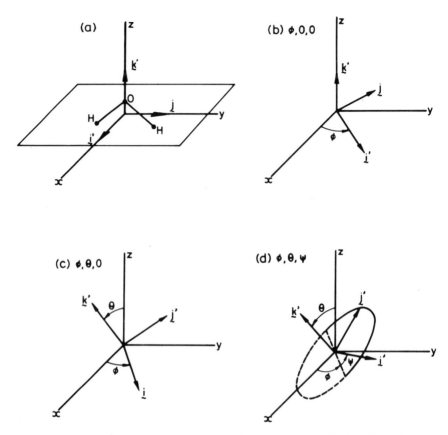

Fig. 8. (a) A coordinate system for the nonlinear water molecule. The Cartesian axes x, y, z represent a fixed laboratory frame (the box frame). The orthogonal unit vectors i', j', k' rotate with the molecule and define the molecular Cartesian frame $x'y'z'$. (b), (c), and (d) show the Euler angle convention for the water molecules. Figure 11 shows the BNS molecule at $\varphi = \theta = \psi = 0$.

frame x', y', and z' (see Fig. 8)

$$\mathbf{M}_i' = (M_{ix'}, M_{iy'}, M_{iz'}) \tag{58}$$

The calculation of $\mathbf{M}_i$ is dependent upon the actual "force point" structure of the model water molecule (i.e., points at which forces act); the components of the force acting at each force point are determined, the x', y', z' coordinates of this point are transformed into the x, y, z frame by the rotation matrix, whence the component in (58) can be calculated additively for each force point. Some details of this calculation are given in Section 4 for a rigid point mass and point charge model of the water molecule. If $\omega_{ix'}$, $\omega_{iy'}$, and $\omega_{iz'}$, are the angular velocity components in the fixed molecular frame x', y', z', then the Euler equations are the following six equations per molecule:

$$I_1\dot{\omega}_{ix'} - \omega_{iy'}\omega_{iz'}(I_2 - I_3) = M_{ix'} \tag{59}$$

$$I_2\dot{\omega}_{iy'} - \omega_{iz'}\omega_{ix'}(I_3 - I_1) = M_{iy'} \tag{60}$$

$$I_3\dot{\omega}_{jz'} - \omega_{ix'}\omega_{iy'}(I_1 - I_2) = M_{iz'} \tag{61}$$

$$\omega_{ix'} = \dot{\varphi}_i \sin\theta_i \sin\psi_i + \dot{\theta}_i \cos\psi_i \tag{62}$$

$$\omega_{iy'} = \dot{\varphi}_i \sin\theta_i \cos\psi_i - \dot{\theta}_i \sin\psi_i \tag{63}$$

$$\omega_{iz'} = \dot{\varphi}_i \cos\theta_i + \dot{\psi}_i \tag{64}$$

A much more powerful, and therefore more time-consuming, numerical scheme is needed to integrate the 3000-or-so simultaneous equations (56)–(64) than the above schemes.[100,879] An extremely comprehensive and powerful numerical integration algorithm has been designed by Gear[346]; this scheme is presently finding a wide application in many fields. The work of Gear[345,346] on numerical integration has been the basis on which an integration of eqns. (56)–(64) has been performed. Gear's algorithm[346] is designed to cope with simultaneous equations in a wide variety of contexts; it is a multivalue predictor–corrector method in which the order of the predictors can be automatically varied to select an optimum step length Δt for a predetermined error bound. A detailed account is given by Gear,[346] who also includes a FORTRAN-coded program of the scheme. The use of the algorithm in its full optimum form would be too time consuming for a full molecular dynamics calculation, but special cases contained in the algorithm can easily be selected after using the fully optimized version for trial experiments, and a suitable procedure is readily obtained. The algorithm is self-starting, the variable time step in the optimized version is especially useful in starting the assembly off; experiments with two molecules [given a $V_e(\mathbf{x}_1, \mathbf{x}_2)$] show that a time step of about 5×10^{-16} sec is about all that can be managed. This is very considerably smaller than the argon calculations. To carry out many computer studies of this type it is unlikely that more than about 5000 time steps for an equilibrium trajectory could be economically justified in terms of computing resources, and this spans only about 10^{-12} sec of real elapsed time. A FORTRAN-coded computer program for the molecular dynamics simulation of liquid water is given in the Appendix; this program includes the fully optimized algorithm of Gear,

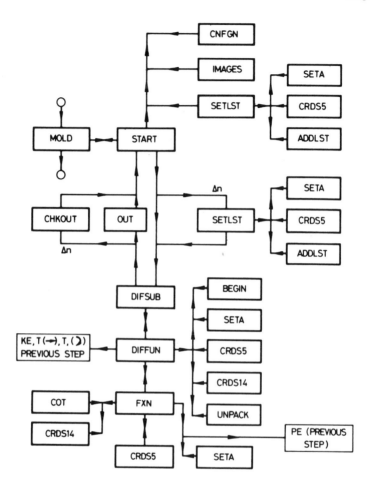

Fig. 9. A flow diagram of the molecular dynamics program to simulate liquid water that is given in the Appendix. The subroutines are described in the Appendix.

and is described by the flow diagram in Fig. 9. The model of the water molecule included in this program is the BNS' model of Section 4.1. Other models can easily be substituted.

Finally we consider how a typical computer run on N-water molecules proceeds. The initial task is to bring the assembly into an "equilibrium" state from an initial $t = 0$ configuration. The dimensions of the box are chosen for a mass density of 1 g/cm³, or the corresponding number density 3.344×10^{22}/cm³. It is impossible to find an initial configuration corresponding to a low value of the total potential energy; instead random center-of-mass positions and random orientations are assigned to each molecule; all molecules are initially set at rest. The initial high value of the potential energy causes rapid angular and translational acceleration to very high values of both the translational and rotational kinetic energies. The temperatures characteristic of any time step configuration are determined by

$$3NkT_t = \sum_{i=1}^{N} mv_i^2 \tag{65}$$

and

$$3NkT_r = \sum_j \mathbf{w}_j \cdot \mathbf{I} \cdot \mathbf{w}_j \tag{66}$$

where T_t and T_r are the translational and rotational temperatures, respectively. In about 10 steps from its initial release the assembly reaches temperatures characteristic of about 10^4–10^5 °K; at this point all the kinetic energy is removed by resetting each molecule to rest. This process of "cooling" is repeated several times to bring the instantaneous temperatures into the required range. If any further temperature control is required, fractional uniform adjustments to all the velocities are made. Initially the rotational motion and kinetic translation are uncoupled and T_t and T_r may be widely separated; the system is allowed to age, a process in which energy modes become coupled. Over this time the potential, rotational, and kinetic energies and the total energy should be monitored and the time-averaged temperatures calculated using (65), (66), and (27). After about 4000–5000 steps with about 200 molecules, $\langle T_t \rangle$ and $\langle T_r \rangle$ should not be separated by more than a few degrees. An example of this aging stage is shown in Fig. 10 for 64 BNS molecules (see Section 4.1) over about 300 time steps. The truncation of the pair potential at a cutoff radius r_c causes the breakdown of energy conservation for noncentral forces; so far experience has shown that this causes only a very small secular rise in temperature over the time intervals covered by the whole calculation. When T_t and T_r are within ± 2°K or so of each other the phase point data $x_i(t)$, $y_i(t)$, $z_i(t)$, $\theta_i(t)$, $\varphi_i(t)$, $\psi_i(t)$, $\dot{x}_i(t)$, $\dot{y}_i(t)$, $\dot{z}_i(t)$, $\dot{\theta}_i(t)$, $\dot{\varphi}_i(t)$, and $\dot{\psi}_i(t)$ are stored for later analysis; it is hoped that 5000 such points might be obtained. Once a system of N molecules has been successfully brought into equilibrium, subsequent variations in temperature, or changes in pair potential functions (i.e., models) can be made without using too much computer time to reestablish equilibrium. A phase point from an equilibrium time step of the original run can be used as the initial configuration for the new system; the assembly clearly feels uncomfortable for some time, but equilibrium is more quickly found than in the original run. So far we have only been concerned with electrically neutral atomic or molecular assemblies; in

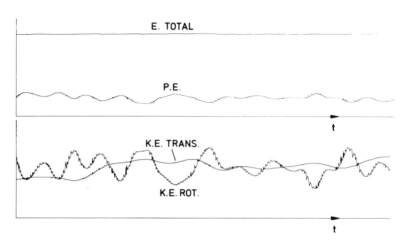

Fig. 10. A plot of the total energy, the potential energy, the translational kinetic energy, and the rotational kinetic energy of an assembly of 64 water molecules (BNS model, Section 4.1) during the aging process. The short period oscillations in the rotational energy are seen to be coupled to the potential energy.

recent years the molecular dynamics method has also been developed to simulate structureless charged particles with Coulombic pair potentials[440,536,554,555]; here Ewald summation techniques[133,554,608,920] for computing the forces acting on point charges are employed ($r_c = \infty$). (See also ionic solutions in Section 8.)

In summary, the molecular dynamics method has become a very powerful tool for investigating both equilibrium structure and kinetic properties in molecular and atomic systems; perhaps the most important single feature of this technique is that it allows the direct determination of any property of the assembly including many significant properties that are *not* measurable by experiment. Thus, if those properties that are comparable with experiment are satisfactory in this comparison, many questions to which *no* experimental answer can be given can be asked, and answered with some confidence.

4. MODELS OF THE WATER MOLECULE

The possibility of simulating liquid water and an aqueous solution has proved to be a great stimulus in the search for an effective water–water intermolecular pair potential function, and also for an effective water–X potential, where X can be anything from a simple ion to the four bases of the DNA molecule! Specific proposals for such functions will be referred to as models. There are essentially two types: these are (a) empirical models with "constructed" functions $V_e(\mathbf{x}_i, \mathbf{x}_j)$ to optimally fit a few experimental parameters, or (b) economized numerical fits to *ab initio* quantum mechanical SCF calculations on the dimer complexes $H_2O–H_2O$ and $H_2O–X$. Here we review the models that have been used in recent molecular dynamics and Monte Carlo simulations.

4.1. The Ben-Naim–Stillinger (BNS) Model

An early point charge model of the water molecule was suggested by Rowlinson,[770] and evaluations of the second and third virial coefficients for water vapor have been made using the Stockmayer potential (50) between point dipoles.[769,771] The recent molecular dynamics calculations on liquid water began with an adaption of Bjerrum's four-point charge model for the water molecule[114] made by Ben-Naim and Stillinger[421] and now known as the BNS model. Ben-Naim and Stillinger adopted the view that the most important property that $V_e(\mathbf{x}_1, \mathbf{x}_2)$ should possess is the preference shown by water molecules for local tetrahedronal coordination, and they chose a model that would favor this structure. The point charge model is shown in Fig. 11; the four charges $\pm\eta e$ are located at the vertices of a regular tetrahedron with the oxygen nucleus at the geometric center of the tetrahedron. The two shielded protons are given a charge of ηe ($<e$) with η a variable parameter and are distant 1 Å from the oxygen atom. The remaining two vertex charges of $-\eta e$ are a crude representation of the unpaired valence shell electrons (the lone pair). The value $\eta = 0.17$ reproduces the dipole moment of the *isolated* molecule. Figure 11 shows the BNS model in the laboratory reference frame x, y, z ($\theta = \varphi = \psi = 0$ in Fig. 8) with the coordinates given in angstroms. The diagonal inertial

Fig. 11. The BNS water molecule; the charges ($\pm$) are located at the vertices of a regular tetrahedron. The OH bond length is 1 Å; the molecule is shown in the standard laboratory frame ($\theta = \varphi = \psi = 0$ in Fig. 8) giving the $x'y'z'$ coordinates of each point charge in (73).

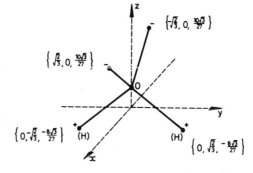

moment tensor is $(13/108, 1/27, 1/12)$ in proton mass Å^2 units, and the center of mass is a little below the geometric center of the tetrahedron.

The potential proposed by BNS[421] is a combination of a central Lennard-Jones potential between oxygen pairs and a noncentral electrostatic potential $V^{(e)}$ between the 16 possible pairs of point charges $\pm\eta e$ on the two molecules. Thus we have initially an expression

$$V_e(\mathbf{x}_i, \mathbf{x}_j) = 4\varepsilon \left[\left(\frac{\sigma}{r_{ij}}\right)^{12} - \left(\frac{\sigma}{r_{ij}}\right)^{6} \right] + (\eta e)^2 \sum_{\alpha_i, \alpha_j} \frac{(-1)^{\alpha_i + \alpha_j}}{d_{\alpha_i \alpha_j}(\mathbf{x}_i, \mathbf{x}_j)} \qquad (67)$$

where the second term is $V^{(e)}$, and $d_{\alpha_i \alpha_j}$ is the distance between point charge number α_i ($\alpha_i = 1, 2, 3, 4$) on the ith molecule and point charge α_j on the jth molecule (negative charges have odd numbers). The Lennard-Jones parameters are chosen to be those of neon, which is isoelectronic with water ($\varepsilon = 5.01 \times 10^{-15}$ erg, later to be scaled to 5.3106×10^{-15} erg, and $\sigma = 2.82$ Å). There is nothing in (67) to prevent the catastrophe $d_{\alpha_i \alpha_j} = 0$ for charges of opposite parity, and this feature is artificially removed by switching out the electronic part $V^{(e)}$ on close approach; (67) is amended to

$$V_e(\mathbf{x}_i, \mathbf{x}_j) = V_{\text{L-J}}(r_{ij}) + S(r_{ij})V^{(e)}(\mathbf{x}_i, \mathbf{x}_j) \qquad (68)$$

where

$$S(r_{ij}) = \begin{cases} 0, & 0 \le r_{ij} \le R_L & (69) \\ (r_{ij} - R_L)^2 (3R_U - R_L - 2r_{ij})/(R_U - R_L)^3, & R_L \le r \le R_U & (70) \\ 1, & R_U \le r_{ij} \le \infty & (71) \end{cases}$$

The switching function smoothly varies between 0 and 1 over the range $r_{ij} = R_L$ to $r_{ij} = R_U$; it is a real part of the potential and is used in forming derivatives to calculate the forces between the oxygen atoms.

The BNS potential contains three variable parameters, η, R_L, and R_U. These were originally fixed by requiring $V_e(\mathbf{x}_i, \mathbf{x}_j)$ to have a minimum corresponding to a nearest-neighbor pair configuration of water molecules occurring in ice; the symmetrical eclipsed pair was chosen (see Fig. 12). Thus the conditions

$$V_e(\mathbf{x}_i, \mathbf{x}_j)\big|_{r_{ij} = 2.76\,\text{Å}} = v_{\min} \qquad (72)$$

and

$$\frac{\partial}{\partial r_{ij}} V_e(x_i, x_j)\big|_{r_{ij} = 2.76\,\text{Å}} = 0$$

determine R_L and R_U, and the value of $v_{\min}$ was chosen to give a good fit

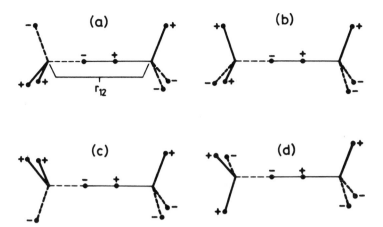

Fig. 12. Nearest-neighbor dimer configurations occurring in the ice lattice, (a) symmetrical eclipsed (SE), (b) nonsymmetrical eclipsed (NSE), (c) symmetrical staggered (SS), and (d) nonsymmetrical staggered (NSS).

to the second virial coefficient $B(T)$ which determined η. The original values quoted by BNS are $\eta = 0.19$, $R_L = 2.0379$ Å, and $R_U = 3.1877$ Å. It is likely that the switching function will seldom be used in the range of (70) in an equilibrium calculation, bearing in mind that the hard core radius σ is 2.82 Å. Figure 13 shows $V_e(\mathbf{x}_i, \mathbf{x}_j)$ as a function of r_{ij} in the four ice lattice configurations shown in Fig. 12.

We can now use the BNS molecule to illustrate the computations needed to set up the differential equations (56)–(64). Let x'_{ni}, y'_{ni}, and z'_{ni} be the coordinates of the nth force point ($n = 5$ is the oxygen atom) of molecule i in the fixed molecular frame shown in Fig. 8; these are fixed and the same for each molecule (Fig. 11). The coordinates of this force point in the molecular dynamics cell are x_{ni}, y_{ni}, and z_{ni}, and these are obtained using the transformation[355]

$$\mathbf{x}_{ni} = \mathbf{x}_{i0} + \mathbf{A}_i^{-1}(\varphi_i, \theta_i, \psi_i)\mathbf{x}'_{ni} \qquad (73)$$

where $\mathbf{x}_{i0}$ are the coordinates of the center of mass, and

$$\mathbf{A}_i = \begin{bmatrix} \cos\psi_i\cos\varphi_i - \cos\theta_i\sin\varphi_i\sin\psi_i & \cos\psi_i\sin\varphi_i + \cos\theta_i\cos\varphi_i\sin\psi_i & \sin\psi_i\sin\theta_i \\ -\sin\psi_i\cos\varphi_i - \cos\theta_i\sin\varphi_i\cos\psi_i & -\sin\psi_i\sin\varphi_i + \cos\theta_i\cos\varphi_i\cos\psi_i & \cos\psi_i\sin\theta_i \\ \sin\theta_i\sin\varphi_i & -\sin\theta_i\cos\varphi_i & \cos\theta_i \end{bmatrix}$$

$$(74)$$

From (73) all the $d_{x_i x_j}$ can be calculated and the forces acting on each point charge $\mathbf{F}_{ni}$ determined; the total force components acting on the molecule

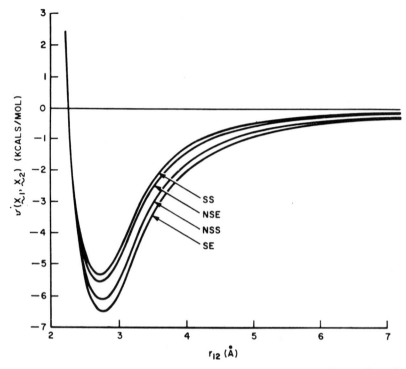

Fig. 13. Potential curves for hydrogen bonding in the four ice lattice configurations.

are calculated for equations (56). The moment vector $\mathbf{M}_i$ in the laboratory frame is given by

$$\mathbf{M}_i = \sum_{j \neq i} \left\{ [\mathbf{\nabla}_i V_{\mathrm{LJ}}(r_{ij}) + V^{(e)}(\mathbf{x}_i, \mathbf{x}_j) \, \mathbf{\nabla}_i S(r_{ij})] \mathbf{A}_i^{-1} \mathbf{x}'_{5i} \right.$$
$$\left. + S(r_{ij}) \sum_{m,n=1}^{4} \mathbf{\nabla}_i V^{(e)}(in, jm) \times \mathbf{A}_i^{-1} \mathbf{x}'_{in} \right\} \qquad (75)$$

where $V^{(e)}(in, jm)$ is obvious notation. The moment components in (58) and (59)–(61) are those in the fixed molecular frame and obtained by using (74) once more:

$$\mathbf{M}'_i = \mathbf{A}_i \mathbf{M}_i \qquad (76)$$

4.2. The ST2 Potential

Initial molecular dynamics calculations using the BNS potential, when compared with experimental results for molecular structure and thermodynamic properties, prompted a revision of the model to improve

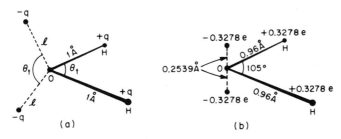

Fig. 14. Point charge tetrads employed in the various water potentials: (a) BNS and ST2; (b) the early model of Rowlinson.[770] For BNS $l = 1$ Å while for ST2 $l = 0.8$ Å. [Reproduced with permission from *J. Chem. Phys.* **60**, 1545 (1974).]

such comparisons. The preference for tetrahedronal order shown by the BNS potential is thought to be too pronounced, and Stillinger and Rahman[831] proposed a modification of the model principally to reduce this tendency. The revised model was encoded the ST2 potential and is geometrically different from the BNS model only in the length l shown in Fig. 14, which is reduced to 0.8 Å. Equation (68) still applies and the parameters are $\varepsilon = 5.2605 \times 10^{-15}$ erg, $\sigma = 3.10$ Å, $q = 0.2357e = 1.13194 \times 10^{-10}$ esu, $R_L = 2.0160$ Å, and $R_U = 3.1287$ Å. The absolute minimum of both BNS and ST2 occur for configurations involving a single hydrogen bond as illustrated in Fig. 15; the respective minima are compared as follows:

BNS: $r_{12} = 2.760$ Å, $\theta = 54.7°$, $\varphi = 54.7°$, $V_e = -6.887$ kcal/mol

ST2: $r_{12} = 2.852$ Å, $\theta = 51.8°$, $\varphi = 53.6°$, $V_e = -6.839$ kcal/mol

The variation of the ST2 potential and the BNS potential with the proton acceptor angle θ at a fixed oxygen–oxygen distance of 2.85 Å in the linear hydrogen bond configuration ($\varphi = 54°44'$) is shown in Fig. 16; the reduced directionality of the ST2 potential is very clear.

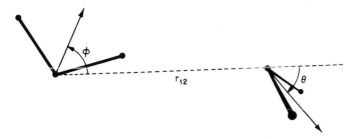

Fig. 15. Geometric coordinates used in the description of the most stable dimer configuration. [Reproduced with permission from *J. Chem. Phys.* **60**, 1545 (1974).]

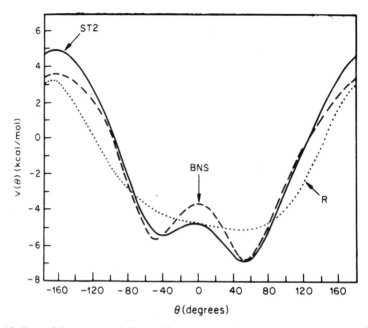

Fig. 16. Potential energy variation with respect to the proton acceptor angle θ at a fixed oxygen–oxygen distance 2.85 Å. The geometry is shown in Fig. 15 with φ selected in each case to yield a linear hydrogen bond. [Reproduced with permission from *J. Chem. Phys.* **60**, 1545 (1974).]

4.3. Hartree–Fock Potential

4.3.1. Hartree–Fock Pair Potentials

The initial simulation of liquid water[732] using the BNS potential (68) carried with it the hope that improvements in the pair potential function would be forthcoming and might ultimately be found in the field of computational quantum mechanics. Some very important developments along these lines have indeed occurred. *Ab initio* quantum mechanical calculations of the ground-state energies of water dimers, trimers, and tetramers have been carried out with the specific objective of converting the raw numerical data into "faithful" and fairly simple functions which can be used in molecular dynamics and Monte Carlo calculations. It should be emphasized that such calculations on the ground-state energy of two water molecules can determine only the pair potential $V_2(x_1, x_2)$ in (7), for a discrete set of configurations x_1, x_2; this will be computed using

$$V_2(x_1, x_2) = U_2(x_1, x_2) - V_1(x_1) - V_2(x_2) \qquad (77)$$

where V_1 is the self-energy of an isolated molecule. These studies still leave us with the problem of converting V_2 into an effective pair potential $V_e(\mathbf{x}_1, \mathbf{x}_2)$.

In the main these SCF calculations have concentrated on a small extension of the configuration space $\mathbf{x}_1, \mathbf{x}_2$ centered on the local minimum energy configuration, subject to prescribed geometrical constraints of particular interest in formation of hydrogen bonding in the neighborhood of such minima. Here we attempt to summarize these calculations, and in particular their use in molecular dynamics and Monte Carlo studies. All of the calculations are dependent upon the particular basis set used, probably more so in $U(\mathbf{x}_1, \mathbf{x}_2)$ than in $V_2(\mathbf{x}_1, \mathbf{x}_2)$; however, many of the current studies possess common characteristics in relation to the formation of a hydrogen bond.

Diercksen[249–251] has performed a very thorough computational investigation of both the isolated water molecule and a hydrogen bonding associated with the most stable configuration of the dimer system. For a single molecule an energy search in varying the HOH bond angle and OH bond length yields a minimum energy configuration OH = 0.9443 Å, and HOH = 105.33°, to be compared with the experimental values of OH = 0.9572 Å, and HOH = 104.52°. On using the experimental values for OH and HOH Diercksen finds a wide-ranging agreement between his orbital model calculation and the experimental values of various properties of the isolated water molecule. From this he concludes that his molecular orbitals are adequate to investigate the dimer system shown in Fig. 17. Diercksen's pair potential energy is shown in Fig. 18 for the three geometries, linear perpendicular, bifurcated perpendicular, and bifurcated planar; it is given as a function of the oxygen–oxygen separation in each case. The absolute minimum energy occurs for the so-called linear hydrogen bond with one hydrogen atom on the line joining the two oxygen atoms. The binding energies (stabilization energies in some quarters) of the minimum energies in Fig. 18 are −4.83 kcal/mol for the linear configuration, −2.59 kcal/mol for the planar bifurcated configuration, and −3.42 kcal/mol for the perpendicular bifurcated configuration. The hydrogen bond in the linear system (2.04 Å) is about 10% longer than the experimental values found for both the cubic and hexagonal ice crystal. This result is a clear indication of the importance of many-body forces in water since we would expect the bond lengths for the dimer system to be smaller than those for the higher polymers, ice being an infinitely large polymer.

A useful summary of *ab initio* SCF calculations on the water molecule and water dimer is given by Del Bene and Pople,[236] who also present their

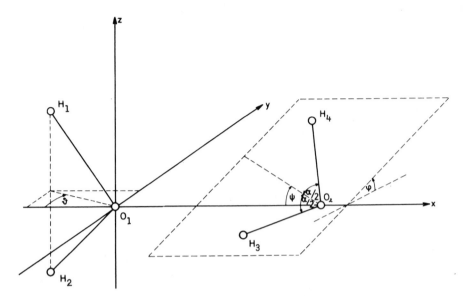

Fig. 17. The coordinate system for the water dimer calculations of Diercksen[250] (see Fig. 18). [Reproduced with permission from *Theoret. Chim. Acta* **21**, 335 (1971).]

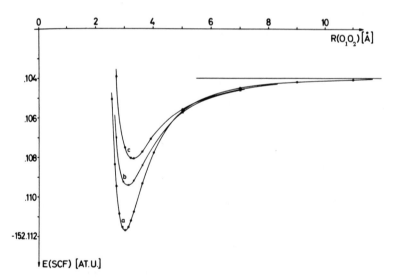

Fig. 18. Total energy variations with the oxygen–oxygen separations obtained in Diercksen's[250] SCF calculations: (a) linear perpendicular geometry, $\varphi = \theta = 0$, $\psi = 52.26°$, $\alpha = 104.52°$; (b) bifurcated perpendicular geometry; $\varphi = \psi = \theta = 0$, $\alpha = 104.52°$, (c) bifurcated planar geometry $\varphi = 90°$, $\psi = \theta = 0$, $\alpha = 104.52°$. [Reproduced with permission from *Theoret. Chim. Acta* **21**, 335 (1971).]

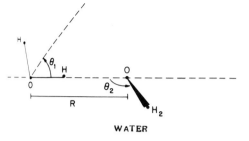

Fig. 19. The geometry of the water dimer used in the calculation of Table I. [Reproduced with permission from *J. Chem. Phys.* **58**, 3605 (1973).]

own calculations to gauge the effects of size and type of basis set. All these studies agree on predicting a linear hydrogen bond, where the plane of the proton donor molecule is a reflection plane, and the nonbonded hydrogens of the proton donor and acceptor molecules have a *trans* configuration as illustrated in Fig. 19. The comparisons between recent calculations shown in Table I are on the linear system shown in Fig. 19; the calculated dipole moments of the single molecule also shown in Table I show considerable variations with the basis set employed.

The calculations shown in Table I are still far removed from obtaining an effective pair potential for use in statistical mechanics; firstly many more points in the configuration space need to be sampled and secondly in molecular dynamics (but not Monte Carlo) the forces and torques in (56), (59), (60), and (61) require all the partial derivatives of $V_2(\mathbf{x}_1, \mathbf{x}_2)$

TABLE I. A Comparison of the Properties of a Single Water Molecule Obtained in Various *ab initio* SCF Calculations[a]

Authors	Total energy (a.u.)	θ_{HOH} (deg)	R_{O-H} (Å)	μ (D)
Bene and Pople[236]	−75.5001	101.1	0.9915	1.82
	−75.8791	111.3	0.9618	2.66
	−75.9086	111.2	0.9506	2.49
Morokuma and Pedersen[631]	−75.5494	113	0.959	
Kollmann and Allen[507]	−75.9757	110	0.967	2.48
Morokuma and Winick[632]	−75.7050	101.1	0.982	
Diercksen[249,250]	−76.0533	105.3	0.944	
Hankins *et al.*[379]	−76.0416	106	0.945	2.19
Experiment	−76.481	104.5	0.957	1.86

[a] Reproduced with permission from *J. Chem. Phys.* **58**, 3605 (1973).

to be defined everywhere in the configuration space x_1, x_2. The significant feature of the above calculations in this respect is likely to be the determination of qualitative characteristics in $V_2(x_1, x_2)$ such as the location of translational minima, and the magnitude of rotational barriers,[250] which can be incorporated as "local forms" into some form of economized model functional expression for $V_2(x_1, x_2)$. The first practical steps in this direction (for water) were taken by Popkie *et al.*[712] and have since been developed by Clementi and co-workers in the form of a computational onslaught on the equilibrium structure of both liquid water and aqueous solutions of ions and complex organic molecules, *ab initio*. The pair potentials for water–X systems are discussed in Section 5.

Popkie *et al.*[712] obtained the Hartree–Fock energies $U_2(x_1, x_2)$ and $V_2(x_1, x_2)$ for 190 configurations of the water dimer; the types of configurations considered are shown in Fig. 20. In addition 26 points close to the absolute minimum energy configuration were obtained using a larger basis set yielding energies close to the Hartree–Fock limit; these calculations were centered on the linear hydrogen-bond system in Fig. 19. These authors have put forward an analytical expression for $V_2(x_1, x_2)$ by numerically fitting the 216 points to a point charge model of the molecule. The point

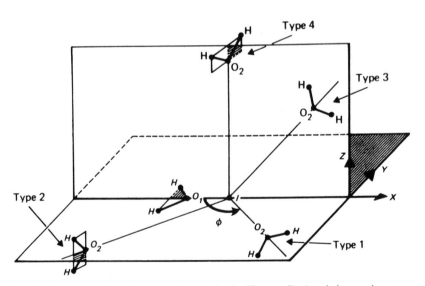

Fig. 20. Nuclear configurations used to obtain the Hartree–Fock pair interaction energy between two water molecules[712] at 190 configurations. The coordinate origin is labelled I. One molecule is kept fixed in the x–y plane; the other molecule is moved such that its symmetry axis points towards the origin. [Reproduced with permission from *J. Chem. Phys.* **59**, 1325 (1973).]

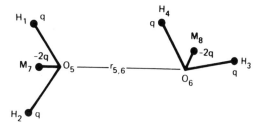

Fig. 21. The Hartree–Fock point charge model used in the analytical expression for the pair potential (78). [Reproduced with permission from *J. Chem. Phys.* **59**, 1325 (1973).]

charge model is only a useful way of choosing a form for the expression—the values of the parameters have no physical significance. The point charge model is shown in Fig. 21 and the best fit is given by

$$V_2(\mathbf{x}_1, \mathbf{x}_2) = q^2 \left(\frac{1}{r_{13}} + \frac{1}{r_{14}} + \frac{1}{r_{23}} + \frac{1}{r_{24}} \right)$$
$$+ \frac{4q^2}{r_{78}} - 2q^2 \left(\frac{1}{r_{18}} + \frac{1}{r_{28}} + \frac{1}{r_{37}} + \frac{1}{r_{47}} \right)$$
$$+ a_1 e^{-b_1 r_{12}} + a_2 (e^{-b_2 r_{13}} + e^{-b_2 r_{14}} + e^{-b_2 r_{23}} + e^{-b_2 r_{24}})$$
$$+ a_3 (e^{-b_3 r_{16}} + e^{-b_3 r_{26}} + e^{-b_3 r_{35}} + e^{-b_3 r_{45}}) \tag{78}$$

where in atomic units $a_1 = 582.277054$, $a_2 = 0.143789$, $a_3 = 5.470184$, $b_1 = 2.520593$, $b_2 = 1.221756$, $b_3 = 1.936626$, $q^2 = 0.449387$, and the $O_5 M_7$ ($O_6 M_8$) distance is 0.436 au. The geometry of the HOH system is the experimental geometry (OH = 0.9572 Å, HOH = 104.52°). The equation (78) is the first serious attempt to provide a link between the two fields of computational quantum mechanics and computational statistical mechanics and could readily be incorporated into the equations of motion (56)–(61); indeed, the overall expression is no more complicated than the BNS potential (68). The fidelity of the fitting achieved by (78) has been biased towards the attractive domain of the potential; this is the important region for Monte Carlo and molecular dynamics calculations.

Interesting comparisons between the Hartree–Fock potential (78) and the BNS potential (68) are shown in Figs. 22, 23, and 24 in the form of equipotential curves. The configurations of the dimer in Fig. 22 are as illustrated; the planes of the two molecules are mutually perpendicular *and* the symmetry axis of molecule 2 points towards the oxygen nucleus of molecule 1 (an orientational constraint). In these configurations the hydrogen bond of the BNS model is seen to be "bent" and much less pronounced than the linear Hartree–Fock bond. The orientational constraint in Fig. 22 is relaxed in Figs. 23 and 24; here for a given oxygen–oxygen separation the

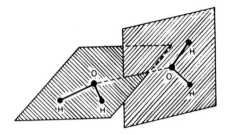

H$_2$O–H$_2$O Total Energy Equipotentials

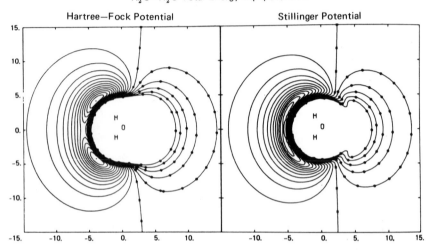

Fig. 22. A comparison of the Hartree–Fock potential (78) and the BNS potential (68). The point charge models are shown in Figs. 11 and 21. The equipotentials are plotted for the dimer configurations shown, where the molecular planes are perpendicular, and the symmetry axis of one molecule is constrained to point towards the oxygen atom of the other molecule. The contour interval is 0.0005 a.u.; solid lines represent negative (attractive) energies, lines with dots represent positive (repulsive) energies, and the line with crosses is the nodal contour of zero energy. The scale is in atomic units with the oxygen atom at the origin. [Reproduced with permission from *J. Chem. Phys.* **59**, 1325 (1973).]

symmetry axis of molecule 2 is positioned such that the dimer energy $U_2(\mathbf{x}_1, \mathbf{x}_2)$ is a minimum (a minimum over the angle α shown in Fig. 23). Now both potentials show a linear hydrogen bond and the Hartree–Fock potential is not as deep as the BNS potential. Kistenmacher *et al.*[490] have developed the original calculations of Popkie *et al.*[712] in a number of directions. Firstly by performing further calculations on 13 additional dimer configurations for the pair potential V_2 a revised version of the Hartree–Fock potential was obtained in a more faithful representation of the SCF data; the point charge model is identical (Fig. 21), but the numerical fitting constants are different, the revised values are (in a.u.)

$$a_1 = 113.996996, \qquad a_2 = 1.242839, \qquad a_3 = 6.508362,$$

$$b_1 = 2.100760, \qquad b_2 = 1.653882, \qquad b_3 = 2.071387, \qquad (79)$$

$$q^2 = 0.419427, \qquad OM = 0.427 \text{ a.u.}$$

Kistenmacher et al.[490] also reconsider the equipotential curves in Figs. 22 and 23. The previous calculations imposed some geometrical constraints on the "searched" dimer configurations. Kistenmacher et al. removed all geometrical constraints in producing minimal energy contour maps;

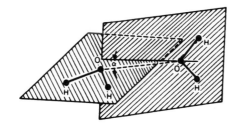

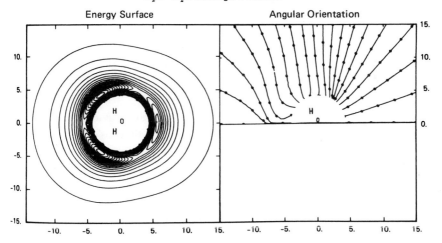

Fig. 23. The Hartree–Fock model potential surface for the illustrated configuration type as a function of the oxygen–oxygen separation (left) and the optimal angle α between the symmetry axis of molecule 2 and the line joining the two oxygen nuclei. For the plot on the left the notation and intervals are as in Fig. 22. For the plot on the right the contour interval is 10°, the horizontal line to the left of the origin is the 0° contour. [Reproduced with permission from J. Chem. Phys. **59**, 1325 (1973).]

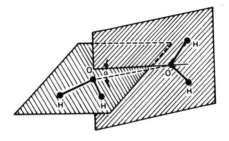

H₂O–H₂O Hartree–Fock Potential

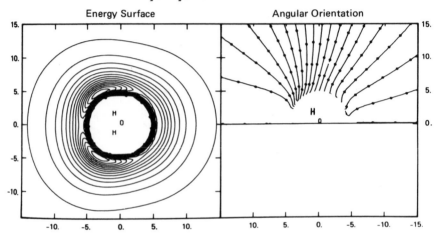

Fig. 24. The BNS model potential surface in exactly the same scheme as described in Fig. 23. [Reproduced with permission from *J. Chem. Phys.* **59**, 1325 (1973).]

molecule 2 in Figs. 22 and 23 is now free to assume any orientation at a given oxygen–oxygen separation, and the minimum energy over all orientations is found using the analytical expression (78). Molecule 1 is kept fixed at the origin of the coordinate system, and is shown in Fig. 25, where the minimal equipotential curves on five z planes are displayed. These authors used the analytical expression (78) and (79) to compute the total intermolecular interaction energies of small clusters of water molecules, $(H_2O)_n$ systems with values of n up to 8. These calculations searched millions of configurations in the various polymers to determine the most stable configuration of a given water polymer. Using only the pair potential for these calculations the authors are of course assuming that the total interaction energy of n-water molecules is pairwise additive. For the trimer and tetramer systems the authors also performed direct Hartree–Fock calculations of $U_3(\mathbf{x}_1, \mathbf{x}_2, \mathbf{x}_3)$ and $U_4(\mathbf{x}_1, \mathbf{x}_2, \mathbf{x}_3, \mathbf{x}_4)$ [see (7)] to check the validity

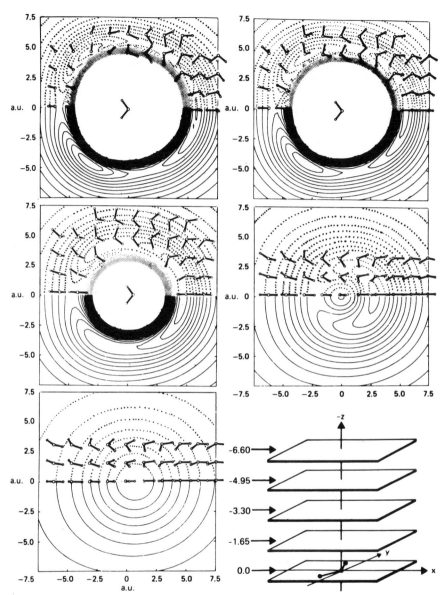

Fig. 25. Minimal energy contour maps for the Hartree–Fock pair potential represented by (79) for the point charge model of Fig. 21. One molecule is kept fixed at the origin. The second molecule has its oxygen nucleus constrained to lie in the plane parallel to the x–y plane at $z = 0$ (top left), $z = -1.65$ a.u. (top right), $z = -3.3$ a.u. (middle left), $z = -4.95$ a.u. (middle right), $z = -6.6$ a.u. (bottom left). The contour interval is 0.0005 a.u. The lowest contour (minimum) is at -4.4 kcal/mol, with the exception of the bottom left insert, where the minimum is -2.8 kcal/mol. The orientations of the second molecule are optimal and some of these orientations are given in a perspective view with the O–H distances being given only in half their actual length and the viewing line being perpendicular to the planes. [Reproduced with permission from *J. Chem. Phys.* **61**, 546 (1974).]

TABLE II. An Analysis of the Energy Partitioning in the Scheme of Eqn. (7) Found by Kistenmacher et al.[490] in Direct Hartree–Fock Calculations on the Most Stable Configurations of the Water Trimer and Tetramer Shown in Fig. 26 and Obtained under the Assumption of Pairwise Additive Forces

Trimer		Tetramer			
Term	Value (a.u.)	Term	Value (a.u.)	Term	Value (a.u.)
$V_2(1, 2)$	-6.67×10^{-3}	$V_2(1, 2)$	-7.14×10^{-3}	$V_3(1, 2, 3)$	-1.05×10^{-3}
$V_2(1, 3)$	-6.71	$V_2(1, 3)$	-2.10	$V_3(1, 2, 4)$	-1.05
$V_2(2, 3)$	-6.10	$V_2(1, 4)$	-7.10	$V_3(1, 3, 4)$	-1.05
$V_3(1, 2, 3)$	-1.80	$V_2(2, 3)$	-7.13	$V_3(2, 3, 4)$	-1.06
		$V_2(2, 4)$	-2.11	$V_4(1, 2, 3, 4)$	-0.31
		$V_2(3, 4)$	-7.14		
Stabilization energy	-19.48×10^{-3}	Total stabilization energy		-37.24×10^{-3}	

of the pairwise additivity assumption at specific three- and four-body configurations. The results of these calculations are shown in Table II; the configurations at which these calculations were performed are the most stable three- and four-molecule configurations predicted by using the analytical expression (78), and are shown in Fig. 26. Probably the most accurate calculations of pair interaction energies at specific configurations x_1, x_2 are the recent calculations of Diercksen et al.[252] using the configura-

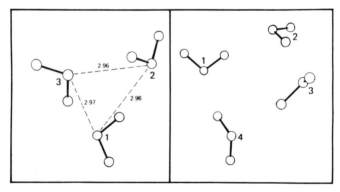

Fig. 26. The most stable configurations of the trimer and tetramer systems of water molecules under the Hartree–Fock potential represented by (78) and (79). [Reproduced with permission from J. Chem. Phys. **61**, 546 (1974).]

tion interaction method. These calculations will serve as a few useful reference points for approximate forms such as (78) and (79). A refinement of (78) is reported by Clementi[184] and is based upon yet more calculations of $U_2(\mathbf{x}_1, \mathbf{x}_2)$ using the configuration interaction method. The point charge model used to obtain an analytical expression is the same as in the previous calculations (Fig. 20); the form of $V_2(x_1, x_2)$ is the same as (78) except that the following term is added:

$$-a_4(e^{-b_4 r_{16}} + e^{-b_4 r_{26}} + e^{-b_4 r_{35}} + e^{-b_4 r_{45}}) \tag{80}$$

The authors of this potential function claim that it yields a more accurate representation of the quantum mechanical potential surface; the constants have been supplied by Matsuoka et al.[184,598] and are (in Å and kcal/mol)

$$a_1 = 1088.213, \quad a_2 = 666.3373, \quad a_3 = 1455.427$$
$$a_4 = 273.5954, \quad b_1 = 5.152712, \quad b_2 = 2.760844 \tag{81}$$
$$b_3 = 2.961895, \quad b_4 = 2.233264, \quad q^2 = 170.9389$$

The hydrogen-bond energies of the potential (81) and (80) are -5.6 kcal/mol (2.98 Å), -4.9 kcal/mol (2.87 Å), and -4.2 kcal/mol (3.01 Å) for the linear, cyclic, and bifurcated dimer configurations, respectively; oxygen–oxygen distances are shown in parentheses.

4.3.2. Effective Potentials Using the Hartree–Fock Potentials

The Hartree–Fock calculations above ignore the electron–electron contributions in the Hamiltonians of both the dimer and monomer systems, the so-called correlation energy contributions to $U_2(\mathbf{x}_1, \mathbf{x}_2)$. These of course are present over the whole range of the H_2O–H_2O interaction. Clementi[184] and co-workers have considered semiempirical corrections to the "bare" Hartree–Fock potentials of Section 4.3.1, the objective being to produce an effective pair potential which can be used in Monte Carlo or molecular dynamics calculations. Kistenmacher et al.[495] have considered making such corrections to deal separately with the intermediate range and long range in terms of the oxygen–oxygen separation r. If $V_{c2}(\mathbf{x}_1, \mathbf{x}_2)$ is the missing electron correlation energy in (77) then

$$V_{c2}(\mathbf{x}_1, \mathbf{x}_2) = U_{c2}(\mathbf{x}_1, \mathbf{x}_2) - U_{c1}(\mathbf{x}_1) - U_{c1}(\mathbf{x}_2) \tag{82}$$

where U_{c2} and U_{c1} are the correlation contributions to the dimer and monomer systems. At an intermediate range Kistenmacher et al.[495] propose that an empirical fitting can be made to the expression for the electron correlation energy originally obtained by Wigner[912] for the N-electron gas, which is

$$E_c = \int a_1 \varrho^{4/3}(a_2 + \varrho^{1/3})^{-1} d\mathbf{r} \quad (a_1 \text{ and } a_2 \text{ constant}) \tag{83}$$

where ϱ is the electron density function of the N-electron system. Adapting (83) for use in (82) for the "electron gas" of the dimer and monomer systems we obtain

$$V_{c2} = \int [a_1 \varrho_d^{4/3}/(a_2 + \varrho_d^{1/3}) - 2a_1 \varrho_m^{4/3}/(a_2 + \varrho_m^{1/3})] d\mathbf{r} \tag{84}$$

where now ϱ_d and ϱ_m are the electron densities in the dimer and monomer, respectively. Kistenmacher et al.[495] used (84) to evaluate V_{c2} for a number of dimer configurations, and the results were numerically fitted to the expression

$$V_{c2}(r) = -ce^{-\delta r} \quad (\delta = 0.9752 \text{ Å}, \ c = 64.62 \text{ kcal/mol}) \tag{85}$$

The long-range corrections for V_c employ the usual London dispersion form for the induced dipole–dipole interaction

$$V_{c2}(r) = -A/r^6 \tag{86}$$

Two values for A are proposed,[495] the Kirkwood–Muller value $A = 1222.2$ kcal/mol and the London value $A = 676.5$ kcal/mol. Kistenmacher et al.[495] make comparisons amongst the following effective pair potentials:

$$V_e(\mathbf{x}_1, \mathbf{x}_2) = V_{HF}(\mathbf{x}_1, \mathbf{x}_2) \qquad\qquad \text{(HF)} \tag{87}$$

$$V_e(\mathbf{x}_1, \mathbf{x}_2) = V_{HF}(\mathbf{x}_1, \mathbf{x}_2) - \frac{1222.2}{r^6} \qquad \text{(HF + K)} \tag{88}$$

$$V_e(\mathbf{x}_1, \mathbf{x}_2) = V_{HF}(\mathbf{x}_1, \mathbf{x}_2) - \frac{676.5}{r^6} \qquad \text{(HF + L)} \tag{89}$$

$$V_e(\mathbf{x}_1, \mathbf{x}_2) = V_{HF}(\mathbf{x}_1, \mathbf{x}_2) - ce^{-\delta r} \qquad \text{(HF + W)} \tag{90}$$

$$V_e(\mathbf{x}_1, \mathbf{x}_2) = V_{HF}(\mathbf{x}_1, \mathbf{x}_2) - ce^{-\delta r} - \frac{1222.2}{r^6} \quad \text{(HF + W + K)} \tag{91}$$

$$V_e(\mathbf{x}_1, \mathbf{x}_2) = V_{HF}(\mathbf{x}_1, \mathbf{x}_2) - ce^{-\delta r} - \frac{676.5}{r^6} \quad \text{(HF + W + L)} \tag{92}$$

The most significant finding of Kistenmacher *et al.*[495] in studies of the above effective potentials is that a preference for tetrahedronal coordination can be seen in the potential. Minimal equipotential surfaces for the potentials HF, HF + L(K) are shown in Fig. 27; the interpretation of these is given in the figure caption. The tetrahedronal coordination is most clearly visible in the HF + L potential, where in the $z = 0$ plane (the plane of the central molecule) two minima associated with two linear hydrogen bonds are clearly present; a third such minimum "grows" with increasing z (see $z = 3.3$ a.u.) and since $V_e(z) = V_e(-z)$ this gives us the four local minima for tetrahedronal coordination.

4.4. Point Charge Models for Monomers

The electrostatic interactions between two water molecules in both the BNS and Hartree–Fock potentials are represented as pairwise electrostatic terms between point charges. Recently Hall,[375] Tait and Hall,[848] and Shipman[804,806] have proposed schemes for obtaining point charge models of molecules, but only for SCF calculations carried out on a single molecule. Such models obtain a representation of the purely electrostatic potential for a point charge sensor in the neighborhood of a single molecule, and are essentially population analyses of the electron density functions of isolated molecules. This class of point charge models originates from attempts to represent the wave functions, and calculate the one-electron properties, of isolated molecules using a basis set of the normalized spherical Gaussian functions

$$\varphi_s(r) = \left(\frac{\pi}{\alpha_s}\right)^{1/4} e^{-\alpha_s(r-r_s)^2} \tag{93}$$

characterized by a center at r_s and an exponent α_s. Both r_s and α_s are chosen optimally in SCF calculations using such basis sets, and yield an SCF wave function

$$\psi(r) = \sum_s c_s \varphi_s(r) \tag{94}$$

and thus a charge density function

$$\varrho(r) = \sum_{st} p_{st} \varphi_s(r)\varphi_t(r) = \sum_{st} p_{st}\Phi_{st}(r) \tag{95}$$

where $\Phi_{st}(r)$ is also a spherical Gaussian function with a center r_{st}

$$r_{st} = (\alpha_s r_s + \alpha_t r_t)/(\alpha_s + \alpha_t) \tag{96}$$

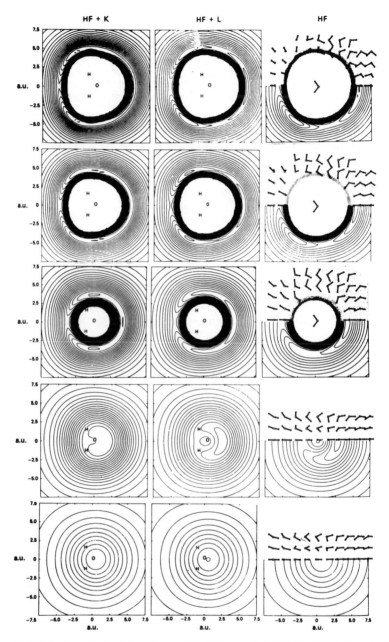

Fig. 27. Minimal equipotential surfaces of the HF potential (87), the HF + L potential (89), and the HF + K potential (88) shown in five z planes at (from top to bottom) $z = 0$, 1.65 a.u., 3.3 a.u., and 4.95 a.u., and 6.6 a.u. The scheme of these plots is as described in Fig. 25. The value of the lowest contour in the first four planes is -6.59 (HF + K), -5.33 (HF + L), and -4.39 (HF) kcal/mol. [Reproduced with permission from *J. Chem. Phys.* **60**, 4455 (1974).]

and an exponent

$$\alpha_{st} = \alpha_s + \alpha_t \qquad (97)$$

The total number of electrons in the molecule is given by

$$n = \int \varrho(r)\, dr = \sum p_{st} S_{st} \qquad (98)$$

and similarly the dipole moment

$$\mu = \int r\varrho(r)\, dr = \sum p_{st} S_{st} r_{st} \qquad (99)$$

We see that in (98) and (99) the system behaves *exactly* as a set of point charges of magnitude p_{ss} at r_s, and $2p_{st}S_{st}$ at r_{st} lying on the line between r_s and r_t. This is the point charge model proposed, having an electron density function

$$\varrho(r) = \sum_{st} p_{st} S_{st}\, \delta(r - r_{st}) \qquad (100)$$

This is not a modeling of the interaction potential between two molecules, although the method could be extended towards finding a symmetrized point charge model of the dimer system. Tait and Hall[848] compare the electrostatic potential of a point charge sensor in the field of a single molecule obtained by using both (95) and (100); other molecular properties are also examined. In the case of water seven Gaussians constitute a minimal-sized basis set and give rise to the 27 point charges shown in Table III. A feature of these models is the appearance of anomalously large charges of opposite parity in close proximity, which would produce a strong angular dependence in the potential at close range. In the case of water we find charges of -197 corresponding to each of the lone pairs, which are neutralized by a charge of 398 on the oxygen nucleus. The form of the electronic pair potential has not yet been examined in any detail, and an obvious difficulty in employing these models in a molecular dynamics or Monte Carlo simulation is the large number of charge pairs required to compute potentials, forces, and torques. Recent work by Shipman[804] and Shipman and Scheraga[806] has concentrated on attempting to reduce the number of point charges in these models.

4.5. Central-Force Models

We have seen that applications of "mathematical statistical mechanics" toward obtaining a theory of the molecular liquid state are completely

TABLE III. A Point Charge Model of a Single Water Molecule Obtained Using a Basis Set of Seven Spherical Gaussians in (94)[a]

Charge	x	y	z
3.97673 (+2)	0.00000 (0)	0.00000 (0)	0.00000 (0)
1.00000 (0)	1.43928 (0)	0.00000 (0)	1.10440 (0)
1.00000 (0)	−1.43928 (0)	0.00000 (0)	1.10440 (0)
6.98333 (−5)	3.98460 (−3)	0.00000 (0)	3.05749 (−3)
−8.91195 (−5)	2.63915 (−2)	0.00000 (0)	2.02509 (−2)
−4.24031 (0)	7.88531 (−1)	0.00000 (0)	6.05061 (−1)
6.98333 (−5)	−3.98460 (−3)	0.00000 (0)	3.05749 (−3)
−8.91195 (−5)	−2.63915 (−2)	0.00000 (0)	2.02509 (−2)
3.56622 (−1)	0.00000 (0)	0.00000 (0)	6.05061 (−1)
−4.24031 (0)	−7.88531 (−1)	0.00000 (0)	6.05061 (−1)
1.44684 (−2)	0.00000 (0)	0.00000 (0)	−1.30342 (−3)
2.49613 (−1)	0.00000 (0)	0.00000 (0)	−8.62759 (−3)
4.56760 (0)	3.89932 (−1)	0.00000 (0)	1.71639 (−1)
4.57670 (0)	−3.89932 (−1)	0.00000 (0)	1.71639 (−1)
−6.19396 (0)	0.00000 (0)	0.00000 (0)	−2.52360 (−1)
−5.43072 (−3)	0.00000 (0)	5.50543 (−4)	0.00000 (0)
−5.16374 (−2)	0.00000 (0)	3.52766 (−3)	0.00000 (0)
−1.85512 (−1)	2.47025 (−1)	3.43364 (−2)	1.89549 (−1)
−1.85512 (−1)	−2.47025 (−1)	3.43364 (−2)	1.89549 (−1)
3.75393 (−2)	0.00000 (0)	3.40990 (−2)	−8.02557 (−2)
−1.96987 (+2)	0.00000 (0)	5.00000 (−2)	0.00000 (0)
−5.43072 (−3)	0.00000 (0)	−5.50543 (−4)	0.00000 (0)
−5.16374 (−2)	0.00000 (0)	−3.52766 (−3)	0.00000 (0)
−1.85512 (−1)	2.47025 (−1)	−3.43364 (−2)	1.89549 (−1)
−1.85512 (−1)	−2.47025 (−1)	−3.43364 (−2)	1.89549 (−1)
3.75393 (−2)	0.00000 (0)	−3.40990 (−2)	−8.02557 (−2)
−1.96987 (+2)	0.00000 (0)	−5.00000 (−2)	0.00000 (0)

[a] Reproduced with permission from *Theor. Chim. Acta* **31**, 311 (1973).

inhibited by an angular dependent pair potential [see eqns. (25) and (26)]. In as far as such mathematical methods can be used for central forces acting between structureless particles they are known to be both reliable and efficient in the liquid density range. We have also seen how an angular-dependent potential greatly complicates molecular dynamics calculations. An attempt to remove these obstacles for a molecular liquid was initiated by Stillinger,[828,829] who has proposed a phenomenological model in which the interaction between two molecules can be represented as a linear combination of central potentials which act between the possible pairs of nuclear species on each molecule. Thus an assembly of N k-atomic molecules becomes a mixture of the nuclear species in the molecule as far as the total potential function is concerned. Indeed in this approach the molecule loses its rigidity, and so in addition to achieving a greater simplicity in $V_e(\mathbf{x}_1, \mathbf{x}_2)$, the advantages of the automatic inclusion of vibrational modes, and the possibility of ionic dissociation (in the case of water $H_2O \rightleftharpoons H^+ + OH^-$) occurring are simultaneous features of these models.

For an N-water "molecule" assembly we have a *mixture* of $2N$ hydrogen atoms and N oxygen atoms; all the interactions envisaged, both within and between molecules, are spherically symmetric pair potential functions. Thus in general, for a k-atomic molecular assembly the total potential of atom i of type α is

$$V_{i_\alpha} = \sum_{\beta \neq \alpha} V(|\mathbf{r}_{i\alpha} - \mathbf{r}_{i\beta}|) + \sum_{j \neq i} \sum_{\beta} V(|\mathbf{r}_{i_\alpha} - \mathbf{r}_{j\beta}|) \qquad (101)$$

where the first term is the intramolecular potential, which clearly must possess a pronounced minimum centered on the equilibrium geometry of a single molecule. For a nuclear mixture the action of (101) would promote the spontaneous formation of N molecules which will remain bound thereafter, apart from possible dissociation effects. In this scheme the equilibrium geometry of the single molecule is now both density and temperature dependent; the nuclear distortions that will occur as the molecules interact provide a natural source of nonadditive interactions between molecules. All in all, this scheme of Stillinger's seems almost perfect from the viewpoint of statistical mechanics: it establishes a "simple" representation of the total intermolecular potential which also takes an automatic account of several important effects, some of which are not treated in the previous models above.

Lemberg and Stillinger[543] have considered the detailed construction of (101) for water. It is required that the centers of mass and charge coincide, thus charges of $2q$ and $-2q$ are placed on the hydrogen and oxygen nuclei, respectively, at the equilibrium positions shown in Fig. 28. The charge q is determined using the experimental dipole moment ($\mu = 1.860$ D) with $r_{OH} = 0.9584$ Å and $\theta = 104.45°$; this yields a screening effect in the value $q = 0.32983e$. To obtain the internal OH bond the oxygen–hydrogen potential $V_{OH}(r)$ must exhibit an attractive well, with $V_{OH}(0.9584)$ being the absolute minimum. Short-range repulsive forces and the long-range attrac-

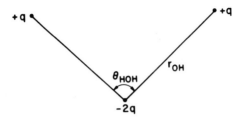

Fig. 28. Point charge representation of the water molecule which corresponds to the central force model of Lemberg and Stillinger.[543] A value of $0.32983e$ for the effective charge q yields a dipole moment in agreement with experiment for the measured parameters $r_{OH} = 0.9534$ Å, $\theta_{HOH} = 104.45°$. [Reproduced with permission from *J. Chem. Phys.* **62**, 1677 (1975).]

tive force are combined in the expression

$$V_{OH}(r) = \frac{2q^2}{r_{OH}} \left[\frac{1}{n} \left(\frac{r_{OH}}{r} \right)^n - \left(\frac{r_{OH}}{r} \right) \right] \tag{102}$$

with n ($n > 1$) to be chosen. Similarly the $V_{HH}(r)$ term is initially conceived in the form

$$V_{HH}(r) = \frac{q^2}{r} + f(r) \tag{103}$$

where $f(r)$ must be such as to have a local minimum at the single molecule HH distance, $r_{HH} = 1.5151$ Å, and also to prevent the opening of $\theta_{HOH} \rightarrow 180°$. The exponent n, and the second derivatives of $f(r)$ are deduced using the experimentally observed normal modes of vibration of a single molecule. Lemberg and Stillinger[543] propose

$$V_{OH}(r) = \frac{2.66366}{r^{14.9797}} - \frac{72.269}{r} \tag{104}$$

and

$$V_{HH}(r) = \frac{36.1345}{r} + \frac{30}{1 + \exp[21.9722(r - 2.125)]} - 26.51983 \exp[-4.728281(r - 1.4511)^2] \tag{105}$$

where the units are Å and kcal/mol. The numerical forms (104) and (105) are of course only one of an infinity of representations that will satisfy the imposed requirements. The remaining nuclear pair potential $V_{OO}(r)$ must be purely repulsive thus preventing the formation of an OH bond with the hydrogens of another molecule; but this potential should also include forces that will encourage hydrogen bonding between neighboring molecules. Lemberg and Stillinger[543] propose

$$V_{OO}(r) = \frac{144.538}{r} + \frac{1.69712 \times 10^6}{r^{12}} - \frac{4.03939 \times 10^3}{r^6} \tag{106}$$

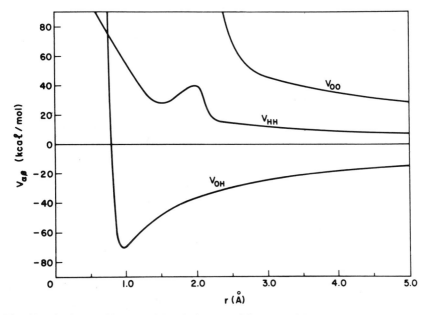

Fig. 29. The interaction potentials of the central-force model, V_{OO}, V_{OH}, and V_{HH} given in (106), (104), and (105), respectively. [Reproduced with permission from *J. Chem. Phys.* **62**, 1681 (1975).]

which has been made to yield a pair potential energy of -6.5 kcal/mol for two molecules placed in the minimal hydrogen bonding configuration of the ST2 potential shown in Fig. 15. The three nuclear pair potentials (104), (105), and (106) are illustrated in Fig. 29 and show the absolute and local minima for the formation of the OH bond and the stabilization of the HH distance.

As a first step in avoiding brute force computer simulations in studies of the molecular liquid state Lemberg and Stillinger[544] have applied the hypernetted chain integral equations (26) to the central force model potentials (104)–(106), but as yet only at high temperatures (1900°K) and low densities (0.32 g/cm³), where a fairly rapid convergence of the numerical iterations can be achieved. A comparison of the most stable dimer configurations obtained using the ST2 potential (See 4.2) and the above central force model is shown in Fig. 30. Lemberg and Stillinger[544] also considered minimal energy configurations of trimers obtained by using the central force potential; the lowest-energy nonplanar configuration is found to be in good agreement with the results of Kistenmacher *et al.*[490] shown in Fig. 26.

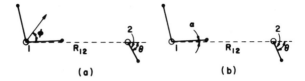

(a)   (b)

Fig. 30. (a) Most stable dimer of the ST2 potential, an SE configuration with $R_{12} =$ 2.852 Å, $\theta = 51.8°$, and $\varphi = 53.6°$. (b) Most stable dimer for central-force model potentials in Fig. 29, allowing for unconstrained bond lengths and angles. The optimized parameter values for this SE arrangement are $R_{12} = 2.865$ Å, $\theta = 60.7°$, and $\alpha = 0.96°$. Bond angles are $\theta_1 = 102.2°$, $\theta_2 = 105.0°$; bond lengths from left to right are 0.9602, 0.9714, 0.9602, and 0.9602 Å. [Reproduced with permission from *J. Chem. Phys.* **62**, 1681 (1975).]

5. POTENTIALS FOR AQUEOUS SOLUTIONS

An important next step after a successful computer simulation of pure water is the inclusion of solute atoms, ions, or molecules in the assembly of water molecules. Such calculations open the way to a direct study of solvent–solute structure in aqueous solutions. Already some stimulating progress has been made in this direction; clearly for either Monte Carlo or molecular dynamics simulations a representation of the water–solute interaction potential is required, and if the solute molecules are to be allowed to move in the molecular dynamics calculation the solute–solute potential is also needed. Here we review some recent attempts to formulate these additional potentials for use in computer simulation.

5.1. Solutions of Inert Atoms

Both the BNS and ST2 potentials (Sections 4.1 and 4.2) can immediately be adapted to form a phenomenological model of an aqueous solution of uncharged structureless particles. An appropriate number of water molecules are turned into structureless particles by setting all the charges ($\pm\eta e$) on these molecules to zero. These molecules can now only interact with the neighboring water molecules (or other solute atoms) via the Lennard-Jones term in (68); the water molecules continue to interact with the full expression (68). In this way one creates several "neonlike" atoms in the presence of water molecules.

On the *ab initio* front, Losonezy *et al.*[571] have computed the ground-state potential of the neon–water pair for many configurations in the Hartree–Fock approximation using a large Gaussian basis set [eqns. (93)

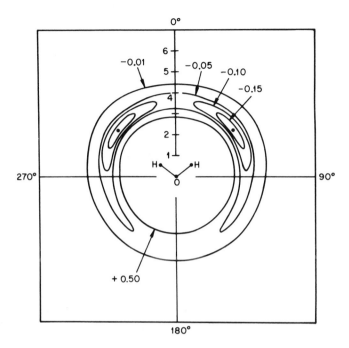

Fig. 31. An equipotential plot of the neon–water interaction potential, where the neon atom is in the molecular plane of the water molecule. The energy identifications are in kcal/mol. [Reproduced with permission from *J. Chem Phys.* **59**, 3264 (1973).]

and (94)]. The equipotential contours in the plane of the water molecule are shown in Fig. 31; a pair of equivalent and distinctive minima are located along the OH directions. These are identified by Losonczy *et al.* as indicators of weak hydrogen bonds occurring at 3.63 Å from the oxygen nucleus, and are computed to have an energy of −0.17 kcal/mol compared to infinite separation.

5.2. Ionic Solutions

Phenomenological models of aqueous ionic solutions can again be obtained by making simple modifications to the BNS and ST2 potentials (Sections 4.1 and 4.2). In this scheme several structureless charged particles are created having the same "size" and mass as the water molecules. Consider a pair of such water molecules selected for modification; the Euler angles and fixed molecular frame coordinates of the point charges (see Fig. 11) of each molecule are set to zero. On each molecule the point charges of $\pm\eta e$ are given the same sign, thus creating a total charge of $\pm0.9428e$ (for ST2) on each molecule, which are now charged mass points in an aqueous solution. The pair potentials now acting are

$$V(H_2O-H_2O), \qquad \text{eqn. (68), (ST2 or BNS)} \qquad (107)$$

$$V(H_2O-\text{ion}) = V_{L-J} \pm 0.9428q \sum_{\alpha=1}^{4} \pm 1/d_{\alpha I} \qquad (108)$$

$$V(\text{ion}-\text{ion}) = V_{L-J} \pm (0.9428e)^2/r_{II} \qquad (109)$$

where $d_{\alpha I}$ is the distance between the point charges $\pm q$ and the ion. The switching function (70) is ignored in (108) and (109).

The above scheme has retained the same value of σ throughout (107)–(109), thus creating ions of the same size as the water molecule. Heinzinger and Vogel[393,394] and Vogel and Heinzinger[883,884] have attempted to create realistic models of ionic solutions of LiCl, CsCl, LiI, NaCl, and CsF by altering the Lennard-Jones parameters σ and ε to model the various ion–water and ion–ion interactions, retaining the ST2 potential for the H_2O-H_2O interaction. All the possible pair potentials are in the form

$$V_{ij}(r, d_{\alpha\beta}) = V_{LJ}(i, j) + V_{ij}^{(e)}(r, d_{\alpha\beta}) \qquad (110)$$

where $V_{ij}^{(e)}$ is the electrostatic contribution, r is the distance between the Lennard-Jones centers (the center of mass for an ion, but the oxygen nucleus for the water molecule), i and j refer to a $\pm$ ion or a water molecule, and $d_{\alpha\beta}$ is the appropriate set of distances between point charges. The Lennard-Jones parameters used by Heinzinger and Vogel are listed in Table IV; the six possible electrostatic contributions proposed are

$$V_{WW}^{(e)} = S(r)q^2 \sum_{\alpha_1, \alpha_2=1}^{4} (-1)^{\alpha_1+\alpha_2}/d_{\alpha_1\alpha_2} \quad [S(r) \text{ as in (70) for ST2]} \qquad (111)$$

$$V_{\substack{++\\--\\(+-)}}^{(e)} = \genfrac{}{}{0pt}{}{+}{(-)} e^2[100S_I(r) + 1]^{-1}r^{-1} \qquad (112)$$

and

$$V_{\substack{+W\\(-W)}}^{(e)} = \genfrac{}{}{0pt}{}{-}{(+)} \sum_{\alpha=1}^{4} (-1)^{\alpha}qe/d_{\substack{+\alpha\\(-\alpha)}} \qquad (113)$$

where $S_I(r)$ is a new switching function of the same form as (70) with $R_L = 9.3678$ Å and $R_U = 28.1034$ Å, introduced by Heinzinger and Vogel

TABLE IV. The Lennard-Jones Parameters σ (Å) and ε (10^{-16} erg) Used in Eqns. (111)–(113) for the Ion–Water Interactions ($+w$, $-w$) and the Ion–Ion Interaction ($+-$). Other Parameters Are the Density d (g/cm³), Box Length l (Å), The Number of Equilibrium Time Steps Developed N, and the Total Elapsed Time (10^{-13} sec)[a]

Parameter	LiI	LiCl	NaCl	CsCl	CsF
σ_{+-}	3.36	2.98	3.09	3.65	3.46
ε_{+-}	35.4934	28.9742	46.7208	128.935	52.8779
σ_{++}		2.37	2.73	3.92	
ε_{++}		24.7996	31.1472	185.435	
σ_{+w}		2.78	2.93	3.53	
ε_{+w}		31.048	38.878	84.2381	
σ_{--}	3.92		3.36		2.73
ε_{--}	185.435		97.7878		31.1472
σ_{-w}	3.58		3.24		2.93
ε_{-w}	84.2381		70.0009		38.878
σ_{ww}			3.10		
ε_{ww}			52.605		
d	1.19	1.05	1.08	1.25	1.27
l	18.681	18.397	18.420	18.737	18.476
N	3291	3398	3443	8000	3588
t	3.587	3.704	3.753	8.72	3.911

[a] Reproduced with permission from *Z. Naturforsch.* **31a**, 463 (1976).

to artificially attenuate the ion–ion potential near the cutoff distance r_c (see Section 3) employed in molecular dynamics calculations. The potentials (110) for LiCl solution are shown in Fig. 32; the extremely artificial force field created by $S_I(r)$ is very evident.

A truly gargantuan computer study of the total ground-state energies and interaction energies between a single water molecule and the ions Li⁺, Na⁺, K⁺, F⁻, and Cl⁻ has been undertaken by Kistenmacher and co-workers using Hartree–Fock calculations.[187,491–494] These authors have calculated the Hartree–Fock energies (and interaction potentials) at 126, 49, 36, 71, and 47 configurations of the ion–water complexes for the Li⁺, Na⁺, K⁺, F⁻, and Cl⁻ ions, respectively. Initially these authors[187,491,492] obtained analytical fitting expressions to the total and interaction energies, and considered computer graphic solutions to the energy surfaces, an

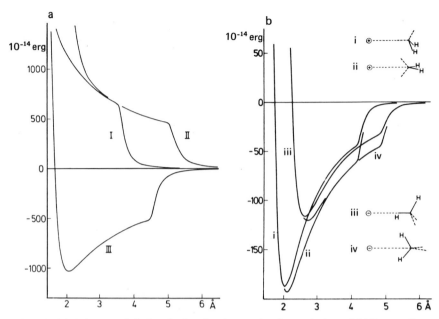

Fig. 32. (a) Pair potentials for the ion–ion interaction in a solution of LiCl according to (109). Curve I is $Li^+ - Li^+$; II: $Cl^- - Cl^-$; and III: $Li^+ - Cl^-$. The strong change in slope between 3.5 and 5.5 Å is due to the switching function in (112). (b) Pair potentials for the ion–water interaction according to (113). For both ions the potentials for two different orientations are given by the inscribed figures. The effects of the switching function are clearly evident. The circle gives the position of the minimum of the water–water interaction in a favorable orientation. [Reproduced with permission from *Z. Naturforsch.* **29a**, 1164 (1974).]

example of which is shown in Fig. 33 and suggests that the form of the cation–water interaction is significantly dependent upon the cation. The analytical expressions fitted to the Hartree–Fock interaction energies of the water–ion complex are based on a simple point charge model of the pair and are thus obtained in exactly the same way as the water–water potentials (78) and (79). The geometry and parameters of these ion–water point charge models are given by Kistenmacher *et al.*[493] and are reproduced here in Fig. 34 and Table V. Two models are presented, the simple model with a potential function

$$V(\mathrm{H_2O\text{--}ion}) = Q\left(\frac{1}{R_{\mathrm{IH_1}}} + \frac{1}{R_{\mathrm{IH_2}}} - \frac{2}{R_{\mathrm{IM}}}\right) - Q_m\left(\frac{1}{R_{\mathrm{IM_1}}} + \frac{1}{R_{\mathrm{IM_2}}} - \frac{2}{R_{\mathrm{IO}}}\right)$$
$$+ A_1[\exp(-a_3 R_{\mathrm{IH_1}}) + \exp(-a_3 R_{\mathrm{IH_2}})] + A_2 \exp(-a_4 R_{\mathrm{IO}})$$

$$(114)$$

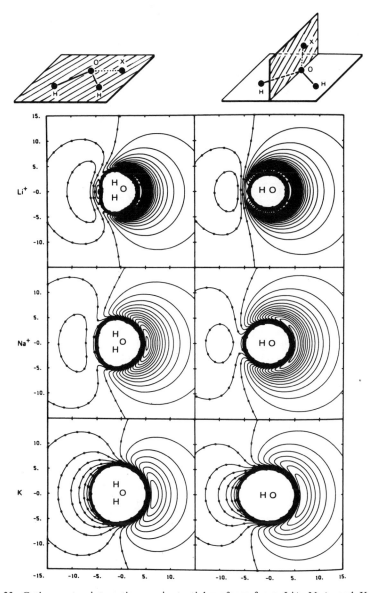

Fig. 33. Cation–water interaction equipotential surfaces for a Li$^+$, Na$^+$, and K$^+$ approaching a rigid H$_2$O molecule in the H$_2$O plane, and in a plane perpendicular to the molecular plane containing the symmetry axis as illustrated. The contour interval is 0.0025 a.u. Solid lines represent negative (attractive) contours, lines with circles represent positive (repulsive) contours, and the line with crosses is the nodal contour of zero interaction energy. The oxygen atom is at the origin. [Reproduced with permission from *J. Chem. Phys.* **58**, 1689 (1973).]

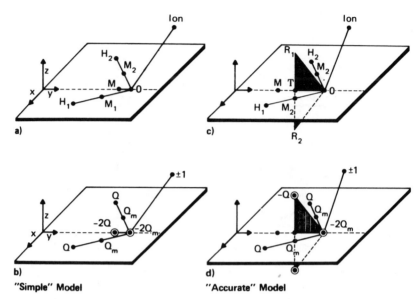

Fig. 34. The point charge models used to establish the potential function expressions (114) and (115) for the ion–water interaction potential obtained in SCF calculations on the ion–water pair. The two figures on the top, (a) and (c), indicate the geometrical relative positions of the charges; the two figures (b) and (d) give the value and sign of the point charges. The numerical values of positional and charge parameters for Li$^+$, Na$^+$, K$^+$, F$^-$, and Cl$^-$ are given in Table V. [Reproduced with permission from *J. Chem. Phys.* **59**, 5842 (1973).]

and an accurate model,

$$V(\text{H}_2\text{O–ion}) = Q\left(\frac{1}{R_{\text{IH}_1}} + \frac{1}{R_{\text{IH}_2}} - \frac{1}{R_{\text{IR}_1}} - \frac{1}{R_{\text{IR}_2}}\right)$$

$$- Q_m\left(\frac{1}{R_{\text{IM}_1}} + \frac{1}{R_{\text{IM}_2}} - \frac{2}{R_{\text{IO}}}\right)$$

$$+ A_1[\exp(-a_3 R_{\text{IH}_1}) + \exp(-a_3 R_{\text{IH}_2})] + A_2 \exp(-a_4 R_{\text{IM}})$$

$$(115)$$

Notice that the point charge model of the water molecule differs from the corresponding water–water potential in Fig. 21.

The Hartree–Fock water–water potential (79) and the corresponding water–ion potentials (115) were employed by Kistenmacher *et al.*[494] to obtain the zero-temperature minimal-energy configurations of clusters of water molecules surrounding positive and negative ions; examples of these minimal energy structures for Li$^+$ and F$^-$ are shown in Figs. 35 and 36.

TABLE V. The Parameters of Charge and Distance in Atomic Units in the Point Charge Models of Fig. 34 Chosen to Represent the Interaction Potentials of Various Water–Ion Systems.[493] [a]

Parameter	Li$^+$	Na$^+$	K$^+$	F$^-$	Cl$^-$
Q	0.63939	0.59222	0.59564	−0.65248	−0.61465
Q_m	−0.06192	−0.06568	−0.05095	−0.02890	0.12698
a_1	7.55251	187.59619	4.45471	4.27508	6.06616
a_2	66.78278	196.08435	373.78784	210.04810	69.61395
a_3	3.55194	3.94727	2.64194	1.99112	1.63899
a_4	2.37924	2.36244	2.16348	2.32209	1.74914
OM	0.03937	0.03385	0.07975	0.15439	0.28188
OT	0.26056	0.26443	0.35249	0.42294	0.46735
$TR_1 = TR_2$	0.00558	0.18066	0.013476	0.28521	0.21344

[a] The corresponding analytical expressions for the ion–water potential are given in eqns. (114) and (115). Here we include only the parameters for the so-called accurate model. [Reproduced with permission from *J. Chem. Phys.* **59**, 5842 (1973).]

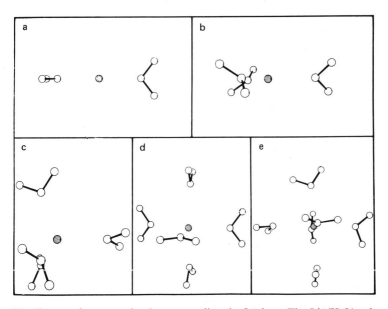

Fig. 35. Clusters of water molecules surrounding the L$^+$ ions. The Li$^+$(H$_2$O)$_n$ clusters shown correspond to the minimum energy structures. The complexes are drawn to scale, the dashed circle represents the Li$^+$ ion. [Reproduced with permission from *J. Chem. Phys.* **61**, 799 (1974).]

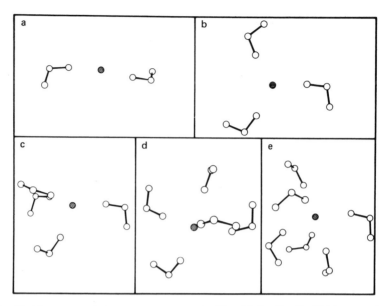

Fig. 36. The same minimum structures as in Fig. 35 for the $F^-(H_2O)_n$ complexes. [Reproduced with permission from *J. Chem. Phys.* **61**, 799 (1974).]

Kress et al.[517] have extended the above calculations in a study of the Hartree–Fock energies of the three-body system H_2O–Li–F. These very elaborate calculations when combined with the ion–water pairs above accomplish two main objectives. The first objective is to express the interaction energy of the three-body system in the pairwise additive approximation using the previous results and the two-body interaction for the Li–F system. The ion–ion interaction was computed at 23 separations and fitted to the expression

$$V(\text{Li–F}) = \sum_{i=1}^{5} \frac{a_i}{R_{\text{LiF}}} + a_6 e^{a_7/R_{\text{LiF}}} + 106.69458 \qquad (116)$$

The constants obtained by Kress et al.[517] are $a_1 = 183.747778$, $a_2 = 161.30035$, $a_3 = 79.10556$, $a_4 = 30.43912$, $a_5 = -7.56712$, $a_6 = -106.69458$, and $a_7 = 1.73191$ (a.u. throughout). The pairwise additive interaction energy of the H_2O–Li–F system is now

$$U_3(H_2O, Li^+, F^-) \simeq V_2(H_2O, Li^+) + V_2(H_2O, F^-) + V_2(Li^+, F^-) \qquad (117)$$

[see (7)], where (115) and (116) are employed.

The second objective is to obtain a direct analytic fit to the Hartree–Fock interaction energies at 250 configurations of the three-body system

H_2O–Li–F for possible use in computer simulations, and also to enable comparisons to be made with the pairwise additive form and assess the importance of the three-body potential $V_3(H_2O, F, Li)$ [see (7)] in

$$U_3(H_2O, Li^+, F^-) = V_2(H_2O, Li^+) + V_2(H_2O, F^-) + V_2(Li^+, F^-) \\ + V_3(H_2O, Li^+, F^-) \tag{118}$$

These authors again employ computer graphic solutions using the analytic expressions obtained as best fits to their data but also quote extensive numerical data on the energy partitioning in the scheme (118). They conclude overall that significant departures from the pairwise additive assumption (117) occur. A comparison of contour maps between (118) and (117) is shown in Figs. 37 and 38. The analytic expression obtained for the three-body interaction potential (118) uses the same point charge model as shown in Fig. 34 (the accurate model); the expression is

$$U_3(H_2O, Li^+, F^-) = Q^F\left(\frac{1}{R_{H_1F}} + \frac{1}{R_{H_2F}} - \frac{1}{R_{FR_1}} - \frac{1}{R_{FR_2}}\right) \\ - Q_m^{\,F}\left(\frac{1}{R_{M_1F}} + \frac{1}{R_{M_2F}} - \frac{2}{R_{FO}}\right) \\ + A_1^{\,F}[\exp(-a_3^{\,F}R_{H_1F}) + \exp(-a_3^{\,F}R_{H_2F})] \\ + A_2^{\,F}\exp(-a_4^{\,F}R_{FM}) \\ + Q^{Li}\left(\frac{1}{R_{H_1Li}} + \frac{1}{R_{H_2Li}} - \frac{1}{R_{LiR_1}} - \frac{1}{R_{LiR_2}}\right) \\ - Q_M^{\,Li}\left(\frac{1}{R_{M_1Li}} + \frac{1}{R_{M_2Li}} - \frac{2}{R_{LiO}}\right) \\ + A_1^{\,Li}[\exp(-a_3^{\,Li}R_{H_1Li}) + \exp(-a_3^{\,Li}R_{H_2Li})] \\ + A_2^{\,Li}\exp(-a_4^{\,Li}R_{LiM}) \\ + \frac{b_1}{R_{LiF}} + \frac{b_2}{R_{LiF}} + b_3\exp(b_4R_{LiF}) \tag{119}$$

where the parameters are given in Table VI; again no physical significance can be attached to the geometry of the point charges.

5.3. Organic Molecules and Water

The technique of fitting analytic expressions to SCF calculations in quantum mechanics has recently been applied towards the most ambitious

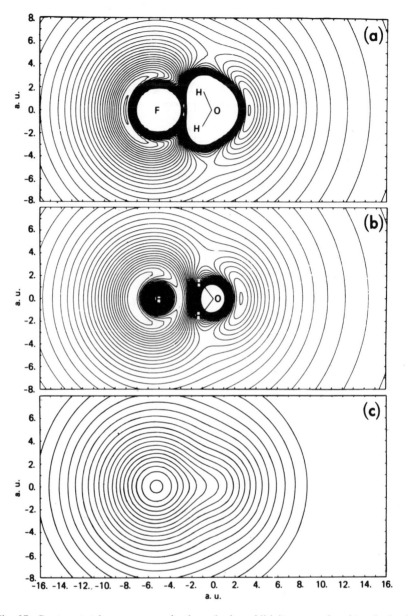

Fig. 37. Contour total energy maps in the pairwise additivity approximation obtained by Kress *et al.*[517] for the Li–F–H$_2$O three-body complex. The fluorine nucleus is at $x = -5.15$ a.u. and the water molecule is fixed as indicated. The three maps correspond to the lithium nucleus in the planes $z = 0$ (a), $z = 2.0$ a.u. (b), and $z = 4.0$ a.u. (c). The interval between contours is 5.0 kcal/mole. Two clear minima are evident in the $z = 0$ and $z = 2$ a.u. planes. [Reproduced with permission from *J. Chem. Phys.* **63**, 3907 (1975).]

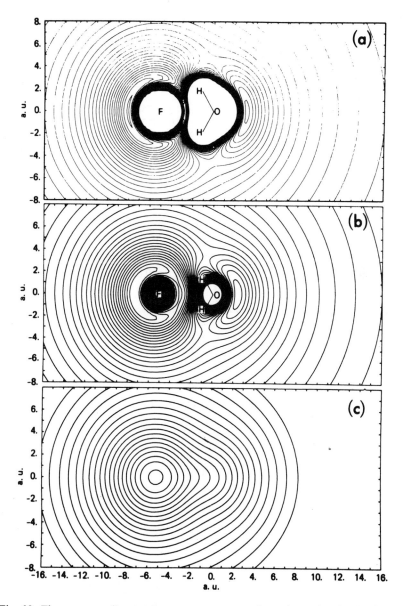

Fig. 38. The corresponding total energy contours to those shown in Fig. 37, but now obtained from the analytical fit (119) to the Hartree–Fock calculations of the three-body system, and therefore not based upon the assumption of pairwise additivity. Although the qualitative features are the same as in Fig. 37, there are quantitative differences, for example in the well depths. [Reproduced with permission from *J. Chem. Phys.* **63**, 3907, (1975).]

TABLE VI. The Constants Used in the Hartree–Fock Point Charge Model of the Triplet System F⁻–Li⁺–H₂O Given in the Analytical Expression (119)[a]

Constant	F	Li	LiF
Q	−0.53544971	0.54877581	
Q_m	−0.05638953	0.01021204	
A_1	0.02382738	8.62957004	
A_2	449.999935	119.999996	
a_3	1.15802317	3.12848282	
a_4	2.52354884	2.76618096	
b_1			−1.05672429
b_2			1.03197828
b_3			39.5052792
b_4			−2.36780241
$x(R_1)$	−0.257415782	$x(M) = -0.0308938$	
$y(R_1)$	0	$y(M) = 0.0$	
$z(R_1)$	0.001385405	$z(M) = 0.0$	

[a] The point charge model itself is the one shown in Fig. 34 (the accurate model) for the ion–water pair, and the charge and length parameters have the same geometry here (atomic units throughout).

program yet conceived. This is to determine the interaction potential between a water molecule and a complex organic molecule. Clementi and co-workers[129,186,188,796] have performed SCF–LCAO–MO calculations on 1690 points in the configuration space of the H₂O–M system where M is one of twenty one amino acids. The interaction potential at these points has been fitted to the atom–atom expression

$$V(\text{H}_2\text{O–M}) = \sum_{i,j} \left(-\frac{A_{ij}^{ab}}{r_{ij}^6} + \frac{B_{ij}^{ab}}{r_{ij}^{12}} + \frac{C_{ij}^{ab} q_i q_j}{r_{ij}} \right) \quad (120)$$

where i, j represent two atoms, one in the amino acid and one in the water molecule, r_{ij} is the distance between the atoms, q_i and q_j are point charges placed on nuclei i and j; the indices a, b serve to distinguish between either H or O on H₂O, atoms in M, and even atomic groups in M with the same Z values, and finally A, B, and C are fitting constants. Clementi and co-workers[129,186,188,796] have provided all the numerical data required to construct the function (120) for 21 amino acids, the four bases of DNA, and lysozyme. To illustrate the use of these data we will consider the glycine molecule shown in Fig. 39. In Table VII the nuclear charges in (120) are

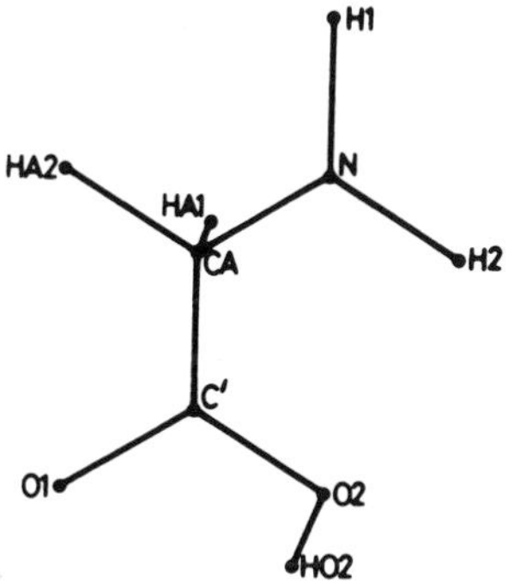

Fig. 39. An illustration of the data reported by Clementi et al.[186] on the interaction between water and the amino acids, here shown for the glycine molecule. Each amino acid is described twice, first to identify each atom with the same alphanumeric code adopted as shown in Table VII, and second to give each atom a class number, which is also listed in Table VII. The potential parameters in (120) are classified according to the class number. [Reproduced with permission from *J. Amer. Chem. Soc.* **99**, 5531 (1977).]

listed for each atom or atomic group in glycine, the corresponding nuclear co-ordinates are given, and a class number labels each group. The class number identifies the incides a, b in (120); there are 23 class numbers for the 21 amino acids considered. From the class number the appropriate constants A, B, and C are selected from the 46 sets of three, A_{ij}, B_{ij}, and C_{ij}. In the case of glycine there are 60 fitting constants in (120)! For all the molecules considered so far Clementi et al. have used the expressions (120) to produce contour maps of the interaction potential, giving an initial idea of how these large molecules are "seen" by a water molecule; also a rough idea of the excluded volume surrounding each molecule, and where the water molecules will tend to place themselves. An example of these maps is shown in Fig. 40 for the DNA base cytosine.

TABLE VII. The Point Charge Model of Glycine (Fig. 39)[a]

Atom	Class	x	y	z	Charge
O2	9	2.835	2.101	1.389	−0.50
O1	10	−1.320	3.200	1.631	−0.38
N	11	2.700	2.032	−3.749	−0.55
C	5	0.673	2.824	0.416	0.46
CA	8	0.580	3.276	−2.431	−0.33
H1	1	2.597	2.270	−5.723	0.24
H2	1	4.433	2.698	−3.125	0.25
HA1	2	0.689	5.303	−2.786	0.18
HA2	2	−1.194	2.587	−3.213	0.22
HO2	4	2.607	1.920	2.160	0.40

[a] The atoms and atomic groups are classified by a class number, from which the interaction constants in eqn. (120) are determined by Clementi and co-workers.[186]

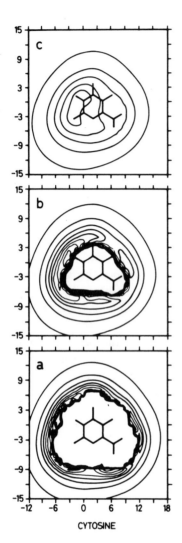

CYTOSINE

Fig. 40. The minimal contour energy maps of the interaction energy between cytosine and a water molecule as obtained by Scordamaglia *et al.*[796] The cases (a), (b), and (c) identify the plane of the cytosine and two parallel planes at heights of 4 and 6 a.u., respectively, above the molecular plane. The interval between contours is 1 kcal/mol. [Reproduced with permission from *J. Amer. Chem. Soc.* **99**, 5545 (1977).]

6. MOLECULAR DYNAMICS STUDIES OF PURE WATER

An extensive study of pure water using the molecular dynamics method (Section 3) has been carried out by Rahman and Stillinger[732–734,830–832]; these authors have employed the phenomenological models above, namely, the BNS, ST2, and central force models (Sections 4.1, 4.2, and 4.5, respectively). In massive computational work of this type it is desirable to have independent calculations performed; such calculations have recently been

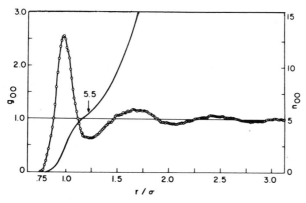

Fig. 41. The oxygen–oxygen pair correlation functions $g_{OO}(r)$ obtained for the BNS model. The monotonically rising curve n_{OO} shows the average number of neighboring oxygens within any radial distance r; $T = 34.3°C$, $\varrho = 1$ g/cm³. [Reproduced with permission from *J. Chem. Phys.* **55**, 3336 (1971).]

carried out by Fox and Wood* on the BNS model, and Tucker and Wood* have performed molecular dynamics calculations using the Hartree–Fock model potential (78). As far as this author is aware these are the only molecular dynamics calculations on pure water at the present time.

6.1. Equilibrium Structure

The outstanding feature of the molecular dynamics method is its ability to provide information that is unavailable through measurement; many of the results reported below illustrate this aspect of the method. The atomic pair correlation functions $g_{\alpha\beta}(r)$ [eqns. (13), (14), (46), and (47)] for water were first determined for the BNS model potential (Section 4.1) using an assembly of 216 molecules at an equilibrium temperature of 34.3°C, and density of 1 g/cm³: they are shown in Figs. 41, 42, and 43.[732] Figure 41 also shows the running co-ordination number $n_{OO}(r)$ (14), which shows that on average there are 5.5 neighboring molecules in a range up to the first minimum in $g_{OO}(r)$, where $r = 1.22\sigma$. It is immediately clear that the $g_{OO}(r)$ curve of the BNS model differs drastically from the corresponding atom–atom pair correlation functions of atomic liquids; the pair correlation function obtained for liquid argon is shown in Fig. 4 and reveals an average coordination of 12.2 neighbors up to the first minimum. The broad second peak in $g_{OO}(r)$ gives a value of 1.69 for the ratio of second

* Unpublished calculations.

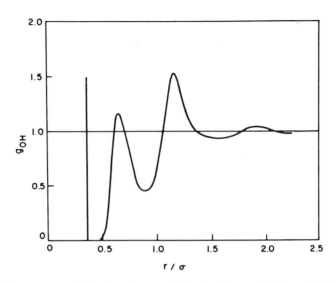

Fig. 42. The cross-correlation function $g_{OH}(r)$ obtained for the BNS model. The vertical line indicates the intramolecular O–H bond length; $T = 34.3°C$, $\varrho = 1$ g/cm³. [Reproduced with permission from *J. Chem. Phys.* **55**, 3336 (1971).]

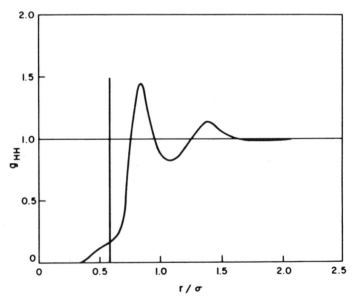

Fig. 43. The proton pair correlation function $g_{HH}(r)$ obtained for the BNS model. The vertical line indicates the intramolecular H–H distance; $T = 34.3°C$, $\varrho = 1$ g/cm³. [Reproduced with permission from *J. Chem. Phys.* **55**, 3336 (1971).]

peak to first peak distances, which is close to that observed for the ideal ice structure $[2(2)^{1/2}/(3)^{1/2} = 1.633]$. The second peak in liquid argon is very different in this respect. In the solid state argon forms an fcc lattice structure; if a remnant of this structure was to persist into the liquid phase on melting, the ratio of the second peak to first peak distances would be centered on the value $2^{1/2}$, and this does not occur. Clearly the disordering effect of the melting transition is very different in the two cases; for water we are seeing the strong effects of directionality in the molecular interactions upon the local order in the liquid state. The first two prominent peaks in the $g_{OH}(r)$ function arise from neighboring water molecules which are in hydrogen-bonded configurations, the first peak occurring for the proton and the acceptor oxygen in the linear hydrogen bond configurations; the second peak averages over the remaining O–H pairs in bonded pairs of molecules. Similarly the first peak in $g_{HH}(r)$ is probably the average proton pair separation in bonded pairs of molecules. The prominent short-range "shoulder" of $g_{HH}(r)$ in the region $r \sim 0.5\sigma$ is probably an artifact of the BNS model since it does not appear again in other improved models. The independent calculations of Fox and Wood* (using a version of the general program in the Appendix) obtained the same overall features in the three correlation functions, including the short-range hump in $g_{HH}(r)$; for comparison with Figs. 42 and 43 the correlation functions $g_{HH}(r)$ and $g_{OH}(r)$ obtained by these authors for a much smaller assembly of 64 molecules at the higher temperature of 75°C are shown in Figs. 44 and 45.

We have already noted the very small time steps Δt allowed in the numerical integration of the Newton–Euler equations (56)–(64) for an assembly of water molecules. The principal reason for this is very high angular velocities resulting from the small mass of the proton. In an attempt to overcome this Fox and Wood* carried out a mass-scaling experiment in which the proton mass was increased by factors of up to 10. Such scaling certainly yields "smoother" dynamics, and a larger span of "molecular time" per unit of computing time is achieved. It is conjectured that the equilibrium structure in the liquid (but clearly not the kinetic properties) will be essentially invariant to substantial mass scaling of this type, since it will serve mainly to eliminate the rapid angular oscillations. The three correlation functions $g_{OO}(r)$, $g_{OH}(r)$, and $g_{HH}(r)$ obtained for 64 BNS molecules at 31°C using a scaling factor of 10 are shown in Figs. 46, 47, and 48, where they are compared with the results of Rahman and Stillinger using 216 molecules at 34.3°C in Figs. 41, 42, and 43. It is clear that the

---

* Unpublished calculations.

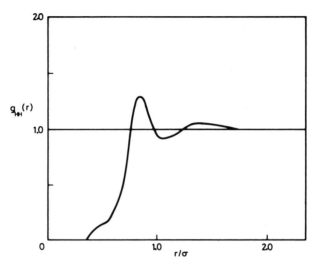

Fig. 44. The proton pair correlation function $g_{HH}(r)$ obtained for 64 BNS molecules (Fox and Wood, unpublished calculations); $T = 75°C$, $\varrho = 1$ g/cm³.

equilibrium structure has been preserved; about 1/20th of the computing time used for the unscaled molecules was required in this calculation.

Stillinger and Rahman[830] also used the BNS molecule to study the breakdown of local hydrogen-bonding order which occurs on increasing

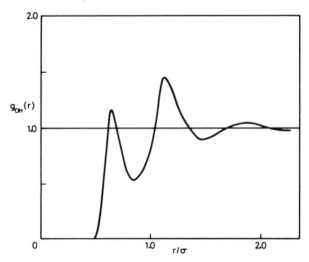

Fig. 45. The cross-correlation function $g_{OH}(r)$ obtained for 64 BNS molecules (Fox and Wood, unpublished calculations); $T = 75°C$, $\varrho = 1$ g/cm³.

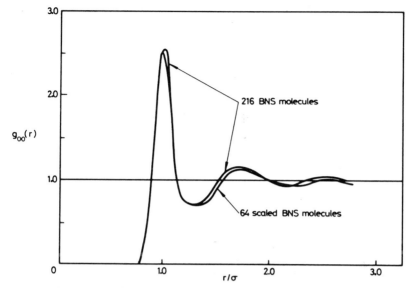

Fig. 46. The oxygen–oxygen pair correlation function $g_{OO}(r)$ obtained by Fox and Wood (unpublished calculations) for 64 BNS water molecules in which the proton mass is scaled by a factor of 10; $T = 31°C$, $\varrho = 1$ g/cm³. The results of Rahman and Stillinger[732] in Fig. 41 are shown for comparison.

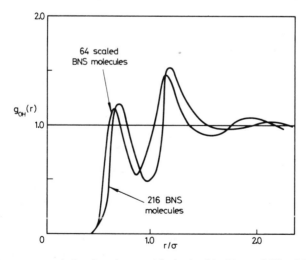

Fig. 47. The cross-correlation function $g_{OH}(r)$ obtained by Fox and Wood (unpublished calculations) for 64 BNS water molecules in which the proton mass is scaled by a factor of 10; $T = 31°C$, $\varrho = 1$ g/cm³. The results of Rahman and Stillinger[732] in Fig. 42 are shown for comparison.

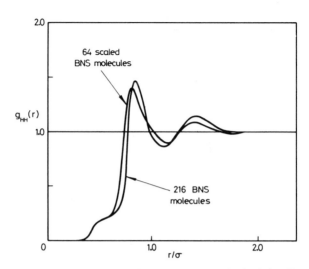

Fig. 48. The proton-pair correlation function $g_{HH}(r)$ obtained by Fox and Wood (unpublished calculations) for 64 BNS water molecules in which the proton mass is scaled by a factor of 10; $T = 31°C$, $\varrho = 1$ g/cm³. The results of Rahman and Stillinger[732] in Fig. 43 are shown for comparison.

the temperature. Calculations were performed at two additional temperatures, $-8.2°C$ and $314.8°C$ corresponding to a super-cooled sample and a sample under a pressure of about 6 kbar, respectively; the density in each case is 1 g/cm³. The three correlation functions $g_{OO}(r)$, $g_{OH}(r)$, and $g_{HH}(r)$ each present an independent view of the liquid structure; on increasing the temperature at a fixed density each shows the same trend of a significant reduction in local molecular correlation. Here we are seeing the decreasing ability of the directional molecular interactions to establish a local tetrahedronal order. The $g_{OO}(r)$ curves through $n_{OO}(r)$ show that the numbers of neighboring molecules up to the first minimum are 5.2 ($-8.2°C$), 5.5 (34.3°C), and 8.0 (314.8°C); the ratio of second peak and first peak distances similarly increase: 1.63 ($-8.2°C$), 1.69 (34.3°C), and 1.83 (314.8°C). At high temperatures any residual ice structure would completely disappear. The $g_{OH}(r)$ functions at the three temperatures are shown in Fig. 49.

The ST2 model (Section 4.2) was introduced by Stillinger and Rahman[831] to improve the overall fidelity of the basic point charge model in areas where comparisons with experimental data are possible. The experimental X-ray scattering experiments of Narten and Levy[644] are used in relation to liquid structure; the "experimental" $g_{OO}(r)$ can only be obtained under the assumption that spherical scattering by the heavy oxygen nucleus dominates the X-ray intensities. A novel feature of the molecular dynamics calculation is that the real scattering experiment can be put into reverse; that is, the process of getting from X-ray or neutron scattering intensities to spatial correlation functions can be reversed. Following Narten,[641] the X-ray scattering intensity has the form

$$I_x(\varkappa) = [\varkappa/(2f_H + f_O)^2]\{f_O^2\gamma_{OO} + 4f_Hf_O[\gamma_{OH} + j_0(\varkappa r_{OH})]$$
$$+ 2f_H^2[2\gamma_{HH} + j_0(\varkappa r_{HH})]\} \tag{121}$$

where f_O and f_H are the $\varkappa$-dependent atomic structure factors, j_0 is the spherical Bessel function, $r_{OH} = 1$ Å, $r_{HH} = 1.63299$ Å, and the γ's are the Fourier transforms of the correlation functions

$$\gamma_{\alpha\beta} = \varrho \int e^{i\mathbf{k}\cdot\mathbf{r}}[g_{\alpha\beta}(r) - 1] \, dr \tag{122}$$

which are inverted in the molecular dynamics calculations. In this way Stillinger and Rahman have calculated $I_x(\varkappa)$ for ST2 water (216 molecules) at 10°C and compared the function with the X-ray data obtained for water at 4°C; this comparison is shown in Fig. 50. A similar comparison with the neutron scattering intensities for $D_2O^{[642]}$ at 25°C is shown in Fig. 51. In all, the agreement achieved between the theoretical and experimental

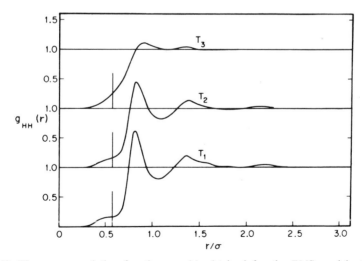

Fig. 49. The cross-correlation function $g_{OH}(r)$ obtained for the BNS model at three different temperatures, $T_1 = -8.2°C$, $T_2 = 34.3°C$, and $T_3 = 314.8°C$. The vertical line indicates the intramolecular OH distance. [Reproduced with permission from *J. Chem. Phys.* **57**, 1281 (1972).]

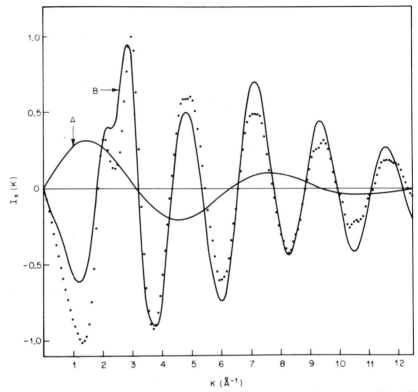

Fig. 50. A comparison of X-ray scattering intensities measured for water at 4°C, with the corresponding intensity derived from a molecular dynamics simulation using the ST2 potential, at 10°C. The latter uses independent atomic scattering factors (121). Experimental results are denoted by discrete points. Curves A and B are the theoretical results, with A just the intramolecular scattering, while B is the complete scattering prediction. [Reproduced with permission from *J. Chem. Phys.* **60**, 1545 (1974).]

results is really very remarkable. In Table VIII specific parameters in the $g_{OO}(r)$ function are compared for various models and temperatures; the experimental results are included. The change from BNS to ST2 has brought the molecular dynamics results into closer agreement with experiment. Stillinger and Rahman[832] used the ST2 potential to study liquid water under high compression at a density of 1.346 g/cm³ and three temperatures, 57, 97, and 148°C. The density was chosen to be that which occurs at the ice-VI–ice-VII–liquid triple point, where $T = 81.6$°C and the pressure is 22 kbar. The pressure developed in the molecular dynamics calculations is much less than the experimental value; in terms of the dimensionless pressure measure $(P/\varrho kT) - 1$ the pressures developed were 3.7 (57°C),

3.9 (97°C), and 4.0 (148°C), compared with the experimental value of 8.98. The discrepancy may be a measure of the neglect of three- and four-body potentials in the ST2 model. The known crystal structures of ice VI and VII provide a natural point of comparison for equilibrium structure obtained in the calculations. The oxygen–oxygen correlation function $g_{OO}(r)$ obtained for the sample under the highest degree of compression (148°C) is shown in Fig. 52, and the liquid has clearly developed an argonlike structure (Fig. 4) with about 11 nearest-neighbor molecules up to the first minimum. The first peak also shows a small degree of hydrogen bond compression (~0.03 Å). Major structural shifts are evident in these calculations and represent the outcome of competition between the strongly directional forces trying to establish an open hydrogen-bonded structure, and the forces of external compression to establish an efficient packing. If fragments of ice VI and VII crystals are present in the liquid state, one would expect to see evidence of minor peaks in $g_{OO}(r)$ at the characteristic second- and third-neighbor distances in the corresponding crystals. These distances are not distinguished in $g_{OO}(r)$. Such low-density clusters have often been invoked to explain the properties of liquid water; they are conjectured to be large

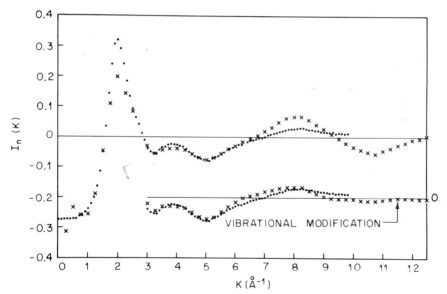

Fig. 51. A comparison of experimental neutron diffraction intensities for D_2O at 25°C (dots), with the corresponding intensity derived from a molecular dynamics simulation using the ST2 potential, at 10°C. The semiempirical vibrational modification is discussed by Stillinger and Rahman.[831] [Reproduced with permission from *J. Chem. Phys.* **60**, 1545 (1974).]

TABLE VIII. A Numerical Comparison of the Oxygen–Oxygen Pair Correlation Function Obtained at Various Temperatures Using the Molecular Dynamics Method on Several Model Potential Functions[a]

Model	T (°C)	R_1 (Å)	R_2 (Å)	R_3 (Å)	$r(M_1)$ (Å)	M_1	$r(M_2)$ (Å)	M_2	$r(m_1)$ (Å)	m_1
ST2	−3	2.63	3.13	4.08	2.83	3.32	4.55	1.2	3.43	0.60
ST2	10	2.63	3.17	4.17	2.84	3.16	4.65	1.18	3.53	0.63
ST2	41	2.63	3.21	4.31	2.86	3.02	4.74	1.08	3.53	0.75
ST2	118	2.63	3.34	4.86	2.86	2.64	5.29	1.03	3.74	0.84
BNS	8	2.45	3.12	4.01	2.73	2.98	4.49	1.28	3.37	0.55
BNS	53	2.47	3.14	4.09	2.76	2.57	4.72	1.18	3.45	0.64
BNS with mass scaling	31	2.42	3.21	4.34	2.80	2.55	4.79	1.22	3.77	0.70
HF	68	2.72	3.72	5.4	2.95	2.32	—	—	4.10	0.88
Exptl.	4	2.64	3.15	4.05	2.85	2.38	4.65	1.17	3.50	0.80
Exptl.	50	2.64	3.31	4.21[b]	2.85	2.28	4.75[b]	1.11[b]	3.64[b]	0.94[b]
Exptl.	100	2.64	3.52	4.10	2.88	1.86	4.50[b]	1.04[b]	3.36	0.90

[a] The experimental results are those of Narten and Levy.[644,831] The R_j are zeros of $g_{OO}(r) - 1$; the M_j are maxima, and m_1 is the minimum of $g_{OO}(r)$. In all cases the density is 1 g/cm³.
[b] Estimated.[831]

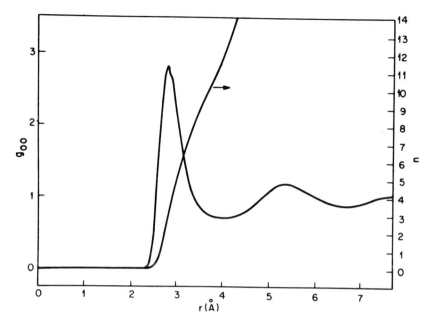

Fig. 52. The oxygen–oxygen pair correlation function $g_{OO}(r)$ obtained for a sample of ST2 water under high compression at $T = 148°C$ and $\varrho = 1.346$ g/cm³. [Reproduced with permission from *J. Chem. Phys.* **61**, 4973 (1974).]

low-density clusters persisting into the liquid state. In the molecular dynamics studies to date no evidence for the presence of such clusters has been found.

A molecular dynamics simulation of liquid water using the Hartree–Fock point charge model potential (78) has been carried out by Tucker and Wood.* This calculation employed only 64 molecules to develop 4000 phase points at 68°C and a density of 1 g/cm³. The Hartree–Fock molecules need a slightly smaller time step for the numerical integration compared to the BNS molecule. In this calculation the electron–electron correlation energy has been totally ignored, so there are no empirical dispersion terms (87)–(92) added to the pair potential. The molecular dynamics results for the three correlation functions are shown in Figs. 53–55, where they are compared with a sample of BNS water at 75°C. The pure Hartree–Fock potential fails to give a second-neighbor peak in $g_{OO}(r)$; a similar effect was found by Popkie et al.[712] in a Monte Carlo study (see Section 8). The first minimum in $g_{OO}(r)$ is much shallower than found in the empirical

* Unpublished calculations.

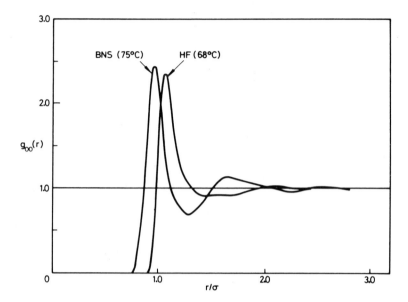

Fig. 53. The oxygen–oxygen pair correlated function $g_{OO}(r)$ obtained by Tucker and Wood (unpublished calculations) using the Hartree–Fock potential (78); $T = 68°C$, $\varrho = 1$ g/cm³.

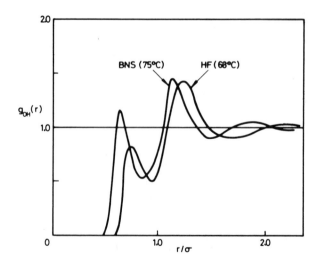

Fig. 54. The cross-correlation function $g_{OH}(r)$ obtained by Tucker and Wood (unpublished calculations) using the Hartree–Fock potential (78); $T = 68°C$, $\varrho = 1$ g/cm³.

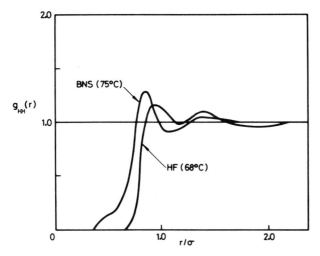

Fig. 55. The proton pair correlation function $g_{HH}(r)$ obtained by Tucker and Wood (unpublished calculations) using the Hartree–Fock potential (76); $T = 68°C$, $\varrho = 1$ g/cm³.

models above and in this the result seems closer to the experimental behavior (see Table VIII); the same is true of the height of the first peak. The clear absence of any neighboring peaks is presumably due to the neglect of dispersion terms in the potential. The first peak in $g_{OH}(r)$ corresponding to the hydrogen bonding is less prominent in the Hartree–Fock model than in the empirical models; however, it is still very evidently present and bears out the potential studies represented in Fig. 27.

Perhaps the most fascinating view of equilibrium structure in pure water appears in an analysis of the formation of hydrogen-bond networks present in the liquid state, and in a corresponding analysis of the partitioning of pair interaction energies over the whole energy range. The hydrogen-bond counter (20), (21), and (22), and the continuous interaction density function $p(V)$ (24) are used in this analysis.[355,830–832] Typical of these studies is the interaction density $p(V)$ obtained by Stillinger and Rahman[831] for the ST2 model at four temperatures, which is shown in Fig. 56. The divergence at $V = 0$ simply reflects the large number of molecular pairs at large separations; the most striking feature of $p(V)$ is the temperature-invariant point at $V \sim -4.0$ kcal/mol. Stillinger and Rahman interpret this feature to evidence a dynamic equilibrium process in which the formation and rupture of hydrogen bonds between neighboring molecules just above and below $V = -4$ kcal/mol is continuously occurring. An analysis of the hydrogen bonding itself requires the specification of a hydrogen-bond

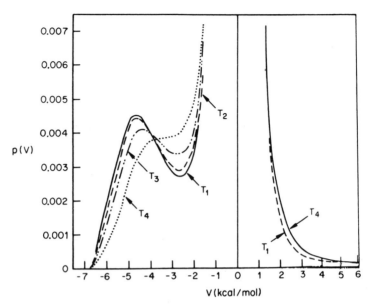

Fig. 56. The pair interaction distribution function $p(V)$ (24) obtained for the ST2 model at four temperatures; $T_1 = -3°C$, $T_2 = 10°C$, $T_3 = 41°C$, and $T_4 = 118°C$; and $\varrho = 1$ g/cm³. For positive V, the curves cluster closely together, so only the extreme temperature cases are shown. [Reproduced with permission from *J. Chem. Phys.* **60**, 1545 (1974).]

energy V_{HB} with the accompanying definition

$$V_e(i, j) < V_{HB} \qquad (i, j \text{ hydrogen bonded}) \qquad (123)$$

$$V_e(i, j) \geq V_{HB} \qquad (i, j \text{ not hydrogen bonded}) \qquad (124)$$

Rahman and Stillinger[56] cover a wide range of bonding schemes by using

$$V_{HB} = V_j = -0.577(j - 1) \text{ kcal/mol}, \qquad j = 1, 2, \ldots, 10 \qquad (125)$$

Their distributions showing the concentrations of molecules engaging in different numbers of hydrogen bonds simultaneously are shown in Fig. 57. The most significant feature of these histograms is the appearance of a single maximum in each case which is evidence against the so-called two-state theories of liquid water.[323,652] In a two-state structure one would expect to see two maxima centered on $n_{HB} = 0$ and $n_{HB} = 4$. This theme is again emphasized by Rahman and Stillinger[733] in an even more detailed account of hydrogen bonding; here an analysis of *networks* present in the liquid is given. The ST2 potential is used and a sample of 216 molecules at 10°C is used to count the average relative numbers of short circuit polygons which are defined in Fig. 58. Four values of V_{HB} are used, -2.121, -3.030, -3.939, and -4.848 kcal/mol to define a hydrogen bond. Hydrogen-bonded pairs of molecules are linked by the solid lines shown schematically in Fig. 58; the resulting linkage patterns form the basis of a network analysis in which average values of configurational parameters are determined over many such patterns. The results are shown in Table IX, and the number of polygons of various sizes that are obtained cannot be reconciled with the view that the liquid state of water is structurally similar to a disordered crystal. The picture of water that emerges from all these studies on equilibrium structure is one of a random network of hydrogen bonds uniformly filling configuration space.

Rahman et al.[734] have performed molecular dynamics calculations using a version of the central force model (Section 4.5). The molecules are no longer rigid, and the assembly consists of 648 particles (432 H and 216 O). An earlier equilibrium configuration of 216 rigid molecules was selected to start the assembly off, thus 216 water molecules are obtained initially by construction. The assembly now evolves under the three independent pair potentials $V_{OO}(r)$, $V_{OH}(r)$, and $V_{HH}(r)$. The equilibrium

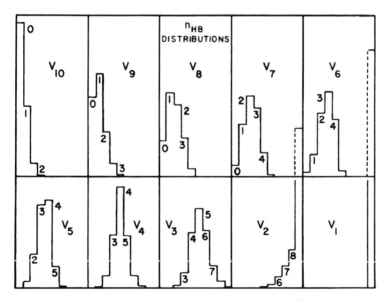

Fig. 57. Distribution of molecules according to the number of hydrogen bonds n_{HB} in which they engage. The set of cutoff energies V_{HB} which are used as alternative hydrogen-bond definitions are given in (125). [Reproduced with permission from J. Chem. Phys. 55, 3336 (1971).]

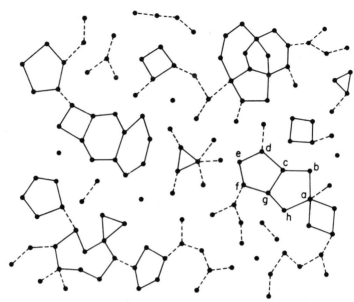

Fig. 58. Schematic hydrogen-bond network. The non-short-circuited polygons have been identified with solid lines; they are polygons with three or more edges and no pair of vertices are linked by a hydrogen-bond path with fewer edges than the minimal path within the polygon itself. Each molecule is denoted by a single vertex. Notice that some of the molecules may participate in no hydrogen bonds, while others can engage simultaneously in more than four. Polygon abcdefgh is short-circuited by bond cg and would not be counted, whereas polygons abcgh and cdefg would be counted. [Reproduced with permission from *J. Amer. Chem. Soc.* **95**, 7943 (1973).]

temperature obtained in this study was 22°C. In view of the long-range potentials now present because of the permanently charged particles Rahman *et al.*[734] consider it necessary to include particle–particle interactions with all orders of periodic images, not just the nearest-neighbor image cells.[570] This is in effect equivalent to an infinite cutoff radius (Section 3). The three spatial correlation functions obtained from about 2000 phase points are shown in Figs. 59–61, but now the *intramolecular* structure is included in $g_{OH}(r)$ and $g_{HH}(r)$ since protons in the same molecule are counted. The running coordination numbers n_{OH} and n_{HH} take the precise values of 2 and 1 at the intramolecular peak, demonstrating that all the "molecules" have remained intact and nonlinear throughout the elapsed time. Tetrahedronal bonding is again evident in $g_{OO}(r)$, and intramolecular peaks apart, the qualitative features are very similar to the earlier studies using rigid molecules. Rahman *et al.* again used the molecular dynamics correlation functions to compute the X-ray scattering intensities (121),

TABLE IX. Parameters Characterizing the Hydrogen-Bond Pattern in Liquid Water at 10°C and 1 g/cm[a]

V_{HB}, kcal/mol	-2.121	-3.030	-3.939	-4.848
$\langle b \rangle$	3.88	3.14	2.26	1.18
n_0	0	0.00331	0.0410	0.249
n_1	0.0026	0.029	0.18	0.415
C_3	0.05952	0.002976	0	0
C_4	0.1564	0.04663	0.007606	0
C_5	0.3459	0.1362	0.03406	0.001323
C_6	0.3548	0.1306	0.02447	0.0006614
C_7	0.3320	0.1280	0.01687	0
C_8	0.2715	0.1045	0.0234	0
C_9	0.1971	0.09854	0.01224	0
C_{10}	0.1118	0.09292	0.01422	0
C_{11}	0.07573	0.08664	0.005952	0

[a] The mean number of hydrogen bonds terminating at a molecule is $\langle b \rangle$, n_0 is the fraction of unbonded molecules, and n_1 is the fraction with precisely one bond. C_j stands for the number of non-short-circuited polygons per molecule with j edges (see Fig. 58). [Reproduced with permission from *J. Amer. Chem. Soc.* **95**, 7943 (1973).]

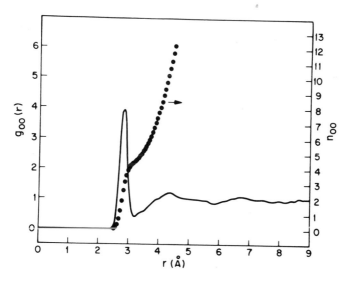

Fig. 59. The oxygen–oxygen pair correlation function $g_{OO}(r)$ obtained for the central force model (104)–(106), and the running coordination number $n_{OO}(r)$; $T = 22°C$, $\varrho = 1$ g/cm³. [Reproduced with permission from *J. Chem. Phys.* **63**, 5223 (1975).]

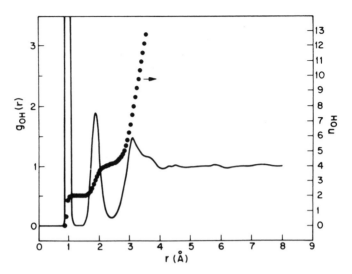

Fig. 60. The cross-correlation function $g_{OH}(r)$ obtained for the central-force model (104)–(106); $T = 22°C$, $\varrho = 1$ g/cm³. [Reproduced with permission from *J. Chem. Phys.* **63**, 5223 (1975).]

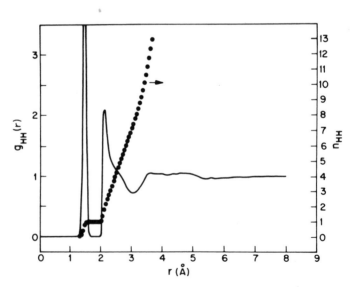

Fig. 61. The proton pair correlation function $g_{HH}(r)$ obtained for the central-force model (104)–(106); $T = 22°C$, $\varrho = 1$ g/cm³. [Reproduced with permission from *J. Chem. Phys.* **63**, 5223 (1975).]

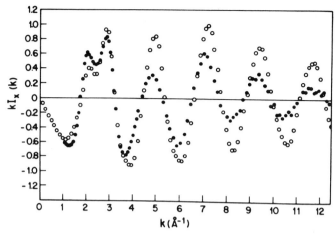

Fig. 62. A comparison between experimental (black circles) and theoretical (open circles) X-ray diffraction intensities (121) using the central-force model (104)–(106). [Reproduced with permission from *J. Chem. Phys.* **63**, 5223 (1975).]

and a comparison with Narten's data[641] at 20°C is shown in Fig. 62. It is particularly satisfying that the central-force model reproduces the double peak at 2.5 Å^{-1} which is a characteristic of X-ray diffraction from water.

6.2. Kinetic Properties

Unlike the Monte Carlo method, molecular dynamics can reveal the particle-averaged short time details of all types of molecular relaxation phenomena.[100] The simplest of these is the velocity autocorrelation function (48) and (49), which gives us the averaged view of center-of-mass diffusive motion from an initial position at $t = 0$. Thus in Fig. 3 we observe that for argon the atoms suffer one initial reverse in direction (back-scattering) before the autocorrelation function runs slowly to zero when all memory of the initial state at $t = 0$ is lost. In water the correlations in center-of-mass motion appear to be more structured over short times; Fig. 63 shows this effect found for BNS molecules at 34.3°C, and Fig. 64 shows the same effect observed for the ST2 potential at 10°C. The molecules are clearly executing pronounced oscillatory motion. At high temperatures (Fig. 65) all trace of these oscillations has disappeared. This phenomenon is consistent with the structural view that emerged in Section 6.1; at low temperatures the molecules act in a fairly rigid network of hydrogen bonds where a molecule is held in position by its neighbors, while at high temperatures the hydrogen-bond network is much less rigid and diffusion is

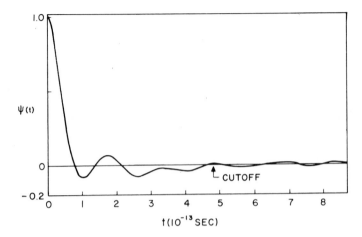

Fig. 63. The center-of-mass velocity autocorrelation function (48) obtained for the BNS model; $T = 34.3°C$, $\varrho = 1$ g/cm³. The cutoff locates the point beyond which statistical noise dominates the autocorrelation function. [Reproduced with permission from *J. Chem. Phys.* **55**, 3336 (1971).]

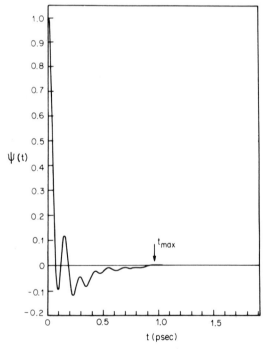

Fig. 64. The center-of-mass velocity autocorrelation function (48) obtained for the ST2 model; $T = 10°C$, $\varrho = 1$ g/cm³. [Reproduced with permission from *J. Chem. Phys.* **60**, 1545 (1974).]

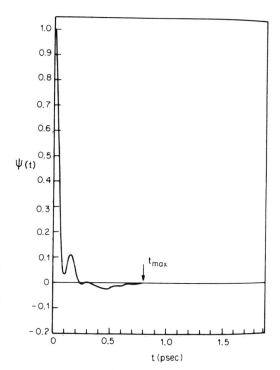

Fig. 65. The center-of-mass velocity autocorrelation function (48) obtained for the ST2 model; $T = 118°C$, $\varrho = 1$ g/cm³. [Reproduced with permission from *J. Chem. Phys.* **60**, 1545 (1974).]

allowed to proceed largely unchecked. The calculation of Tucker and Wood* on the pure Hartree–Fock potential (78) for water at 68°C yields a very different picture of short time mass diffusion (Fig. 66); here no evidence of rapid oscillations was found and probably reflects the rather weak hydrogen-bonding minima in this potential (See Fig. 27) which are unable to "hold" a molecule in position.

Although the self-diffusion coefficient may be determined from the velocity autocorrelation function via the relation

$$D = \frac{1}{3} \int_0^\infty \langle v_j(0) \cdot v_j(t) \rangle \, dt \qquad (126)$$

the short elapsed times that are available in the molecular dynamics data ($\sim 10^{-12}$ sec) make it generally more reliable to compute the asymptotic slope [at times for which $\langle v_j(0) \cdot v_j(t) \rangle$ has decayed to zero] of the mean square displacement

$$D \sim \langle [\Delta r_j(t)]^2 \rangle / 6t \qquad (127)$$

* Unpublished calculations.

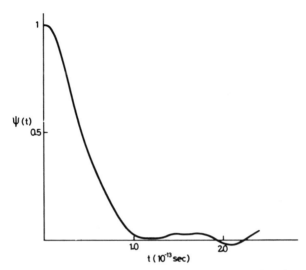

Fig. 66. The velocity autocorrelation function (48) obtained for the Hartree–Fock model potential (76) (Tucker and Wood, unpublished calculations); $T = 68°C$, $\varrho = 1$ g/cm³.

Some estimates of D obtained from molecular dynamics calculations using (127) are given in Table X. In the central-force model calculations[734] the translational autocorrelation function is resolved into nuclear components

$$A_\alpha(t) = \langle \mathbf{v}_{\alpha_j}(0) \cdot \mathbf{v}_{\alpha_j}(t) \rangle / \langle v_{\alpha_j}(0)^2 \rangle \tag{128}$$

where α denotes the nuclear species; the results for $A_O(t)$ and $A_H(t)$ are shown in Figs. 67 and 68. The corresponding diffusion coefficients for oxygen and hydrogen are slightly different, $D_H = 0.87 \times 10^{-5}$ cm²/sec, and $D_O = 0.73 \times 10^{-5}$ cm²/sec. If the molecules remain intact (which they do; see Figs. 60 and 61) the two coefficients should be identical.[734] The experimental value of D at this temperature (22.2°C) is 2.1×10^{-5} cm²/sec, and clearly the central force molecules are diffusing too slowly, unlike the rigid molecules for which the value of D exceeds the experimental values.

The directional relaxation of the water molecule dipoles $\boldsymbol{\mu}_j$ has also been examined. Hydrogen bonding forces will act so as to prevent a rapid tumbling of the molecules. The rotational retardation can be studied by defining an autocorrelation function for the *direction* of the dipole moment $\Gamma(t)$

$$\Gamma(t) = \langle \hat{\boldsymbol{\mu}}_j(0) \cdot \hat{\boldsymbol{\mu}}_j(t) \rangle$$
$$= \langle \cos \theta_j(t) \rangle \tag{129}$$

TABLE X. The Self-Diffusion Constant D Obtained Using Molecular Dynamics, at Various Temperatures, and for Various Model Potentials[a]

Model	T (°C)	D from molecular dynamics	D (exptl.) at 68°C
BNS	−8.2	1.5	0.74
BNS	34.3	4.2	2.9
BNS	314.8[b]	23	24[b]
ST2	−3	1.3	1.00
ST2	10	1.9	1.55
ST2	41	4.3	3.32
ST2	118	8.4	—
HF[c]	68	∼7.5	

[a] The units are 10^{-5} cm²/sec.
[b] Estimated.[830]
[c] Tucker and Wood, unpublished calculations.

where $\theta_j(t)$ is the angle through which the dipole direction of molecule j turns in a time t. The function $\Gamma(t)$ is a key function in the theory of dielectric relaxation.[647] This autocorrelation function obtained for the BNS model at three different temperatures (−8.2, 34.3, and 314.8°C) is shown

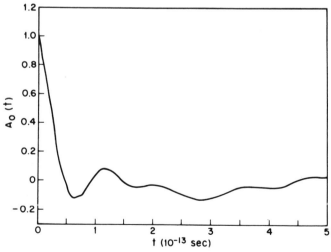

Fig. 67. The velocity autocorrelation function (48) for the oxygen atoms obtained for the central-force model (104)–(106); $T = 22$°C, $\varrho = 1$ g/cm³. [Reproduced with permission from *J. Chem. Phys.* **63**, 5223 (1975).]

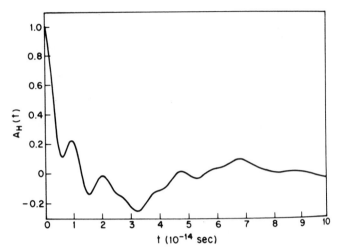

Fig. 68. The velocity autocorrelation function (48) for the hydrogen atoms obtained for the central-force model (104)–(106); $T = 22°C$, $\varrho = 1$ g/cm³. [Reproduced with permission from *J. Chem. Phys.* **63**, 5223 (1975).]

in Fig. 69 and clearly illustrates that rotational motion proceeds much more freely at high temperatures. The net dipole moment of the assembly can be represented by

$$\mathbf{M} = \sum_{i=1}^{N} \hat{\boldsymbol{\mu}}_i \tag{130}$$

the way that the molecular interactions act to quench the assembly's net dipole moment can be seen by computing

$$G_K = \langle \mathbf{M}^2 \rangle / N \tag{131}$$

For completely uncorrelated dipole directions $G_K = 1$, the values of G_K obtained for various models are shown in Table XI; a substantial quenching effect is evident. Stillinger and Rahman[831] have argued that g_k defined by

$$g_k = [(\varepsilon_0 + 2)(2\varepsilon_0 + 1)/9\varepsilon_0]G_K \tag{132}$$

is the factor introduced by Kirkwood in describing orientational correlations, and appearing in the dielectric formula[153,386]

$$\frac{\varepsilon_0 - 1}{\varepsilon_0 + 2} = \frac{\varepsilon_\infty - 1}{\varepsilon_\infty + 2} + \frac{4\pi\varrho\varepsilon_0 g_k \mu_i^2}{kT(2\varepsilon_0 + 1)(\varepsilon_0 + 2)} \tag{133}$$

where ε_∞ is the high-frequency dielectric constant. By using (132) and the

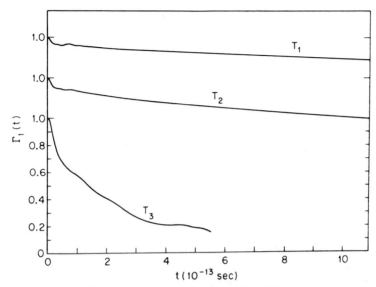

Fig. 69. The dipole direction correlation function $\Gamma(t)$ (129) obtained for the BNS model at the three temperatures $T_1 = -8.2°C$, $T_2 = 34.3°C$, and $T_3 = 314.8°C$, and $\varrho = 1$ g/cm³. [Reproduced with permission from *J. Chem. Phys.* **57**, 1281 (1972).]

experimental values of ε_0 and ε_∞ to obtain g_k, (133) can then be used to compute μ_l, the liquid phase mean dipole moment per molecule, and this can be compared with Onsager's result[672]

$$(\mu_l)_{\text{Onsager}} = [(2\varepsilon_0 + 1)(\varepsilon_\infty + 2)/3(2\varepsilon_0 + \varepsilon_\infty)]\mu_v \qquad (134)$$

TABLE XI. Dielectric Properties Obtained in Molecular Dynamics Calculations Using the ST2 Potential (One HF Result Included[a]) and the Relations (135) and (136)

Model	T (°C)	G_K	$\varepsilon_0{}^b$	$\varepsilon_\infty{}^b$	g_k	$(\mu_l)_{\text{md}}$	$(\mu_l)_{\text{Onsager}}$
ST2[831]	-3	0.18	88.9	1.78	3.66	1.84	2.30
ST2	10	0.15	83.8	1.78	2.88	2.06	2.30
ST2	41	0.16	72.8	1.77	2.68	2.09	2.29
	118	0.21	51.4	1.75	2.52	2.02	2.27
HF[a]	68	0.186	60.4	1.76	2.60	2.02	2.28

[a] Tucker and Wood (unpublished calculations).
[b] The static (ε_0) and optical frequency (ε_∞) dielectric constants are experimental values for 1 g/cm³.

using $\mu_v = 1.830$ D for the mean dipole moment in the vapor phase. These calculations are shown in Table XI, where the values for the Hartree–Fock molecules* are included.

7. MONTE CARLO SIMULATIONS OF PURE WATER

In this article the technical details of the Monte Carlo method are not included. In statistical mechanics this method is essentially synonymous with a specific algorithm first introduced by Metropolis et al.[619] for sampling points in the configuration space of the assembly according to a prescribed probability distribution, this being the classical canonical distribution function itself. The method is restricted to the evaluation of equilibrium properties; the molecular configurations generated by the sampling process do *not* represent "real" instantaneous molecular configurations; they are configurations that make a substantial contribution to the integrals over configuration space in (4).[159,310,376,919]

An early Monte Carlo calculation on *liquid* water at 25°C was reported by Barker and Watts[54] using the first empirical point charge model of liquid water proposed by Rowlinson,[769,770] and experimental properties of ice and steam. The results obtained for the center-of-mass radial correlation functions [roughly $g_{OO}(r)$] were rather poor in comparison with early experimental data by Narten et al.[643] Popkie et al.[712] included a small Monte Carlo simulation (27 molecules) in their work on obtaining the pure Hartree–Fock potential (78). These authors used this very small assembly to obtain the three spatial correlation functions $g_{OO}(r)$, $g_{OH}(r)$, and $g_{HH}(r)$ at 4, 25, and 75°C; at the latter temperature these results are in very close agreement with the correlation functions obtained by Tucker and Wood† in a molecular dynamics calculation using the same potential, and shown in Figs. 53, 54, and 55. Kistenmacher et al.[495] have performed Monte Carlo calculations on up to 125 molecules at 26°C using the Hartree–Fock potential (78) and the empirical corrections to the Hartree–Fock potential HF + K (88) and HF + L (89). The results obtained for the three spatial correlation functions are shown in Figs. 70–72 where the authors compare their Monte Carlo results with "calculations" of $g_{OO}(r)$, $g_{HH}(r)$, and $g_{OH}(r)$ using Narten's experimental X-ray data. These so-called "experimental" curves are, however, clearly incorrect since $g_{HH}(r)$ and $g_{OH}(r)$

---

* Tucker and Wood, unpublished calculations.
† Unpublished calculations.

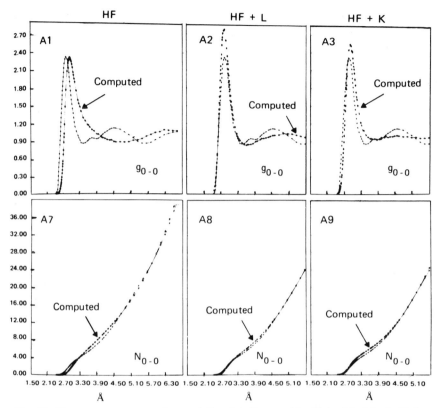

Fig. 70. The oxygen–oxygen pair correlation function $g_{OO}(r)$ obtained in Monte Carlo calculations for the HF, HF + L, and HF + K potentials [(78), (88), and (89)]; $T = 26°C$, $\varrho = 1$ g/cm³. The running coordination number $n_{OO}(r)$ is shown separately. The computed curves are compared with experimental results (see text). [Reproduced with permission from *J. Chem. Phys.* **60**, 4455 (1974).]

show very close H–H and O–H pairs, which contradicts the corresponding $g_{OO}(r)$ curve. These experimental curves are not very meaningful since they themselves require assumptions about liquid water structure. The pure Hartree–Fock potential fails to give the correct local order in water; the structure is in fact more akin to an atomic liquid (Fig. 5), the number of neighboring water molecules up to the first minimum being about 12 as in argon.* The empirically corrected versions HF + K and HF + L both show a significant effect in reducing the coordination to about 5 up to the first minimum and reflects the stronger directionality of these potentials

* The authors[495] report about 4.5 nearest neighbors, but this is using the experimental curve, not the Monte Carlo results to locate the first minimum.

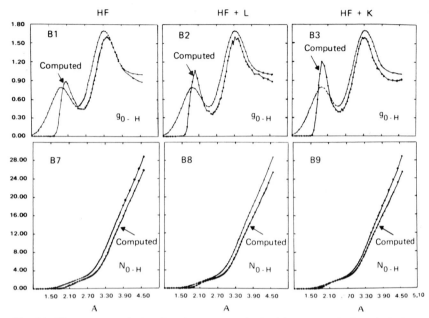

Fig. 71. The cross-correlation function $g_{OH}(r)$ obtained in Monte Carlo calculations. The details are as in Fig. 70. [Reproduced with permission from *J. Chem. Phys.* **60**, 4455 (1974).]

as evidenced in Fig. 27. It is rather disappointing that the HF + K and HF + L results for $g_{OO}(r)$ are not in closer agreement with experiments. In an attempt to improve the fidelity of the model Lie and Clementi[560] obtained a modified Hartree–Fock potential by using the version (79) coupled with new calculations on the dispersion force corrections, which now take the form

$$V_c(r) = \frac{c_1}{r_{56}^6} - \frac{c_2}{r_{56}^8} + \frac{c_3}{r_{56}^{10}} \qquad (135)$$

where in kcal/mol $c_1 = 922.781$, $c_2 = 17283.5$, and $c_3 = 24119.7$. These modifications parallel the ST2 modifications to the BNS model for the empirical models. The Monte Carlo results obtained using 343 water molecules at 25°C for the oxygen–oxygen correlation function are shown in Fig. 73 and clearly represent a very substantial improvement. These authors also compute the X-ray and neutron diffraction intensities using the Monte Carlo correlation functions and (121) and (122); the agreement between the experimental and theoretical intensities is very satisfactory. Kistenmacher *et al.*[495] and Lie and Clementi have obtained the internal energy and specific heat of the various Hartree–Fock models; the results are shown in Table XII.

Sarkisov et al.[225,780] have estimated an empirical model of the water molecule and performed Monte Carlo calculations on up to 64 molecules at 27, 57, and 87°C. These authors have built an atom–atom potential function with three spherically symmetric components

$$V_{HH}(r) = -57\left(\frac{\sigma}{r}\right)^6 + 4.2 \times 10^4 \exp\left(-4.86\frac{r}{\sigma}\right) \tag{136}$$

$$V_{OO}(r) = -259.4\left(\frac{\sigma}{r}\right)^6 + 7.77 \times 10^4 \exp\left(-4.18\frac{r}{\sigma}\right) \tag{137}$$

$$V_{OH}(r) = D\{1 - \exp[-n(r - r_0)]\}^2 - D \tag{138}$$

where $\sigma = 1$ Å, D is the depth of the potential within the O–H bond, r_0 is the equilibrium OH bond distance of 1.78 Å, and n is a parameter chosen to fit an empirical view of the electrostatic interaction of two water molecules (the units of $V_{\alpha\beta}$ are kcal/mol). Although the above model is of

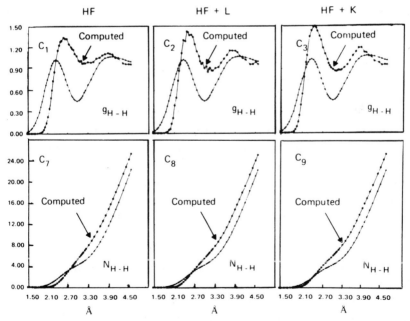

Fig. 72. The proton pair correlation function $g_{HH}(r)$ obtained in Monte Carlo calculations. The details are as in Fig. 70. [Reproduced with permission from J. Chem. Phys. **60**, 4455 (1974).]

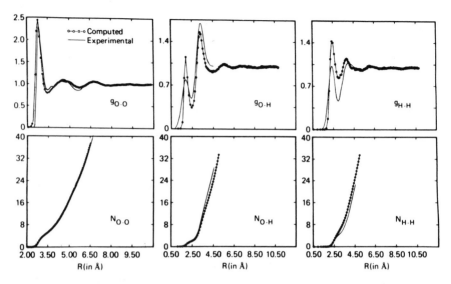

Fig. 73. The pair correlation functions $g_{\alpha\beta}(r)$ and the running coordination $n_{\alpha\beta}(r)$ obtained in a Monte Carlo calculation on 343 water molecules for the improved Hartree–Fock potential (78) and (79) and the dispersion interaction (135); $T = 25°C$, $\varrho = 1$ g/cm³. The theoretical result is compared with the experimental results obtained from X-ray scattering. [Reproduced with permission from *J. Chem. Phys.* **62**, 2195 (1975).]

a rigid molecule, and employs the experimental equilibrium geometry, the construction of (136)–(138) and the choice of n follows arguments similar to those used by Lemberg and Stillinger[544] to establish the $V_{OO}(r)$, $V_{OH}(r)$, and $V_{HH}(r)$ in the nonrigid central-force model (Section 4.5). The Monte Carlo calculations of Sarkisov *et al.* were performed using values of $n = 3$ Å⁻¹ and 5 Å⁻¹. The analysis of spatial structure in H_2O covers only a very small number of canonical configurations and is rather crude, the results being still clearly dependent upon the choice of initial configuration chosen in the Markov chain of canonical configurations. The results obtained for the internal energy, specific heat C_V, and free energy are included in Table XII.

Sections 6 and 7 complete the summary of attempts that have so far been made to simulate the behavior of liquid water; there are clearly two viewpoints represented in all this effort and "expenditure." Firstly the outlook of Kistenmacher, Clementi, and co-workers is to see if the properties of liquid water (and therefore other molecular liquids) can be obtained in a long sequence of very large computer studies which begin *ab initio* with the wave functions of monomer, dimer, and in some cases trimer and

TABLE XII. The Internal Energy and Specific Heat Obtained in Monte Carlo Calculations Using the Hartree–Fock Potential (78) and Empirical Corrections for the Electron Correlation Energy, (87)–(92), at 25°C

Potential	U (kcal/mol)	C_v (cal deg^{-1} mol^{-1})
HF	−6.9	18.0
HF + L	−9.2	15.0
HF + K	−12.2	19.0
HF + W	−8.0	—
HF + W + L	−8.1	—
HF + W + K	−8.6	—
exptl.	−8.8	18.0

The following results have been obtained using the potential (136)–(138):

T (°K)	U (kcal/mol)	U (exptl.)	C_v (cal mol^{-1} deg^{-1})	C_v (exptl.)
300	−9.9	−9.98	20.5	17.95
320	−9.5	−9.65	20.2	17.41
350	−9.0	−9.10	20.0	16.69

tetramer systems. In this work the intermolecular potential function has its origins in quantum mechanics, and the statistical mechanics is equivalent to the molecules moving classically in a quantum mechanical potential. At the end of this enormous use of computing resources stands the Monte Carlo simulation of liquid water by Lie and Clementi[560]; their results clearly show that an accurate thermodynamic account can indeed be obtained from *ab initio* calculations in quantum chemistry. The second approach is embodied in the work of Stillinger and Rahman, which seeks to build a "simple" effective model of a *single* water molecule using available experimental data (instead of Hartree–Fock data) to determine empirical parameters in a model potential function which purports to yield an effective pair potential energy of two water molecules. The molecular dynamics method in which these models have been used is far more searching than the Monte Carlo method in its test of the fidelity of a given model, and here the ST2 potential[831] also supports the claim that a very good fit over a wide range of properties of liquid water can be obtained using phenomenological models.

Both of these viewpoints have not escaped the dismissive description of "number crunching" by many chemists, and there are equally as many researchers ready to criticize such a use of computing facilities as a waste of resources. It should be emphasized that the problem of interpreting the many and diverse experiments on molecular liquids, and indeed on molecular systems in general, is more often than not a formidable theoretical problem. The much favored view that such experimental data can be adequately interpreted using "simple" macroscopic models is, in all probability, simply incorrect. Thus in the present studies no evidence has been found to support the "two state" and "iceberg" type theories of liquid water; in fact, strong evidence against such ideas has been found. It is at least clear that a great deal of experimental effort (and resources) has so far been unable to resolve the controversial and mutually contradicting multiplicity of theories of the equilibrium structure of liquid water and aqueous solutions. In this context the computer studies are surely neither expensive nor wasteful, rather they are an extremely valuable and unbiased source of information originating at a *molecular* level.

8. MOLECULAR DYNAMICS RESULTS FOR AQUEOUS SOLUTIONS

Molecular dynamics calculations on ionic solutions have been carried out from two different standpoints. The ST2 model of pure water[831] does not show large fluctuations to be occurring in the total electric moment of the sample; such fluctuations however exist in real water because of the large dielectric constant. The view expressed by Rahman[731] is that the essential features of dielectric behavior can be seen at the microscopic level over a few psec, thus avoiding the difficulty with macroscopic behavior. The microscopic behavior that is sought in a model calculation is the ability of the water molecules to maintain a spatial separation between two near and oppositely charged ions. The macroscopic view adopted by Heinzinger and Vogel[393, 394, 883, 884] is to introduce a potential model of an ionic solution as realistic as possible with macroscopic ionic concentrations.

Rahman[731] selected two ST2 water molecules (Section 4.2) in an equilibrium sample of 216 molecules and converted them into two oppositely charged ions in the scheme of (107)–(109). The assembly is now allowed to evolve under the new forces acting; the introduction of two charged particles in this way disturbs the equilibrium state and a period of readjustment is required for the system to relax into a new equilibrium state. The ionic separation in units of σ [see (108)] during this initial period of adjustment is shown in Fig. 74, in which the ionic separation oscillates about a distance close to its initial separation, and then shows a subsequent substantial increase in value before leveling off at a new "equilibrium" separation. Rahman repeated this "experiment" at different temperatures and with different initial conditions; the same abrupt rise in the ionic separation occurred in each case. The essential dielectric effect is beautifully illustrated by Rahman, who calculates the scalar products $\mathbf{r} \cdot \mathbf{F}^+$ and $\mathbf{r} \cdot \mathbf{F}^-$ over this time interval, where $\mathbf{r}$ is the ionic separation vector and $\mathbf{F}^+$ and $\mathbf{F}^-$ are the total forces acting on each ion. The ion–ion contribution to $\mathbf{r} \cdot \mathbf{F}^-$ and $\mathbf{r} \cdot \mathbf{F}^+$ is just the ion–ion potential (109); for the electrostatic force $\mathbf{r} \cdot \mathbf{F}^+ = \pm q^2 e^2 / r$. The plots of these scalar products are shown in Fig. 75, where we see how the products fluctuate about zero and how the quenching of the electrostatic attractive force between the ions occurs due to the dynamics of the neighboring water molecules. Rahman carried out a parallel experiment using two uncharged particles in the scheme of Section 5.1. The time dependence of the separation of the inert particles is compared with the ion–ion separations in Fig. 74: the oscillation periods of the inert particles are much longer, the inert molecules appearing to move very sluggishly in a cage of water molecules.

Such a system of two inert particles in water constitutes a useful model for studies of the hydrophobic interaction which plays such an important role in the maintenance of biological structures, e.g., cell membranes, folded globular proteins, etc. The results of the simulation, and also the statistical mechanical calculations described in Chapter 5, demonstrate that the details of this peculiar interaction are very much more complex than is usually allowed for in biochemistry texts.

Heinzinger and Vogel[393,394,883,884] have performed molecular dynamics calculations on aqueous solutions of the alkali halides using the system of pair potentials in (111)–(113), both with and without the switching function $S_{\mathrm{I}}(r)$, the effects of which can be seen in Fig. 32; in all cases these authors have employed a cutoff radius $r_c \sim 2.5$ box length. Most authors employing the molecular dynamics method on particles that experience long-range Coulombic forces find it necessary to use the full Ewald summation over all orders of periodic images to evaluate the force acting upon a single charged particle.[412,440,536,554,555,570,608,920] Thus Rahman and Stillinger included all orders of periodic images in their calculations using the nonrigid central force model of the water molecule.[734] The use of the switching function and the neglect of Ewald summations by Heinzinger and Vogel is a possible source of inaccuracies in their results. In the third paper of their series[883]

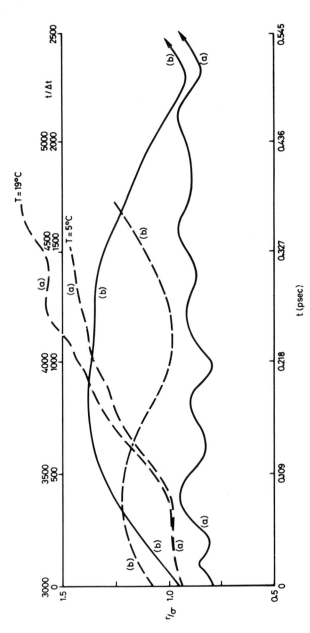

Fig. 74. The time dependence of the separation (a) between two ions in a sample of 216 ST2 water molecules (the ions are "created" at $t = 0$ as described in the text); (b) between two uncharged particles. The solid curves correspond to the range 0–2500 on the $t/\Delta t$ scale; the dashed curves are continuations of the solid curves and correspond to the range 3000–5000 on the $t/\Delta t$ scale. r is the distance between the particles and $\sigma = 2.82$ Å.

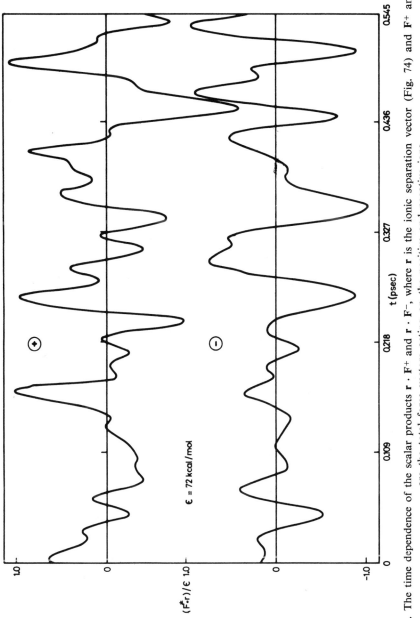

Fig. 75. The time dependence of the scalar products $\mathbf{r} \cdot \mathbf{F}^+$ and $\mathbf{r} \cdot \mathbf{F}^-$, where $\mathbf{r}$ is the ionic separation vector (Fig. 74) and $\mathbf{F}^+$ and $\mathbf{F}^-$ are the total force vectors acting on the positive and negative ions.

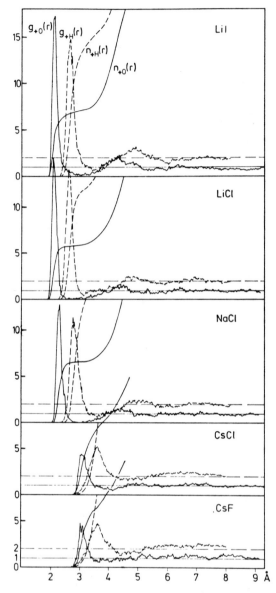

Fig. 76. The cation–oxygen and cation–hydrogen correlation functions, obtained in a molecular dynamics simulation of 2.2 molar solutions of alkali halides, using the potential functions (111)–(113). [Reproduced with permission from *Z. Naturforsch.* **31a**, 463 (1976).]

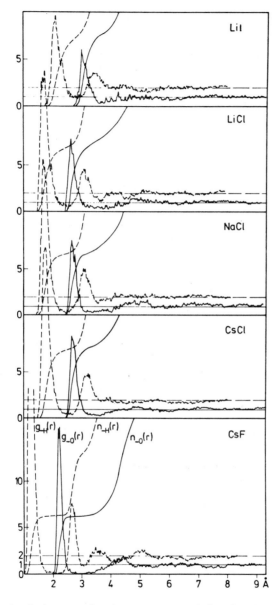

Fig. 77. The anion–hydrogen and anion–oxygen correlation functions corresponding to the curves shown in Fig. 76. [Reproduced with permission from *Z. Naturforsch.* **31a**, 463 (1961).]

Heinzinger and Vogel present a comparative study of 2.2 molar solutions of LiI, LiCl, NaCl, CsCl, and CsF, using 200 ST2 water molecules, eight cations, and eight anions; the cell dimensions and potential parameters employed in these calculations are given in Table IV.

The radial correlation functions g_{+O}, g_{-O}, g_{+H} and g_{-H} are shown in Figs. 76, and 77, where the running coordination numbers $n_{\pm O}$ and $n_{\pm H}$ are also shown [see eqns. (13) and (14)]. Second hydration shells are clearly present for the light nuclei Li$^+$ and F$^-$, but this feature disappears for the heavier ions. In the above calculations of Rahman[731] on water structure neighboring a single ion pair, a pronounced second hydration peak was evident for both ions which in this case were made as large as the water molecules by retaining the same value of the Lennard-Jones parameter σ for all interactions. The hydration numbers, which are defined to be the number of neighboring molecules up to the first minimum in $g_{\pm O}(r)$ are shown in Table XIII; characteristics of the correlation functions shown in Figs. 76 and 77 are given in Table XIV. The positions of first peaks in the $g_{\pm O}$ functions give the average ion–oxygen distances of the first hydration shell; a comparison of these results with Monte Carlo calculations (see below) and with experimental data is given in Table XV. One interesting feature of the correlation functions is that in all cases the distance between the first peaks of $g_{-H}(r)$ and $g_{-O}(r)$ is exactly 1 Å, and this is the distance of the OH bond in the ST2 water molecule, and from this it is concluded that a linear hydrogen bond is being formed with the anion.

TABLE XIII. Hydration Numbers Deduced from Radial Pair Correlation Functions[a]

	F$^-$	Cl$^-$	I$^-$
Li$^+$		7.4 ± 0.4	7.3 ± 0.3
		5.7 ± 0.2	7.1 ± 0.1
Na$^+$		6.7 ± 0.3	
		6.6 ± 0.1	
Cs$^+$	6.3 ± 0.1	7.9 ± 0.3	
	7.3 ± 0.7	8.3 ± 0.8	

[a] The first and second entries for each ion are for anion and cation hydration numbers, respectively. These results are obtained from the molecular dynamics calculations of Heinzinger and Vogel[394] and should be compared with the Monte Carlo results shown in Table XVI. [Reproduced with permission from Z. Naturforsch. **31a**, 463 (1976).]

TABLE XIV. Characteristic Values of Ion–Oxygen Pair Correlation Functions $g_{XO}(r)$ for the Five Different Alkali Halide Solutions Shown in Figs. 76 and 77[a]

X	Solute	R_1 (Å)	r_{M_1} (Å)	$g_{XO}(r_{M_1})$	R_2 (Å)	r_{m_1} (Å)	$g_{XO}(r_{m_1})$	r_{M_2} (Å)	$g_{XO}(r_{M_2})$
Li⁺	LiI	1.97	2.10	17.29	2.47	3.1 ± 0.2	0.12	4.30	2.24
	LiCl	1.95	2.06	15.30	2.34	2.72 ± 0.20	0.04	4.5 ± 0.1	1.6
Na⁺	NaCl	2.14	2.31	12.77	2.57	3.11	0.011	4.5 ± 0.1	1.6
Cs⁺	CsCl	2.92	3.10 ± 0.06	4.38	3.50	4.14	0.43	4.9 ± 0.3	1.3
	CsF	2.93	3.10 ± 0.04	≈ 4.7	3.45 ± 0.05	3.9 ± 0.2	0.5	5.30 ± 0.15	1.4
F⁻	CsF	2.09	2.22	≈15.8	2.49	2.92	0.03	4.46	1.88
	LiCl	2.50	2.68	7.77	3.05 ± 0.05	3.88 ± 0.10	0.25	5.1 ± 0.4	1.5
Cl⁻	NaCl	2.51	2.66	8.0	2.96	3.40 ± 0.10	0.18	5.0 ± 0.2	1.6
	CsCl	2.51	2.66	8.93	3.04	3.55 ± 0.10	0.27	4.85 ± 0.10	1.28
I⁻	LiI	2.88	3.02	≤6.0	3.47 ± 0.04	3.7 ± 0.1	0.30	4.8 ± 0.5	1.4
	LiI	2.54	2.90	2.87	3.37 ± 0.05	3.75 ± 0.15	0.75	5.45 ± 0.10	1.13
	LiCl	2.63	2.90 ± 0.04	2.81	3.33	3.75 ± 0.15	0.77	5.3 ± 0.3	1.15
O	NaCl	2.63	2.90	2.96	3.30 ± 0.04	3.73 ± 0.15	0.74	5.30 ± 0.10	1.13
	CsCl	2.66	2.91 ± 0.04	2.82	3.40	4.08 ± 0.15	0.82	5.40	1.09
	CsF	2.63	2.90	2.87	3.35	3.82	0.78	5.20 ± 0.10	1.07
	Pure water	2.63	2.85	3.13	3.19	3.53	0.72	4.70	1.13

[a] The parameters listed here are in the same scheme as given in Table VIII. The values of pure water are taken from Stillinger and Rahman[(83)] by interpolating between 10°C and 41°C. The uncertainties in R_i, $r(M_i)$, and $r(m_i)$ are less than ±0.02 if not stated otherwise. A general statement about the uncertainties in the values of g_{XO} is not possible; in each case it has been estimated from the statistical noise as shown in Figs. 76 and 77. [Reproduced with permission from Z. Naturforsch. 31a, 463 (1976).]

TABLE XV. Average Ion–Oxygen Distances of the First Hydration Shell Obtained from Molecular Dynamics Calculations, X-Ray Scattering Experiments, Crystal Radii Data, and Monte Carlo Calculations[a] (see Table XVI)

Ion	Molecular dynamics	X-ray[b]	Crystal[c] radii	Monte Carlo
Li+	2.08	1.95, 2.1	2.35	1.75
Na+	2.31	—	2.66	—
Cs+	3.10	3.14	3.25	—
F−	2.22	—	2.64	2.3
Cl−	2.67	3.10, 3.15	—	—
		2.9–3.25	—	—
I−	3.02	3.69	3.46	—

[a] See Refs. 185 and 341.
[b] See Refs. 103, 538, 556, and 646.
[c] See Ref. 767.

9. MONTE CARLO RESULTS FOR AQUEOUS SOLUTIONS

The extensive Hartree–Fock calculations on ion–water, ion–ion, and cation–anion–water systems discussed in Section 5.2 have been used in a Monte Carlo simulation of LiF by Watts et al.[897] and Fromm et al.[341] These two papers differ only in the size of the assembly employed, the latter authors using the larger assembly of 200 water molecules and the single ion pair Li+–F−. In these calculations the ions are held in fixed positions and the equilibrium structure is the structure determined only for the water molecules in the external field of two ions. With this restriction there is, of course, no need to included the ion–ion potential. The potential functions used in these calculations are the Hartree–Fock pair potential (78) and (79), and (135) for the interactions between water molecules, and the Hartree–Fock *triplet* potential (119) for the triplets Li+–F−–H_2O. The water temperature in these calculations is 25°C.

The calculations were performed for three different separations between the Li+ and F− ions, 2, 6, and 10 Å. The equilibrium structure is presented in a nonstandardized fashion. Fromm et al. define simple unnormalized local density functions

$$f_\alpha(x, r) = \langle \varrho_\alpha(x, r) \rangle \tag{139}$$

where $\alpha = 0$, or H and ϱ_α is the density of type-α nuclei in a solid annulus

between r, and $r + \delta r$ ($r^2 = y^2 + z^2$) and x and $x + \delta x$ in a single Monte
Carlo configuration; thus $f_\alpha(x, r)$ is this local density simply averaged over
all the Monte Carlo configurations. The local density functions $f_\alpha(x, r)$
describe the average local density of a cluster of 200 water molecules in the

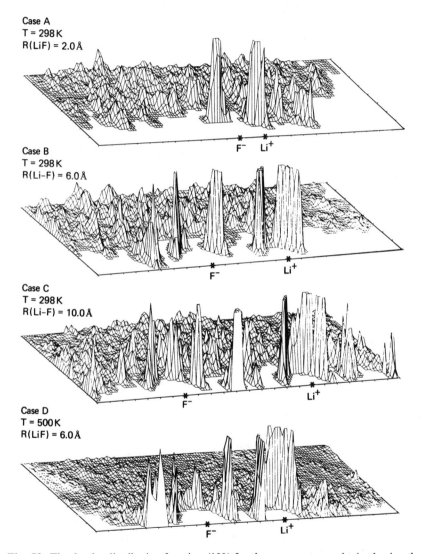

Case A
T = 298 K
R(LiF) = 2.0 Å

F⁻ Li⁺

Case B
T = 298 K
R(Li–F) = 6.0 Å

F⁻   Li⁺

Case C
T = 298 K
R(Li–F) = 10.0 Å

F⁻   Li⁺

Case D
T = 500 K
R(LiF) = 6.0 Å

F⁻   Li⁺

Fig. 78. The density distribution function (139) for the oxygen atoms obtained using the
three-body potention function (119) for 200 water molecules and the ion pair Li⁺–F⁻
at three *fixed* separations. The figure is a three-dimensional representation of the data.
[Reproduced with permission from *J. Chem. Phys.* **62**, 1388 (1975).]

external field of a single ion pair; the oxygen density distributions are rather dramatically illustrated in a three-dimensional form in Fig. 78 which gives an instant visual impression of the hydration shell structure. A first hydration shell is evident in each case, and at 25°C a second shell structure is clearly present. The secondary hydration shell is seen to disappear at the single high-temperature calculation which was performed at $T = 227$°C. The running coordination numbers $n_H(r)$ and $n_O(r)$ of hydrogen and oxygen atoms inside a spherical volume of radius r centered on the positive and negative ions obtained from the distributions in Fig. 78 are shown in Fig. 79. The formation of the first hydration shell is evidenced by the plateau formed in these running coordination numbers.

Identical Monte Carlo calculations on the fixed ion pairs F^--Na^+ and F^--K^+ are reported by Clementi[184] and Clementi et al.[185]; the results for the spatial distribution of the water molecules are qualitatively the same as those for the Li^+-F^- system shown in Figs. 78 and 79. Fromm et al.[341]

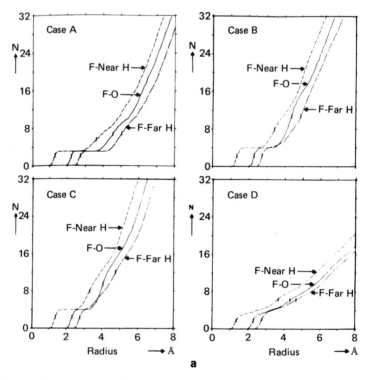

Fig. 79. (a) The running coordination numbers for each of the two hydrogens of H_2O, and for the oxygen centered on the fluorine nucleus, corresponding to the distributions shown in Fig. 78. (b) The running coordination numbers for hydrogen and oxygen atoms

have analyzed the average spatial distributions of the water molecules in the four cases represented in Fig. 78 to put forward estimates of the average radii of the hydration shells and average water molecule coordination within the first and second shell structure. The results are shown in Table XVI, where additional results obtained for the ion pairs F^--Na^+, and F^--K^+ at a separation of 10 Å are included.[184,185]

Dashevsky and Sarkisov[225] have undertaken Monte Carlo calculations on highly phenomenological models of nonpolar solutes in water. The two cases considered are simple hard spheres and a spherically symmetric potential to represent the interaction of two methane molecules, where a system of atom–atom potentials V_{CH}, V_{HH}, and V_{CC} have been averaged over all orientations of two methane molecules.[473] The result is a simple potential function in terms of the center-of-mass separation of two methane molecules. No specific potentials between the solute and water molecules are considered; these are rather unsatisfactorily obtained using the so-called

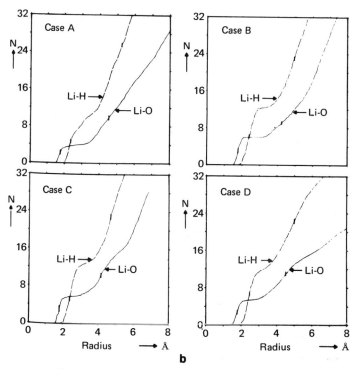

b

centered on the lithium nucleus corresponding to the distributions shown in Fig. 78. Case A: $T = 298°K$, $R(Li–F) = 2.0$ A. Case B: $T = 298°K$, $R(Li–F) = 6.0$ A. Case C: $T = 298°K$, $R(Li–F) = 10.0$ A. Case D: $T = 500°K$, $R(Li–F) = 6.0$ A. [Reproduced with permission from *J. Chem. Phys.* **62**, 1388 (1975).]

TABLE XVI. The Outer Radii (Å) of the First and Second Hydration Shells, and the Radii of the Maximum Density in the Distribution of Hydrogen and Oxygen Atoms around the Ions F$^-$ and Li$^+$ (see Fig. 78)[a]

Case A	Case B	Case C	Case D	Radii
		Fluorine		
1.3	1.3	1.3	1.3	H(1)–F
2.2	2.3	2.3	2.3	O–F
2.7	2.8	2.8	2.8	H(2)–F
2.9 ± 0.1	2.9 ± 0.1	2.8 ± 0.1	3.2 ± 0.1	First shell
3.8 ± 0.2	3.5 ± 0.2	3.4 ± 0.2	3.7 ± 0.2	H(1)–F
4.7 ± 0.2	4.4 ± 0.2	4.2 ± 0.2	4.4 ± 0.2	O–F
5.1 ± 0.3	5.1 ± 0.3	5.0 ± 0.3	4.9 ± 0.3	H(2)–F
5.4 ± 0.5	5.3 ± 0.5	5.2 ± 0.5	5.4 ± 0.5	Second shell
		Lithium		
1.75	1.75	1.75	1.75	O–Li
2.3	2.4	2.4	2.5	H–Li
3.5 ± 0.2	3.2 ± 0.2	3.0 ± 0.2	3.2 ± 0.4	First shell
5.0 ± 0.5	4.3 ± 0.5	4.5 ± 0.5	4.8 ± 0.5	O–Li
5.1 ± 0.5	4.5 ± 0.5	4.8 ± 0.5	5.3 ± 0.5	H–Li
5.5 ± 0.5	5.5 ± 0.5	5.5 ± 0.5	6.8 ± 0.5	Second shell

Coordination and Shell Radii Deduced from Ion Pairs at a Separation of 10 Å

Ion	First shell		Second shell	
	Radius (Å)	Coordination	Radius (Å)	Coordination
Li$^+$	2.7 ± 0.1	5.4 ± 0.7	5.1 ± 0.4	13.9 ± 2.7
Na$^+$	3.4 ± 0.3	6.0 ± 1.1	5.9 ± 0.3	17.1 ± 2.8
K$^+$	4.0 ± 0.3	7.2 ± 1.2	5.4 ± 0.3	12.6 ± 4.0
F$^-$	3.0 ± 0.5	4.5 ± 0.7	4.7 ± 0.4	15.4 ± 1.7
Cl$^-$	3.9 ± 0.5	5.1 ± 0.8	6.2 ± 0.5	17.9 ± 2.4

[a] H(1) designates the hydrogen of H_2O nearer to the F$^-$ (hydrogen bonded to F$^-$), and H(2) designates the hydrogen of H_2O further from F$^-$. Also shown below are the coordination numbers and hydration shell radii obtained for Li$^+$, Na$^+$, K$^+$, F$^-$, and Cl$^-$. (Compare Table XIII). [Reproduced with permission from *J. Chem. Phys.* **62**, 1388 (1975), and *Lecture Notes in Chemistry*, Vol. 2, Springer-Verlag, New York/Berlin (1976).]

"mean geometric rule."[408] The water–water interactions are those given in (136)–(138) between atom pairs OO, OH, and HH. The Monte Carlo calculations are performed on assemblies of 63 water molecules and one solute molecule and 62 water molecules and two solute molecules. In a substantial part of these calculations the periodic boundary conditions were removed to economize on computing time; they were replaced by hard walls. It is likely that this will have significantly influenced the water molecule structure throughout the box volume, although the authors argue that the first hydration sphere is small enough not to be affected by the boundary structuring of the water molecules. Dashevsky and Sarkisov have calculated the partial thermodynamic properties of solvation, these being the internal energy U_{solv}, the Helmholtz free energy A_{solv}, and the entropy S_{solv} defined according to

$$P_{solv} = P_{soln} - \frac{N-k}{N} P_N \qquad (140)$$

where P is the thermodynamic property, P_{soln} is the value of P for a solution of k solutes in $N-k$ solvent molecules, and P_N is the value of P for the N-molecule pure solvent. These quantities are shown in Table XVII for a single hard sphere solute of radius varying between 1–5 Å; the results are plotted in Fig. 80, where they are compared with those predicted by a theory due to Pierotti.[706,707] Here we see a fundamental difference between the Monte Carlo calculations and the theoretical predictions in the behavior of the internal energy of solvation, where the Monte Carlo calculations indicate that cavity formation is energetically favorable.

TABLE XVII. Some Thermodynamic Functions of Solutions Obtained by Dashevsky and Sarkisov[224] Using (140), with $N = 64$, and a Hard-Sphere Solute Particle of Varying Diameter d (Å)[a]

d	U_{64}	F_{64}	U_{soln}	A_{soln} (kcal/mol)	U_{solv}	A_{solv}	S_{solv}
1	−364.5	−204.8	−359.0	−199.7	−0.2	1.9	−5.7
2	−364.5	−204.8	−359.0	−198.5	−0.2	3.1	−9.7
3	−364.5	−204.8	−359.0	−196.8	−0.2	4.8	−15.3
4	−364.5	−204.8	−359.3	−194.7	−0.5	6.9	−21.3
5	−364.5	−204.8	−359.3	−191.8	−0.5	9.8	−31.0

[a] Reproduced with permission from *Molec. Phys.* **27**, 1271 (1974).

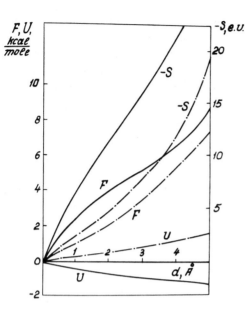

Fig. 80. The thermodynamic functions of solvation (140) obtained by Dashevsky and Sarkisov[225] for 63 water molecules interacting via the atom–atom potential scheme (136)–(138) and one hard-sphere solute molecule of varying diameters. [Reproduced with permission from *Mol. Phys.* **27**, 1271 (1974).]

APPENDIX

The computer program below is illustrated by the flow diagram in Fig. 9, and is here coded throughout in CDC-7600 FORTRAN. The program is for a molecular dynamics simulation of an assembly of BNS water molecules (see Section 4.1), although, as will be evident below, the program can be very simply altered to deal with any point charge model of a rigid molecule. The program is due to Dr. P. F. Fox and is illustrated in modular form in Fig. 9. The subroutines in Fig. 9 are briefly described below.

MOLD: Molecular assembly defined, and constants set for the integration algorithm.

START: Initiates the integration of the differential equations over a new time step, and calls on three subroutines: CONFGN, IMAGES, and SETLST.

CONFGN: Supplies a previous phase point of the assembly or a random generator to establish an initial zero temperature control.

IMAGES: Sets up the array IMCUBE; a 27×6 array with special word packing (available on CDC-7600). Each row in this array is a 6-vector with 0 or 1 elements, to denote the shift operation on the coordinates of the central cell to produce the corresponding image cell coordinates. Thus $\left(\begin{smallmatrix} x & y & z \\ 0 & 0 & 0 \end{smallmatrix} \quad \begin{smallmatrix} -x & -y & -z \\ 1 & 0 & 0 \end{smallmatrix}\right)$ shifts the x coordinate by $-L$, the box length.

SETLST: Sets up the neighborhood lists of each molecule (see below). Each molecule has an assigned list of the neighboring molecules that reside inside a spherical volume of radius RB. This listing for all molecules is updated every Δn time steps, where Δn (see Fig. 9) is an estimate of the number of time steps over which molecules can traverse the skin depth RB − RC (RC = cut off radius). SETLST calls on three subroutines, SETA, CRD55, and ADDLST.

SETA: Sets up the rotation matrix and its transpose for a given set of Euler angles φ_i, θ_i, ψ_i.

CORDS5: Determines the laboratory coordinates of the oxygen atom [$n = 5$ in (73)].

ADDLST: Adds a molecule to the bubble lists in SETLST if molecule is inside the bubble radius RB.

The main force loop is now entered, here all the forces and torques are determined to provide the right hand sides of (59)–(61) and the forces $\mathbf{F}_i$ in (56). Program enters DIFSUB which is the general algorithm given by Gear[346] and calls on its main subroutine DIFFUN.

DIFFUN: Sets up input data needed for the numerical integration; calls on BEGIN.

BEGIN: Provides partial phase point data which is already known, namely, $\mathbf{r}_i$ and $\mathbf{v}_i$.

SETA: (See above.)

CORDS5: (See above.)

CRDS14: Evaluates the coordinates of the point charges 1, 2, 3, and 4 in (73) in the laboratory frame.

UNPACK: Works out the neighborhood list molecules from the special word packing used in SETLST (CDC-7600).

Program enters FXN.

FXN: Evaluates the forces and torques for each molecule using those molecules inside the neighborhood list of a given molecule. Program leaves DIFSUB after successful integration step and returns to START via OUT.

OUT: An output channel for new phase point.

CHKOUT: Checks for escapes from box; in effect the box dimensions are reset after every Δn steps, (see SETLST).

Program Variables

XLAB:	Laboratory coordinates.
XLOC:	Coordinates in molecular frame (see Fig. 11).
XPCON:	Standard configuration coordinates, $\theta = \varphi = \psi = 0$ (see Fig. 11).
XIX:	Moment tensor (I_1, I_2, I_3).
ECH:	Charge on the electron.
RU, RL:	BNS model parameters (see Section 4.1).
T:	Initial time.
H, Δt, HMIN, HMAX, EPS:	Parameters in Gear's algorithm.[346]
BOXL:	Box length.
BOXL2:	1/2 box length.
RB:	Bubble radius.
RC:	Cutoff radius.
Y, DY, SAVE, ERROR, AVE, YMAX:	Parameters in Gear's algorithm.
LIST:	Array for SETLST.
NMOL:	Number of molecules.
NSTEPS:	Number of steps to be run.
NGAP:	Δn.
MAXDER:	A parameter in Gears Algorithm.[346]
VB, AR:	Average velocity (estimate) to determine Δn.
NPRINT:	Print out (if required) every N steps.

The symmetry of the periodic boundary conditions is employed in the preparation of the neighborhood lists in SETLST, and in the additive calculations of forces and torques in the main force loop $F \times N$. The use of symmetry is briefly described in Fig. A.1, which is based upon the periodic system shown in Fig. 1 and here illustrated in a two-dimensional scheme. The cells are arranged in the following complementary pairs:

[0, 0]   [1, 5]   [2, 6]   [3, 7]   [4, 8]   [5, 1]   [6, 2]   [7, 3]   [8, 4]

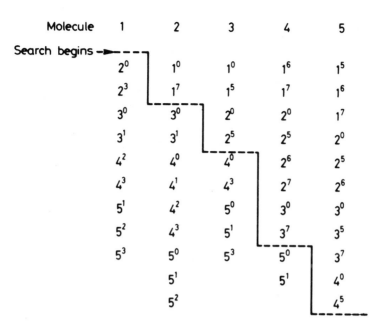

Fig. A.1.

The notation adopted in Fig. 1 is such that 4^7 is the image of particle 4 in the image cell 7. The neighborhood lists (inside a bubble radius RB) for the system in Fig. 1 is shown in Fig. A.1. Only the elements in the lists below the dotted line need be determined by searching; the elements above this line are related to those below. Thus if 3^1 is in the neighborhood list of molecule number 1; 1^5 is in the neighborhood list of molecule number 3, 5 being the complementary image cell to image cell 1. The central cell is self-complementary. The general rule is, if n^m is in the neighborhood list of molecule j then $j^{m'}$ is in the list of molecule n, where m' is the complementary cell to m.

```
PRUGRAM MOLD(INPUT,OUTPUT,DSTUP5,DSTRT5,TAPE5=INPUT,
1TAPE6=OUTPUT,TAPE7=DSTUP5,TAPE3=DSTRT5)
COMMON/GEOM/XLAB(2,5,3),XLOC(2,5,3),XPCON(5,3)
COMMON/CONMOL/XIX(3),ECH,RU,RL
COMMON/PARAMS/T,H,HMIN,HMAX,EPS
COMMON/CONSYS/BOXL,BOXL2,RB,RBSQ,RANGE
COMMON/VARBLS/Y,DY
COMMON/WORKRS/SAVE,ERRUR,AVE,YMAX
COMMON/LISTER/LIST,ITRAK,INT1,INT2
COMMON/ENEHGY/PE,ETRAN,EROI,PEIJ
DIMENSION Y(8,1500),DY(1500)
DIMENSION SAVE(10,1500),ERROR(1500),AVE(1500),YMAX(1500)
DIMENSION LIST(6250),ITRAK(125),INT1(125),INT2(250)
```

```
      LEVEL 2,Y,DY
      LEVEL 2,SAVE,ERROR,AVE,YMAX
      LEVEL 2,LIST,ITRAK,INT1,INT2
      READ(5,1)NMOL,NSTEPS,NGAP,MAXDER
      READ(5,2)T,H,HMIN,HMAX,EPS
      READ(5,3)BOXL,RANGE,VBAR
      READ(5,4)NPRINT,NTAKOF
    1 FORMAT(4I6)
    2 FORMAT(5E10.1)
    3 FORMAT(3E10.1)
    4 FORMAT(2I6)
      H0=H
      WRITE(6,9000)T,H,HMIN,HMAX,EPS
 9000 FORMAT(1X,5(1X,E15.7))
      NMOL=115
 5000 NMOL=NMOL+10
      IF(NMOL.GT.125) GO TO 5001
      T=0.0
      H=H0
      NMOL50=NMOL*50
      NMOL2=NMOL*2
      SKIN=0.0
      RB=RANGE+SKIN
      RBSQ=RB*RB
      BOXL2=BOXL/2.0
      NEQNS=NMOL*12
      XTEMP2=SQRT(2.0)
      XTEMP3=SQRT(3.0)
      XTRAT=XTEMP2/XTEMP3
      XREP3=1.0/XTEMP3
      XPCON(1,1)=-XTRAT/2.82
      XPCON(1,2)=0.0
      XPCON(1,3)=(10.0*XTEMP3)/(27.0*2.82)
      XPCON(2,1)=0.0
      XPCON(2,2)=XPCON(1,1)
      XPCON(2,3)=-1.0*(8.0*XTEMP3)/(27.0*2.82)
      XPCON(3,1)=-XPCON(1,1)
      XPCON(3,2)=0.0
      XPCON(3,3)=XPCON(1,3)
      XPCON(4,1)=0.0
      XPCON(4,2)=XPCON(3,1)
      XPCON(4,3)=XPCON(2,3)
      XPCON(5,1)=0.0
      XPCON(5,2)=0.0
      XPCON(5,3)=XTEMP3/(27.0*2.82)
      XIX(1)=(8.0*13.0)/(9.0*108.0*2.82*2.82)
      XIX(2)=8.0/(9.0*27.0*2.82*2.82)
      XIX(3)=8.0/(9.0*12.0*2.82*2.82)
      SEPSIG=SQRT(1.41282)
      ECM=(16.021*2.998)/SEPSIG
      RU=3.1877/2.82
      RL=2.0379/2.82
      CALL START(NMOL,NMOL50,NMOL2,NEQNS,NSTEPS,NGAP,MAXDER,Y,DY,SAVE,
     1ERROR,AVE,YMAX,LIST,ITRAK,INT1,INT2,NPRINT,BOXL,NTAKOF)
      GOTO 5000
 5001 CONTINUE
      STOP
      END




      SUBROUTINE START(NMOL,NMOL50,NMOL2,NEQNS,NSTEPS,NGAP,MAXDER,Y,DY,
     1SAVE,ERROR,AVE,YMAX,LIST,ITRAK,INT1,INT2,NPRINT,BOXL,NTAKOF)
      COMMON/PARAMS/T,H,HMIN,HMAX,EPS
      COMMON/IPATRN/IMCUBE(27,6)
```

```
      DIMENSION Y(8,NEQNS),DY(NEQNS)
      DIMENSION SAVE(10,NEQNS),ERROR(NEQNS),AVE(NEQNS),YMAX(NEQNS)
      DIMENSION LIST(NMOL50),ITRAK(NMOL),INT1(NMOL),INT2(NMOL2)
      LEVEL 2,Y,DY
      LEVEL 2,SAVE,ERROR,AVE,YMAX
      LEVEL 2,LIST,ITRAK,INT1,INT2
      CALL CNFGIN(NEQNS,Y,BOXL,NMOL,NTAKOF)
C*****
C*****
      DO 200 I=1,NMOL
      IMBLK=(I-1)*12
      IMBLK1=IMBLK+1
      IMBLK6=IMBLK+6
      IMBLK7=IMBLK+7
      IMBLKT=IMBLK+12
      WRITE(6,201)I,(Y(1,J),J=IMBLK1,IMBLK6)
  200 WRITE(6,201)I,(Y(1,J),J=IMBLK7,IMBLKT)
  201 FORMAT(1X,I4,1X,6(E15.7))
C*****
C*****
      CALL IMAGES
C*****
C*****
      DO 100 K=1,NEQNS
  100 YMAX(K)=1.0
      JSTART=0
      MF=0
      NNEXT=NGAP
      NPNEXT=NPRINT
      DO 10 ISTEP=1,NSTEPS
      WRITE(6,900)
  900 FORMAT(1X,11HINTO DIFSUB)
      NRGY=0
      CALL DIFSUB(NEQNS,NMOL,NMOL50,NMOL2,Y,SAVE,ERROR,AVE,YMAX,LIST,
     2ITRAK,INT1,INT2,T,H,HMIN,HMAX,EPS,MF,KFLAG,JSTART,MAXDER,IFORBD,
     3NRGY)
      WRITE(6,901)
  901 FORMAT(1X,13HOUT OF DIFSUB)
      IF(KFLAG.NE.1.OR.IFORBD.NE.0) GO TO 20
      IF(ISTEP.EQ.NNEXT) GO TO 30
      IF(ISTEP.NE.1.AND.ISTEP.NE.NPNEXT) GO TO 50
      CALL OUT(ISTEP,T,NEQNS,NMOL,Y)
      NPNEXT=NPNEXT+NPRINT
   50 CONTINUE
      GO TO 10
   30 CALL CHKOUT(ISTEP,T,NEQNS,NMOL,Y)
      NNEXT=NNEXT+NGAP
   10 CONTINUE
      GO TO 40
   20 WRITE(6,21)KFLAG,IFORBD
   21 FORMAT(1X,6HKFLAG=,I6,5X,7HIFORBD=,I6)
      GO TO 41
   40 CONTINUE

      CALL LPOINT(T,NMOL,Y,NEQNS)
   41 CONTINUE
      RETURN
      END




      SUBROUTINE CNFGIN(NEQNS,Y,BOXL,NMOL,NTAKOF)
      DIMENSION Y(8,NEQNS)
      LEVEL 2,Y
      PI=3.1415926535900
      IF(NTAKOF.NE.0) GO TO 200
```

```
C*****
C*****
      DO 2 I=1,NEQNS
    2 Y(1,I)=0.0
      I=1
      IBLK=(I-1)*12
    1 XX=10.0+BOXL*RANF(1)
      YY=10.0+BOXL*RANF(1)
      ZZ=10.0+BOXL*RANF(1)
      IF(I.EQ.1) GO TO 222
      IM=I-1
      DO 3 J=1,IM
      JBLK=(J-1)*12
      DSQ=(XX-Y(1,JBLK+1))**2+(YY-Y(1,JBLK+2))**2+(ZZ-Y(1,JBLK+3))**2
      IF(DSQ.LE.0.745) GO TO 1
    3 CONTINUE
  222 Y(1,IBLK+1)=XX
      Y(1,IBLK+2)=YY
      Y(1,IBLK+3)=ZZ
      I=I+1
      IF(I.GT.NMOL) GO TO 4
      IBLK=(I-1)*12
      GO TO 1
    4 I=1
      IBLK=(I-1)*12
   22 Y(1,IBLK+7)=2.0*PI*RANF(1)
      Y(1,IBLK+9)=2.0*PI*RANF(1)
   11 Y(1,IBLK+8)=2.0*PI*RANF(1)
      RTHETA=Y(1,IBLK+8)/PI
      FTHETA=Y(1,IBLK+8)-PI*AINT(RTHETA)
      IF(ABS(FTHETA).LE.0.03) GO TO 11
      I=I+1
      IF(I.GT.NMOL) GO TO 33
      IBLK=(I-1)*12
      GO TO 22
   33 WRITE(6,100)
  100 FORMAT(1X,16HALL MOLECULES IN)
C*****
      GO TO 201
  200 READ(3,302)T
  302 FORMAT(E22.1)
      DO 300 I=1,NMOL
      IRB=(I-1)*12
      READ(3,301)Y(1,IRB+1),Y(1,IRB+2),Y(1,IRB+3)
      READ(3,301)Y(1,IRB+4),Y(1,IRB+5),Y(1,IRB+6)
      READ(3,301)Y(1,IRB+7),Y(1,IRB+8),Y(1,IRB+9)
      READ(3,301)Y(1,IRB+10),Y(1,IRB+11),Y(1,IRB+12)
  301 FORMAT(E23.1,2E22.1)
  300 CONTINUE
      IF(NIAKOF.NE.2) GO TO 250
      WRITE(6,101)
  101 FORMAT(1X,31HALL MOLECULES IN NORMAL RESTART)
      GO TO 201
  250 DO 400 I=1,NMOL
      IRB=(I-1)*12
      DO 401 I1=4,6
  401 Y(1,IRB+I1)=0.0
      DO 402 I1=10,12
  402 Y(1,IRB+I1)=0.0
  400 CONTINUE
      WRITE(6,102)
  102 FORMAT(1X,29HALL MOLECULES IN ZERO RESTART)
  201 CONTINUE
      RETURN
      END
```

```
      SUBROUTINE IMAGES
      COMMON/IPATRN/IMCUBE(27,6)
      IMCUBE(1,1)=IMCUBE(1,2)=IMCUBE(1,3)=0
      IMCUBE(1,4)=IMCUBE(1,5)=IMCUBE(1,6)=0
      I=1
      III=1
    1 II=2*I+1
      DO 2 I1=1,II
      I2=4-I1
      I3=I2/3
      I4=I2/2
      I5=(I4-1)**2
      DO 2 J1=1,3
      III=III+1
      J2=4-J1
      J3=J2/3
      J4=J2/2
      J5=(J4-1)**2
      IMCUBE(III,6)=I
      IMCUBE(III,3)=0
      IMCUBE(III,5)=I3
      IMCUBE(III,2)=I5
      IMCUBE(III,4)=J3
    2 IMCUBE(III,1)=J5
      I=I-1
      IF(I.LT.0) GO TO 3
      GO TO 1
    3 IMCUBE(14,1)=IMCUBE(14,2)=IMCUBE(14,3)=0
      IMCUBE(14,5)=IMCUBE(14,6)=0
      IMCUBE(14,4)=1
      DO 4 I=2,14
      J=I+13
      DO 4 K=1,3
      IMCUBE(J,K)=IMCUBE(I,K+3)
    4 IMCUBE(J,K+3)=IMCUBE(I,K)
      RETURN
      END


      SUBROUTINE DIFSUB(N,NMOL,NMOL50,NMOL2,Y,SAVE,ERROR,AVE,YMAX,LIST,
     1ITRAK,INT1,INT2,T,H,HMIN,HMAX,EPS,MF,KFLAG,JSTART,MAXDER,IFORBD,
     2NRGY)
      DIMENSION Y(6,N)
      DIMENSION SAVE(10,N),ERROR(N),AVE(N),YMAX(N)
      DIMENSION LIST(NMOL50),ITRAK(NMOL),INT1(NMOL),INT2(NMOL2)
      DIMENSION A(6),PERTST(7,2,3)
      LEVEL 2,Y
      LEVEL 2,SAVE,ERROR,AVE,YMAX
      LEVEL 2,LIST,ITRAK,INT1,INT2
      DATA PERTST /2.0,4.5,7.333,10.42,13.7,17.15,1.0,
     1             2.0,12.0,24.0,37.89,53.33,70.08,87.97,
     2             3.0,6.0,9.167,12.5,15.96,1.0,1.0,
     3             12.0,24.0,37.89,53.33,70.08,87.97,1.0,
     4             1.0,1.0,0.5,0.1667,0.04133,0.008267,1.0,
     5             1.0,1.0,2.0,1.0,0.3157,0.07407,0.0139/
      DATA A(2) /-1.0 /
      IFORBD=0
  900 FORMAT(10X,100(*=*)/45X,*FORBIDDEN ZONE IFORBD=*,I2/100(*=*))
  950 FORMAT(1X,18HENTERING DIFFUN AT,I4)
      IRET=1
      KFLAG=1
      IF (JSTART.LE.0) GO TO 140


  100 DO 110 I=1,N
      DO 110 J=1,K
```

```
110 SAVE(J,I)=Y(J,I)
    HOLD=HNEW
    IF(H.EQ.HOLD) GO TO 130
120 RACUM=H/HOLD
    IRET1=1
    GO TO 750
130 NQOLD=NQ
    TOLD=T
    RACUM=1.0
    IF(JSTART.GT.0) GO TO 250
    GO TO 170
140 IF (JSTART.EQ.-1) GO TO 160
    NQ=1
    IGODIF=1
    WRITE(6,950)IGODIF
    CALL DIFFUN(N,NMOL,NMOL50,NMOL2,Y,AVE,LIST,ITRAK,INT1,INT2,NRGY)
    DO 150 I=1,N
150 Y(2,I)=AVE(I)*H
    HNEW=H
    K=2
    GO TO 100
160 IF (NQ.EQ.NQOLD) JSTART=1
    T=TOLD
    NQ=NQOLD
    K=NQ+1
    GO TO 120
170 IF (MF.EQ.0) GO TO 180
    IFORBD=1
    WRITE(6,900)IFORBD
    RETURN
180 IF (NQ.GT.7) GO TO 190
    GO TO (211,212,213,214,215,216,217),NQ
190 KFLAG=-2
    RETURN
211 A(1)=-1.0
    GO TO 230
212 A(1)=-0.500000000
    A(3)=-0.500000000
    GO TO 230
213 A(1)=-0.41666666666667
    A(3)=-0.750000000
    A(4)=-0.16666666666667
    GO TO 230
214 A(1)=-0.375000000
    A(3)=-0.91666666666667
    A(4)=-0.33333333333333
    A(5)=-0.04166666666667
    GO TO 230
215 A(1)=-0.34861111111111
    A(3)=-1.04166666666667
    A(4)=-0.48611111111111
    A(5)=-0.10416666666667
    A(6)=-0.00833333333333
    GO TO 230
216 A(1)=-0.32986111111111
    A(3)=-1.14166666666667
    A(4)=-0.625000000
    A(5)=-0.17708333333333
    A(6)=-0.025000000
    A(7)=-0.00138888888889
    GO TO 230
217 A(1)=-0.31559193121693
    A(3)=1.235000000
    A(4)=-0.75185185185185
    A(5)=-0.25520833333333
```

```
       A(6)=-0.04661111111111
       A(7)=-0.00486111111111
       A(8)=-0.00019841269841
 230   K=NQ+1
       IDOUB=K
       MTYP=(4-MF)/2
       ENQ2=0.5/FLOAT(NQ+1)
       ENQ3=0.5/FLOAT(NQ+2)
       ENQ1=0.5/FLOAT(NQ)
       PEPSH=EPS
       EUP=(PERTST(NQ,MTYP,2)*PEPSH)**2
       E=(PERTST(NQ,MTYP,1)*PEPSH)**2
       EDWN=(PERTST(NQ,MTYP,3)*PEPSH)**2
       IF (EDWN.EQ.0) GO TO 780
       BND=EPS*ENQ3/FLOAT(N)
 240   IWEVAL=MF
       GO TO (250,680),IRET
 250   T=T+H
       DO 260 J=2,K
       DO 260 J1=J,K
       J2=K-J1+J-1
       DO 260 I=1,N
 260   Y(J2,I)=Y(J2,I)+Y(J2+1,I)
       DO 270 I=1,N
 270   ERROR(I)=0.0
       DO 430 L=1,3
       IGUDIF=2
       WRITE(6,950)IGUDIF
       CALL DIFFUN(N,NMUL,NMOL50,NMUL2,Y,AVE,LIST,ITRAK,INT1,INT2,NRGY)
       IF (IWEVAL.LT.1) GO TO 350
       IFORBD=1
       WRITE(6,900)IFORBD
       RETURN
 350   IF (MF.NE.0) GO TO 370
       DO 360 I=1,N
 360   SAVE(9,I)=Y(2,I)-AVE(I)*H
       GO TO 410
 370   IFORBD=2
       WRITE(6,900)IFORBD
       RETURN
 410   NT=K
       DO 420 I=1,N
       Y(1,I)=Y(1,I)+A(1)*SAVE(9,I)
       Y(2,I)=Y(2,I)-SAVE(9,I)
       ERROR(I)=ERROR(I)+SAVE(9,I)
C      *    *    *    *    *    *    *    *    *    *    *
C      *    *    *    *    *    *    *    *    *    *    *
       XPETE1=SAVE(9,I)
       IF(ABS(XPETE1).LE.(BND*YMAX(I))) NT=NT-1
C      *    *    *    *    *    *    *    *    *    *    *
C      *    *    *    *    *    *    *    *    *    *    *
 420   CONTINUE
       IF(NT.LE.0) GO TO 490
 430   CONTINUE
 440   T=TOLD
       IF ((H.LE.(HMIN*1.00001)).AND.((IWEVAL-MTYP).LT.-1)) GO TO 460
       IF ((MF.EQ.0).OR.(IWEVAL.NE.0)) RACUM=RACUM*0.25
       IWEVAL=MF
       IRET1=2
       GO TO 750
 460   KFLAG=-3
 470   DO 480 I=1,N
       DO 480 J=1,K
 480   Y(J,I)=SAVE(J,I)
       H=HOLD
```

```
      NQ=NQOLD
      JSTART=NQ
      RETURN
490   D=0.0
      DO 500 I=1,N
500   D=D+(ERROR(I)/YMAX(I))**2
      IWEVAL=0
      IF (D.GT.E) GO TO 540
      IF (K.LT.3) GO TO 520
      DO 510 J=3,K
      DO 510 I=1,N
510   Y(J,1)=Y(J,I)+A(J)*ERROR(I)
520   KFLAG=+1
      HNEW=H
      IF (IDOUB.LE.1) GO TO 550
      IDOUB=IDOUB-1
      IF (IDOUB.GT.1) GO TO 700
      DO 530 I=1,N
530   SAVE(10,I)=ERROR(I)
      GO TO 700
540   KFLAG=KFLAG-2
      IF (H.LE.(HMIN*1.00001)) GO TO 740
      T=TOLD
      IF (KFLAG.LE.-5) GO TO 720
550   PR2=(D/E)**ENQ2*1.2
      PR3=1.E+20
      IF ((NQ.GE.MAXDER).OR.(KFLAG.LE.-1)) GO TO 570
      D=0.0
      DO 560 I=1,N
560   D=D+((ERROR(I)-SAVE(10,I))/YMAX(I))**2
      PR3=(D/EUP)**ENQ3*1.4
570   PR1=1.E+20
      IF (NQ.LE.1) GO TO 590
      D=0.0
      DO 580 I=1,N
580   D=D+(Y(K,I)/YMAX(I))**2
      PR1=(D/EDWN)**ENQ1*1.3
590   CONTINUE
      IF (PR2.LE.PR3) GO TO 650
      IF (PR3.LT.PR1) GO TO 660
600   R=1.0/AMAX1(PR1,1.E-4)
      NEWQ=NQ-1
610   IDOUB=10
      IF ((KFLAG.EQ.1).AND.(R.LT.(1.1))) GO TO 700
      IF (NEWQ.LE.NQ) GO TO 630
      DO 620 I=1,N
620   Y(NEWQ+1,I)=ERROR(I)*A(K)/FLOAT(K)
630   K=NEWQ+1
      IF (KFLAG.EQ.1) GO TO 670
      RACUM=RACUM*R
      IRET1=3
      GO TO 750
640   IF(NEWQ.EQ.NQ) GO TO 250
      NQ=NEWQ
      GO TO 170
650   IF (PR2.GT.PR1) GO TO 600
      NEWQ=NQ
      R=1.0/AMAX1(PR2,1.E-4)
      GO TO 610
660   R=1.0/AMAX1(PR3,1.E-4)
      NEWQ=NQ+1
      GO TO 610
670   IRET=2
      R=AMIN1(R,HMAX/ABS(H))
      H=H*R
```

```
        HNEW=H
        IF (NQ.EQ.NEWQ) GO TO 680
        NQ=NEWQ
        GO TO 170
  680   R1=1.0
        DO 690 J=2,K
        R1=R1*R
        DO 690 I=1,N
  690   Y(J,I)=Y(J,I)*R1
        IDOUB=K
  700   DO 710 I=1,N
C       *      *      *      *      *      *      *      *      *      *      *
C       *      *      *      *      *      *      *      *      *      *      *
        XPETE1=YMAX(I)
        XPETE2=Y(1,I)
  710   YMAX(I)=AMAX1(XPETE1,ABS(XPETE2))
C       *      *      *      *      *      *      *      *      *      *      *
C       *      *      *      *      *      *      *      *      *      *      *
        JSTART=NQ
        RETURN
  720   IF (NQ.EQ.1) GO TO 780
        IGODIF=3
        WRITE(6,950)IGODIF
        CALL DIFFUN(N,NMOL,NMOL50,NMOL2,Y,AVE,LIST,ITRAK,INT1,INT2,NRGY)
        R=H/HOLD
        DO 730 I=1,N
        Y(1,I)=SAVE(1,I)
        SAVE(2,1)=HOLD*AVE(I)
  730   Y(2,I)=SAVE(2,1)*R
        NQ=1
        KFLAG=1
        GO TO 170
  740   KFLAG=-1
        HNEW=H
        JSTART=NQ
        RETURN
  750   RACUM=AMAX1( ABS(HMIN/HOLD),RACUM)
        RACUM=AMIN1(RACUM, ABS(HMAX/HOLD))
        R1=1.0
        DO 760 J=2,K
        R1=R1*RACUM
        DO 760 I=1,N
  760   Y(J,I)=SAVE(J,1)*R1
        H=HOLD*RACUM
        DO 770 I=1,N
  770   Y(1,I)=SAVE(1,I)
        IDOUB=K
        GO TO (130,250,640),IRET1
  780   KFLAG=-4
        GO TO 470
        END



        SUBROUTINE SETA(MOL,NUMBER,NEQNS,Y)
        COMMON/XMATRX/AI(3,3),AJ(3,3),AIT(3,3),AJT(3,3)
        DIMENSION Y(8,NEQNS)
        DIMENSION A(3,3)
        LEVEL 2,Y
        MOLBLK=(MOL-1)*12
        PHI=Y(1,MOLBLK+7)
        THETA=Y(1,MOLBLK+8)
        PSI=Y(1,MOLBLK+9)
        A(1,1)=COS(PSI)*COS(PHI)-COS(THETA)*SIN(PHI)*SIN(PSI)
        A(1,2)=COS(PSI)*SIN(PHI)+COS(THETA)*COS(PHI)*SIN(PSI)
```

```
      A(1,3)=SIN(PSI)*SIN(THETA)
      A(2,1)=-SIN(PSI)*COS(PHI)-COS(THETA)*SIN(PHI)*COS(PSI)
      A(2,2)=-SIN(PSI)*SIN(PHI)+COS(THETA)*COS(PHI)*COS(PSI)
      A(2,3)=COS(PSI)*SIN(THETA)
      A(3,1)=SIN(THETA)*SIN(PHI)
      A(3,2)=-SIN(THETA)*COS(PHI)
      A(3,3)=COS(THETA)
      GO TO (1,2),NUMBER
    1 DO 3 I=1,3
      DO 3 J=1,3
      AI(I,J)=A(I,J)
    3 AIT(J,I)=AI(I,J)
      GO TO 4
    2 DO 5 I=1,3
      DO 5 J=1,3
      AJ(I,J)=A(I,J)
    5 AJT(J,I)=AJ(I,J)
    4 CONTINUE
      RETURN
      END



      SUBROUTINE CRDS5(MOL,NUMBER,NEQNS,Y)
      COMMON/XMATRX/AI(3,3),AJ(3,3),AIT(3,3),AJT(3,3)
      COMMON/GEOM/XLAB(2,5,3),XLOC(2,5,3),XPCON(5,3)
      DIMENSION Y(8,NEQNS)
      LEVEL 2,Y
      MBLK=(MOL-1)*12
      DO 20 K=1,3
   20 XLOC(NUMBER,5,K)=0.0
      DO 1 I=1,3
      DO 2 J=1,3
      GO TO (3,4),NUMBER
    3 XLOC(NUMBER,5,I)=XLOC(NUMBER,5,I)+AIT(I,J)*XPCON(5,J)
      GO TO 2
    4 XLOC(NUMBER,5,I)=XLOC(NUMBER,5,I)+AJT(I,J)*XPCON(5,J)
    2 CONTINUE
    1 XLAB(NUMBER,5,I)=XLOC(NUMBER,5,I)+Y(1,MBLK+I)
      RETURN
      END



      SUBROUTINE CRDS14(MOL,NUMBER,NEQNS,Y)
      COMMON/XMATRX/AI(3,3),AJ(3,3),AIT(3,3),AJT(3,3)
      COMMON/GEOM/XLAB(2,5,3),XLOC(2,5,3),XPCON(5,3)
      DIMENSION Y(8,NEQNS)
      LEVEL 2,Y
      MBLK=(MOL-1)*12
      DO 20 N=1,4
      DO 20 K=1,3
   20 XLOC(NUMBER,N,K)=0.0
      DO 10 N=1,4
      DO 1 I=1,3
      DO 2 J=1,3
      GO TO (3,4),NUMBER
    3 XLOC(NUMBER,N,I)=XLOC(NUMBER,N,I)+AIT(I,J)*XPCON(N,J)
      GO TO 2
    4 XLOC(NUMBER,N,I)=XLOC(NUMBER,N,I)+AJT(I,J)*XPCON(N,J)
    2 CONTINUE
    1 XLAB(NUMBER,N,I)=XLOC(NUMBER,N,I)+Y(1,MBLK+I)
   10 CONTINUE
      RETURN
      END
```

```
      SUBROUTINE DIFFUN(NEQNS,NMOL,NMOL50,NMOL2,Y,DY,LIST,ITRAK,INT1,
     1INT2,NRGY)
      COMMON/CONSYS/BOXL,BOXL2,RB,RBSQ,RANGE
      COMMON/IPATRN/IMCUBE(27,6)
      COMMON/GEOM/XLAB(2,5,3),XLOC(2,5,3),XPCON(5,3)
      COMMON/ENERGY/PE,ETRAN,EROT,PEIJ
      COMMON/CONMOL/XIX(3),ECH,RU,RL
      DIMENSION Y(8,NEQNS),DY(NEQNS)
      DIMENSION LIST(NMOL50),ITRAK(NMOL),INT1(NMOL),INT2(NMOL2)
      LEVEL 2,Y,DY
      LEVEL 2,LIST,ITRAK,INT1,INT2
      WRITE(6,100)
100   FORMAT(1X,14HDIFFUN ENTERED)
      CALL BEGIN(NEQNS,NMOL,Y,DY)
      DO 300 I=1,NMOL
      II=2*I
300   INT1(I)=INT2(I)=INT2(II)=0
      I1=1
      J2=2
      PE=0.0
      DO 1 IMOL=1,NMOL
      INT1(IMOL)=1
      CALL SETA(IMOL,I1,NEQNS,Y)
      CALL CRDS5(IMOL,I1,NEQNS,Y)
      CALL CRDS14(IMOL,I1,NEQNS,Y)
      IMOL1=IMOL+1
      IF(IMOL1.GT.NMOL) GO TO 10
      DO 2 JMOL=IMOL1,NMOL
      IF(INT2(IMOL+JMOL).EQ.0) GO TO 3
      IF(INT1(JMOL).EQ.0) GO TO 3
      GO TO 2
  3   INT2(IMOL+JMOL)=1
      CALL SETA(JMOL,J2,NEQNS,Y)
      CALL CRDS5(JMOL,J2,NEQNS,Y)
      DO 4 IMAGE=1,27
      R12SQ=0.0
      DO 20 K=1,3
 20   R12SQ=R12SQ+(XLAB(I1,5,K)-XLAB(J2,5,K)-BOXL
     1*(IMCUBE(IMAGE,K+3)-IMCUBE(IMAGE,K)))**2
      IF(R12SQ.GT.RBSQ) GO TO 4
      CALL FXN(IMOL,JMOL,IMAGE,I1,J2,NEQNS,Y,DY)
      PE=PE+PEIJ
  4   CONTINUE
  2   CONTINUE
  1   CONTINUE
 10   CONTINUE
      IF(NRGY.NE.0) GO TO 11
      ETRAN=0.0
      EROT=0.0
      DO 12 I=1,NMOL
      IBLK=(I-1)*12
      TERMT=Y(1,IBLK+4)**2+Y(1,IBLK+5)**2+Y(1,IBLK+6)**2
      TERMR=XIX(1)*(Y(1,IBLK+10)**2)+XIX(2)*(Y(1,IBLK+11)**2)
     2+XIX(3)*(Y(1,IBLK+12)**2)
      ETRAN=ETRAN+0.5*TERMT
      EROT=EROT+0.5*TERMR
 12   CONTINUE
      NRGY=1
      ETOTAL=PE+ETRAN+EROT
      WRITE(6,30)PE,ETRAN,EROT,ETOTAL
 30   FORMAT(1X,25HENERGIES AT START OF STEP,4(1X,E20.12))
      TEMPTK=(24.19333*ETRAN)/FLOAT(NMOL)
      TEMPRK=(24.19333*EROT)/FLOAT(NMOL)
      WRITE(6,31)TEMPTK,TEMPRK
 31   FORMAT(65X,E20.12,1X,E20.12)
```

```
  11 CONTINUE
     RETURN
     END


     SUBROUTINE FXN(IMOL,JMOL,IMAGE,I1,J2,NEQNS,Y,DY)
     COMMON/GEOM/XLAB(2,5,3),XLOC(2,5,3),XPCON(5,3)
     COMMON/FORCES/FOXGNS(3),FLJ(3),FLABI(3),FLABJ(3),FTEMP(2,4,3),
    8FV(3)
     COMMON/TORQES/XNLABI(3),XNLABJ(3),XNLOCI(3),XNLOCJ(3)
     COMMON/CONMOL/XIX(3),ECM,RU,RL
     COMMON/XMATRX/AI(3,3),AJ(3,3),AIT(3,3),AJT(3,3)
     COMMON/CONSYS/BOXL,BOXL2,RB,RBSQ,RANGE
     COMMON/IPATRN/IMCUBE(27,6)
     COMMON/ENERGY/PE,ETRAN,EROT,PEIJ
     DIMENSION Y(8,NEQNS),DY(NEQNS)
     DIMENSION SHFTBX(3)
     LEVEL 2,Y,DY
C*****
C*****
C*****
C*****
     XIMG=1.0
     IF(IMAGE.EQ.1) XIMG=1.0
     DO 500 K=1,3
 500 SHFTBX(K)=BOXL*(IMCUBE(IMAGE,K+3)-IMCUBE(IMAGE,K))
     DO 600 K=1,3
     FLABI(K)=0.0
     FLABJ(K)=0.0
     XNLABI(K)=0.0
     XNLABJ(K)=0.0
     XNLOCI(K)=0.0
     XNLOCJ(K)=0.0
     FOXGNS(K)=0.0
 600 CONTINUE
     RIJ=0.0
     DO 10 K=1,3
  10 RIJ=RIJ+(XLAB(I1,5,K)-XLAB(J2,5,K)-SHFTBX(K))**2
     RIJ=SQRT(RIJ)
     FLJMOD=24.0*((2.0/RIJ**13)-(1.0/RIJ**7))
     PEIJ=XIMG*4.0*((1.0/RIJ**12)-(1.0/RIJ**6))
     DO 20 K=1,3
     FLJ(K)=FLJMOD*(XLAB(I1,5,K)-XLAB(J2,5,K)-SHFTBX(K))/RIJ
     FLABI(K)=FLABI(K)+FLJ(K)
     FLABJ(K)=FLABJ(K)-FLJ(K)
  20 FOXGNS(K)=FOXGNS(K)+FLJ(K)
     IF(RIJ.LT.RL) GO TO 100
     IF(RIJ.GE.RL.AND.RIJ.LE.RU) GO TO 101
     IF(RIJ.GT.RU) GO TO 102
 100 SRIJ=0.0
     DSRIJ=0.0
     ICASE=1
     GO TO 103
 101 SRIJ=(3.0*RU-RL-2.0*RIJ)*(RIJ-RL)**2/(RU-RL)**3
     DSRIJ=6.0*(RIJ-RL)*(RU-RIJ)/(RU-RL)**3
     ICASE=2
     GO TO 103
 102 SRIJ=1.0
     DSRIJ=0.0
     ICASE=3
 103 CONTINUE
     GO TO (200,201,201),ICASE
 201 CALL CRDS14(JMOL,J2,NEQNS,Y)
     DO 601 IIF=1,2
     DO 601 JJF=1,4
```

```
      DO 601 KKF=1,3
601   FTEMP(IIF,JJF,KKF)=0.0
      FVMOD=0.0
      VEL=0.0
      DO 30 N=1,4
      DO 30 M=1,4
      RNIMJ=0.0
      DO 31 K=1,3
 31   RNIMJ=RNIMJ+(XLAB(I1,N,K)-XLAB(J2,M,K)-SHFTBX(K))**2
      RNIMJ=SQRT(RNIMJ)
      FNIMJ=((-1)**(N+M+1))*((0.19*ECH)**2)*SRIJ/(RNIMJ**2)
      VEL=VEL+((-1)**(N+M))*((0.19*ECH)**2)*SRIJ/RNIMJ
      IF(ICASE.EQ.3) GO TO 400
      FVMOD=FVMOD+((-1)**(N+M))*((0.19*ECH)**2)*DSRIJ/RNIMJ
400   CONTINUE
      DO 32 K=1,3
      FIJNM=-FNIMJ*(XLAB(I1,N,K)-XLAB(J2,M,K)-SHFTBX(K))/RNIMJ
      FTEMP(I1,N,K)=FTEMP(I1,N,K)+FIJNM
 32   FTEMP(J2,M,K)=FTEMP(J2,M,K)-FIJNM
 30   CONTINUE
      PEIJ=PEIJ+XIMG*VEL
      DO 35 NM=1,4
      DO 35 K=1,3
      FLABI(K)=FLABI(K)+FTEMP(I1,NM,K)
 35   FLABJ(K)=FLABJ(K)+FTEMP(J2,NM,K)
      IF(ICASE.EQ.3) GO TO 401
      DO 40 K=1,3
      FV(K)=-FVMOD*(XLAB(I1,5,K)-XLAB(J2,5,K)-SHFTBX(K))/RIJ
      FLABI(K)=FLABI(K)+FV(K)
      FLABJ(K)=FLABJ(K)-FV(K)
 40   FOXGNS(K)=FOXGNS(K)+FV(K)
401   CONTINUE
200   CONTINUE
      XNLABI(1)=XNLABI(1)+FOXGNS(3)*XLOC(I1,5,2)-FOXGNS(2)*XLOC(I1,5,3)
      XNLABJ(1)=XNLABJ(1)-FOXGNS(3)*XLOC(J2,5,2)+FOXGNS(2)*XLOC(J2,5,3)
      XNLABI(2)=XNLABI(2)-FOXGNS(1)*XLOC(I1,5,3)+FOXGNS(3)*XLOC(I1,5,1)
      XNLABJ(2)=XNLABJ(2)-FOXGNS(1)*XLOC(J2,5,3)+FOXGNS(3)*XLOC(J2,5,1)
      XNLABI(3)=XNLABI(3)+FOXGNS(2)*XLOC(I1,5,1)-FOXGNS(1)*XLOC(I1,5,2)
      XNLABJ(3)=XNLABJ(3)+FOXGNS(2)*XLOC(J2,5,1)+FOXGNS(1)*XLOC(J2,5,2)
      GO TO (300,301,301),ICASE
301   DO 50 NM=1,4
      XNLABI(1)=XNLABI(1)+FTEMP(I1,NM,3)*XLOC(I1,NM,2)-FTEMP(I1,NM,2)
     2*XLOC(I1,NM,3)
      XNLABJ(1)=XNLABJ(1)+FTEMP(J2,NM,3)*XLOC(J2,NM,2)-FTEMP(J2,NM,2)
     3*XLOC(J2,NM,3)
      XNLABI(2)=XNLABI(2)+FTEMP(I1,NM,1)*XLOC(I1,NM,3)-FTEMP(I1,NM,3)
     4*XLOC(I1,NM,1)
      XNLABJ(2)=XNLABJ(2)+FTEMP(J2,NM,1)*XLOC(J2,NM,3)-FTEMP(J2,NM,3)
     5*XLOC(J2,NM,1)
      XNLABI(3)=XNLABI(3)+FTEMP(I1,NM,2)*XLOC(I1,NM,1)-FTEMP(I1,NM,1)
     6*XLOC(I1,NM,2)
      XNLABJ(3)=XNLABJ(3)+FTEMP(J2,NM,2)*XLOC(J2,NM,1)-FTEMP(J2,NM,1)
     7*XLOC(J2,NM,2)
 50   CONTINUE
300   DO 60 II=1,3
      DO 60 JJ=1,3
      XNLOCI(II)=XNLOCI(II)+AI(II,JJ)*XNLABI(JJ)
 60   XNLOCJ(II)=XNLOCJ(II)+AJ(II,JJ)*XNLABJ(JJ)
      DO 70 K=1,3
      DY((IMOL-1)*12+K+3)=DY((IMOL-1)*12+K+3)+FLABI(K)
      DY((JMOL-1)*12+K+3)=DY((JMOL-1)*12+K+3)+FLABJ(K)
      DY((IMOL-1)*12+K+9)=DY((IMOL-1)*12+K+9)+XNLOCI(K)/XIX(K)
 70   DY((JMOL-1)*12+K+9)=DY((JMOL-1)*12+K+9)+XNLOCJ(K)/XIX(K)
      RETURN
      END
```

```
      SUBROUTINE BEGIN(NEQNS,NMOL,Y,DY)
      COMMON/CONMOL/XIX(3),ECM,RU,RL
      DIMENSION Y(8,NEQNS),DY(NEQNS)
      LEVEL 2,Y,DY
      DO 1 I=1,NMOL
      IB=(I-1)*12
      THETA=Y(1,IB+8)
      PSI=Y(1,IB+9)
      DO 2 K=1,3
      DY(IB+K)=Y(1,IB+K+3)
    2 DY(IB+K+3)=0.0
      DY(IB+7)=Y(1,IB+10)*SIN(PSI)/SIN(THETA)+Y(1,IB+11)*COS(PSI)
     8/SIN(THETA)
      DY(IB+8)=Y(1,IB+10)*COS(PSI)-Y(1,IB+11)*SIN(PSI)
      DY(IB+9)=Y(1,IB+12)-Y(1,IB+10)*SIN(PSI)*COT(THETA)-Y(1,IB+11)
     9*COS(PSI)*COT(THETA)
      DY(IB+10)=(Y(1,IB+11)*Y(1,IB+12)*(XIX(2)-XIX(3)))/XIX(1)
      DY(IB+11)=(Y(1,IB+12)*Y(1,IB+10)*(XIX(3)-XIX(1)))/XIX(2)
    1 DY(IB+12)=(Y(1,IB+10)*Y(1,IB+11)*(XIX(1)-XIX(2)))/XIX(3)
      RETURN
      END




      FUNCTION COT(X)
      COT=COS(X)/SIN(X)
      RETURN
      END




      SUBROUTINE CHKOUT(ISTEP,T,NEQNS,NMOL,Y)
      COMMON/CONSYS/BOXL,BOXL2,RB,RBSQ,RANGE
      DIMENSION Y(8,NEQNS)
      DIMENSION MARKER(6)
      LEVEL 2,Y
      ORIGIN=10.0+BOXL2
C*****
C*****
      WRITE(6,200)
C*****
  200 FORMAT(1X,14HCHKOUT ENTERED)
C*****
      DO 5 I=1,NMOL
      IBLK=(I-1)*12
      DO 1 K=1,3
      CRD=Y(1,IBLK+K)-ORIGIN
      IF(ABS(CRD).LE.BOXL2) GO TO 2
      IF(CRD.LT.0.0) GO TO 3
      IF(CRD.GT.0.0) GO TO 4
    2 MARKER(K)=MARKER(K+3)=0
      GO TO 1
    3 MARKER(K)=1
      MARKER(K+3)=0
      Y(1,IBLK+K)=Y(1,IBLK+K)+BOXL
      GO TO 1
    4 MARKER(K)=0
      MARKER(K+3)=1
      Y(1,IBLK+K)=Y(1,IBLK+K)-BOXL
    1 CONTINUE
      I1=IBLK+1
      I2=IBLK+2
      I3=IBLK+3
      I7=IBLK+7
      I8=IBLK+8
```

```
      I9=IBLK+9
      IF(IBLK.NE.0) GO TO 1000
      WRITE(6,100)ISTEP,(MARKER(II),II=1,6),T,Y(1,I1),Y(1,I2),Y(1,I3),
     1Y(1,I7),Y(1,I8),Y(1,I9)
  100 FORMAT(1X,I5,1X,6I1,1X,7(E15.7))
      IGOFF=IBLK+1
      ISTOP=IBLK+12
      DO 2000 IIPI=IGOFF,ISTOP
 2000 WRITE(6,2001)Y(1,IIPI)
 2001 FORMAT(1X,E20.12)
 1000 CONTINUE
    5 CONTINUE
      RETURN
      END




      SUBROUTINE OUT(ISTEP,T,NEQNS,NMOL,Y)
      DIMENSION Y(8,NEQNS)
      LEVEL 2,Y
      WRITE(6,100)ISTEP,T,(Y(1,12+J),J=1,3),(Y(1,18+J),J=1,3)
      GO TO 1000
      DO 1 I=1,NMOL
      IBLK=(I-1)*12
      I1=IBLK+1
      I2=IBLK+2
      I3=IBLK+3
      I7=IBLK+7
      I8=IBLK+8
      I9=IBLK+9
      WRITE(6,100)ISTEP,T,Y(1,I1),Y(1,I2),Y(1,I3),Y(1,I7),Y(1,I8)
     1,Y(1,I9)
  100 FORMAT(1X,I5,1X,7(E15.7))
    1 CONTINUE
 1000 CONTINUE
      RETURN
      END




      SUBROUTINE LPOINT(T,NMOL,Y,NEQNS)
      DIMENSION Y(8,NEQNS)
      LEVEL 2,Y
      WRITE(7,9)T
    9 FORMAT(1X,E21.14)
      DO 1 I=1,NMOL
      IBLK=(I-1)*12
      WRITE(7,10)Y(1,IBLK+1),Y(1,IBLK+2),Y(1,IBLK+3)
      WRITE(7,10)Y(1,IBLK+4),Y(1,IBLK+5),Y(1,IBLK+6)
      WRITE(7,10)Y(1,IBLK+7),Y(1,IBLK+8),Y(1,IBLK+9)
      WRITE(7,10)Y(1,IBLK+10),Y(1,IBLK+11),Y(1,IBLK+12)
   10 FORMAT(1X,3(1X,E21.14))
    1 CONTINUE
      RETURN
      END
```

References

1. M. H. Abraham, *Prog. Phys. Org. Chem.* **11**, 1 (1974).
2. M. H. Abraham and P. L. Grellier, *J. Chem. Soc. Perkin Trans.* 2, 1856 (1975).
3. M. H. Abraham, R. J. Irving, and G. F. Johnston, *J. Chem. Soc. A*, 199 (1970).
4. M. H. Abraham and G. F. Johnston, *J. Chem. Soc. A*, 188, 193 (1970).
5. M. H. Abraham and G. F. Johnston, *J. Chem. Soc. A*, 1610 (1971).
6. M. H. Abraham, J. F. C. Oliver, and J. A. Richards, *J. Chem. Soc. A*, 203 (1970).
7. R. J. Abraham and D. Birch, *Mol. Pharmacol.* **11**, 663 (1975).
8. A. S. Adair and M. E. Adair, *Proc. Roy. Soc. B* **120**, 422 (1936).
9. G. E. Adams and J. N. Israelachvili, *Nature* **262**, 774 (1976).
10. G. E. Adams and J. N. Israelachvili, *J. Chem. Soc. Faraday II* **74**, 975 (1978).
11. E. T. Adman, L. C. Sieker, and L. H. Jensen, *J. Biol. Chem.* **248**, 3987 (1973).
12. E. T. Adman, L. C. Sieker, and L. H. Jensen, *Acta Cryst.* **A31** (S3), S34 (1975).
13. E. T. Adman, L. C. Sieker, and L. H. Jensen, *J. Biol. Chem.* **251**, 3801 (1975).
14. E. T. Adman, L. C. Sieker, L. H. Jensen, M. Bruschi, and J. le Gall, *J. Mol. Biol.* **112**, 113 (1977).
15. J. C. Ahluwalia, C. Ostiguy, G. Perron, and J. E. Desnoyers, *Can. J. Chem.* **55**, 3364 (1977).
16. G. Alagona, A. Pullman, E. Scrocca, and J. Tomasi, *Int. J. Peptide Protein Res.* **5**, 251 (1973).
17. J. Alber, G. A. Petsko, and D. Tsernoglou, *Nature* **263**, 297 (1976).
18. W. J. Albery, *Prog. React. Kin.* **4**, 353 (1967).
19. W. J. Albery and B. H. Robinson, *Trans. Faraday Soc.* **65**, 980 (1969).
20. J. N. Albright, *J. Chem. Phys.* **56**, 3783 (1972).
21. R. A. Alden, J. J. Birktoft, J. Kraut, J. D. Robertus, and C. S. Wright, *Biochem. Biophys. Res. Commun.* **45**, 337 (1971).
22. B. J. Alder, *J. Chem. Phys.* **40**, 2724 (1964).
23. B. J. Alder, H. L. Strauss, and J. J. Weis, *J. Chem. Phys.* **59**, 1002 (1973).
24. B. J. Alder and T. E. Wainwright, *J. Chem. Phys.* **27**, 1208 (1957).
25. B. J. Alder and T. E. Wainwright, *J. Chem. Phys.* **31**, 459 (1959).
26. B. J. Alder and T. E. Wainwright, *J. Chem. Phys.* **33**, 1439 (1960).
27. B. J. Alder and T. E. Wainwright, *Phys. Rev. A* **1**, 18 (1970).
28. R. W. Alder, R. Baker, and J. M. Brown, "Mechanism in Organic Chemistry," Wiley, New York, p. 48 (1971).
29. L. C. Allen and P. A. Kollman, *J. Amer. Chem. Soc.* **92**, 4108 (1970).
30. M. Alves Marques and M. I. de Barros Marques, *Proc. K. Ned. Akad. Wet., Ser. B* **77**, 286 (1974).

31. E. S. Amis and J. F. Hinton, "Solvent Effects on Chemical Phenomena," Vol. I, Academic Press, New York (1973).

32. L. Anderson, *J. Mol. Biol.* **94**, 33 (1975).

33. A. Arcoria, V. Librando, E. Maccarone, G. Musumarra, and G. A. Tomaselli, *Tetrahedron* **33**, 105 (1977).

34. E. M. Arnett *in* "Physico-chemical Processes in Mixed Aqueous Solvents" (F. Franks, ed.), Heinemann, London (1967).

35. E. M. Arnett, W. G. Bentrude, J. J. Burke, and P. McC. Duggleby, *J. Amer. Chem. Soc.* **87**, 1541 (1965).

36. E. M. Arnett, P. McC. Duggleby, and J. J. Burke, *J. Amer. Chem. Soc.* **85**, 1350 (1963).

37. E. M. Arnett, F. M. Jones, III, M. Taagepera, W. G. Henderson, J. L. Beauchamp, D. Holtz, and R. W. Taft, *J. Amer. Chem. Soc.* **94**, 4724 (1972).

38. E. M. Arnett and D. R. McKelvey, *Rec. Chem. Progr.* **26**, 184 (1965).

39. E. M. Arnett and D. R. McKelvey *in* "Solute–Solvent Interactions" (J. F. Coetzee and C. D. Ritchie, eds.), M. Dekker, New York (1969).

40. E. M. Arnett, E. T. Mitchell, and T. S. S. R. Murthy, *J. Amer. Chem. Soc.* **96**, 3875 (1974).

41. E. M. Arnett, L. E. Small, R. T. McIver, Jr., and J. S. Miller, *J. Amer. Chem. Soc.* **96**, 5638 (1974).

42. E. M. Arnett, L. E. Small, D. Oancea, and D. Johnston, *J. Amer. Chem. Soc.* **98**, 7346 (1976).

43. A. Arnone, C. J. Bier, F. A. Cotton, E. E. Hazen, D. C. Richardson, J. S. Richardson, and A. Yonath, *J. Biol. Chem.* **246**, 2302 (1971).

44. D. H. Aue, H. M. Webb, and M. T. Bowers, *J. Amer. Chem. Soc.* **98**, 311, 318 (1976).

45. L. Avedikian, J. Juillard, J.-P. Morel, and M. Ducros, *Thermochimica Acta* **6**, 283 (1973).

46. B. T. Baliga and E. Whalley, *J. Phys. Chem.* **71**, 1166 (1967).

47. B. T. Baliga and E. Whalley, *Can. J. Chem.* **48**, 528 (1970).

48. B. T. Baliga, R. J. Withey, D. Poulton, and E. Whalley, *Trans. Faraday Soc.* **61**, 517 (1965).

49. A. Banin and D. M. Anderson, *Nature* **255**, 261 (1975).

50. D. W. Banner, A. C. Bloomer, G. A. Petsko, D. C. Phillips, C. I. Pogson, I. A. Wilson, P. H. Corran, A. J. Furth, J. D. Milman, R. E. Offord, J. D. Priddle, and S. G. Waley, *Nature* **255**, 609 (1975).

51. D. W. Banner, A. C. Bloomer, G. A. Petsko, D. C. Phillips, P. S. Rivers, and I. A. Wilson, *Acta Cryst.* **A31** (S3), S27 (1975).

52. S. H. Banyard, D. K. Stammers, and P. M. Harrison, *Nature* **271**, 282 (1978).

53. J. A. Barker and D. Henderson, *Rev. Mod. Phys.* **48**, 587 (1976).

54. J. A. Barker and R. O. Watts, *Chem. Phys. Lett.* **3**, 144 (1969).

55. R. Barker, "Organic Chemistry of Biological Compounds," Prentice-Hall, Englewood Cliffs, New Jersey, Chap. 2 (1971).

56. P. Barnes, "Report of CECAM Workshop on Molecular Dynamics of Water," CECAM, Orsay, France, p. 77 (1972).

57. P. Barnes *in* "Progress in Liquid Physics" (C. A. Croxton, ed.), John Wiley, Chichester (1978).

58. P. Barnes and J. L. Finney, *in* "Abstr. 4th European Crystallographic Meeting, Oxford," A377 (1977).
59. J. Barojas, D. Levesque, and B. Quentrec, *Phys. Rev. A* **7**, 1092 (1973).
60. R. G. Bates, *in* "Hydrogen Bonded Solvent Systems" (A. K. Covington and P. Jones, eds.), Taylor and Francis, London (1968).
61. C. Battistini, G. Berti, P. Crotti, M. Ferretti, and F. Macchia, *Tetrahedron* **33**, 1629 (1977).
62. R. J. Baxter, *J. Chem. Phys.* **49**, 2770 (1968).
63. J. L. Beauchamp and M. C. Caserio, *J. Amer. Chem. Soc.* **94**, 2638 (1972).
64. J. W. Becker, G. N. Reeke, J. L. Wang, B. A. Cunningham, and G. M. Edelman, *J. Biol. Chem.* **250**, 1513 (1975).
65. S. Bedarkar, W. G. Turnell, T. L. Blundell, and C. Schwabe, *Nature* **270**, 449 (1977).
66. J. L. Beeby, *J. Phys. C.* **6**, 2262 (1973).
67. G. M. Bell, *J. Math. Phys.* **10**, 1753 (1969).
68. G. M. Bell, *J. Phys. A.* **3**, 568 (1970).
69. G. M. Bell, *J. Phys. C. Solid State Phys.* **5**, 889 (1972).
70. G. M. Bell and D. A. Lavis, *J. Phys. A.* **3**, 427, 568 (1970).
71. G. M. Bell and D. W. Salt, *J. Chem. Soc., Faraday Trans. II* **72**, 77 (1975).
72. J. R. Bell, J. L. Tyvoll, and D. L. Wertz, *J. Amer. Chem. Soc.* **95**, 1456 (1973).
73. R. P. Bell, *Chem. Soc. Rev.* 513 (1974).
74. R. P. Bell and B. G. Cox, *J. Chem. Soc. B* 194 (1970).
75. R. P. Bell and B. G. Cox, *J. Chem. Soc. B* 783 (1971).
76. R. P. Bell, J. E. Critchlow, and M. I. Page, *J. Chem. Soc. Perkin Trans. 2*, 66 (1974).
77. R. P. Bell and R. L. Tranter, *Proc. Roy. Soc. A* **337**, 517 (1974).
78. A. Ben-Naim, *Trans. Faraday Soc.* **66**, 2749 (1970).
79. A. Ben-Naim, *J. Chem. Phys.* **54**, 1387 (1971).
80. A. Ben-Naim, *Mol. Phys.* **24**, 705 (1972).
81. A. Ben-Naim, *in* "Water: A Comprehensive Treatise" (F. Franks, ed.), Vol. 2, p. 585, Plenum Press, New York (1973).
82. A. Ben-Naim, "Water and Aqueous Solutions, Introduction to a Molecular Theory," Plenum Press, New York (1974), Chapter 6.
83. A. Ben-Naim, *Biopolymers* **14**, 1337 (1975).
84. A. Ben-Naim, *J. Phys. Chem.* **79**, 1268 (1975).
85. A. Ben-Naim, "Statistical mechanics of aqueous fluids," *in* "Progress in Liquid Physics" (C. A. Croxton, ed.), Wiley, New York (1977).
86. A. Ben-Naim, *J. Phys. Chem.* **82**, 792 (1978).
87. A. Ben-Naim and F. H. Stillinger *in* "Structure and Transport Properties in Water and Aqueous Solutions" (R. A. Horne, ed.), Wiley Interscience, New York (1972).
88. A. Ben-Naim and R. Tenne, *J. Chem. Phys.* **67**, 627 (1977).
89. A. Ben-Naim, J. Wilf, and M. Yaacobi, *J. Phys. Chem.* **77**, 95 (1973).
90. A. Ben-Naim and M. Yaacobi, *J. Phys. Chem.* **78**, 170 (1974).
91. A. Ben-Naim and M. Yaacobi, *J. Phys. Chem.* **79**, 1263 (1975).
92. G. A. Bentley, E. Duee, S. A. Mason, A. C. Nunes, and L. F. Power, *in* "Abstr. 4th European Crystallographic Meeting, Oxford," A356 (1977).
93. H. J. C. Berendsen, *Fed. Proc.* **25**, 971 (1966).
94. H. J. C. Berendsen, *in* "Water: A Comprehensive Treatise" (F. Franks, ed.), Vol. 5, p. 293, Plenum Press, New York (1975).

95. H. J. C. Berendsen and G. A. van der Velde, "Report of CECAM Workshop on Molecular Dynamics of Water," p. 63, CECAM, Orsay, France (1972).
96. J. D. Bernal, *Trans. Faraday Soc.* **33**, 27 (1937).
97. J. D. Bernal, *Proc. Roy. Soc. A* **280**, 299 (1964).
98. J. D. Bernal and J. L. Finney, *Acta Cryst.* **A25** (S3), S89 (1969).
99. J. D. Bernal and R. H. Fowler, *J. Chem. Phys.* **1**, 515 (1933).
100. B. J. Berne and G. D. Harp, *Adv. Chem. Phys.* **17**, 63 (1970).
101. J. Bernhardt and H. Pauly, *J. Phys. Chem.* **79**, 584 (1975).
102. A. Berson, Z. Hamlet, and W. A. Mueller, *J. Amer. Chem. Soc.* **84**, 297 (1962).
103. H. Bertagnolli, J. U. Weidner, and H. W. Zimmerman, *Ber. Bunsenges. Phys. Chem.* **78**, 1 (1974).
104. G. L. Bertrand and T. F. Fagley, *J. Amer. Chem. Soc.* **98**, 7944 (1976).
105. D. L. Beveridge, M. M. Kelly, and R. J. Radna, *J. Amer. Chem. Soc.* **96**, 3769 (1974).
106. D. L. Beveridge, R. J. Radna, G. W. Schnuelle, and M. M. Kelly, *in* "Molecular and Quantum Pharmacology" (E. D. Bergmann and B. Pullman, eds.), pp. 153–178, Reidel Publishing Co., Dordrecht, Holland (1974).
107. D. L. Beveridge and G. W. Schnuelle, *J. Phys. Chem.* **78**, 2064 (1974).
108. B. Bielski and S. Freed, *Biochim. Biophys. Acta* **89**, 314 (1964).
109. R. L. Biltonen, Colloques Int. CNRS No. 246, "L'Eau et les Systèmes Biologiques," p. 13, CNRS, Paris (1975).
110. J. J. Birktoft and D. M. Blow, *J. Mol. Biol.* **68**, 187 (1972).
111. J. J. Birktoft, D. M. Blow, R. Henderson, and T. A. Steitz, *Phil. Trans. Roy. Soc.* **B257**, 67 (1970).
112. W. H. Bishop and F. M. Richards, *J. Mol. Biol.* **33**, 415 (1968).
113. W. H. Bishop and F. M. Richards, *J. Mol. Biol.* **38**, 315 (1968).
114. N. Bjerrum, *Kgl. Danske Videnskab. Selskab. Skr.* **27**, 1 (1951).
115. C. C. F. Blake, *Nature* **255**, 278 (1975).
116. C. C. F. Blake, G. A. Mair, A. C. T. North, D. C. Phillips, and V. R. Sarma, *Proc. Roy. Soc.* **B167**, 365 (1967).
117. C. C. F. Blake and S. J. Oatley, *Nature* **268**, 115 (1977).
118. M. J. Blandamer and J. Burgess, *Chem. Soc. Rev.* **4**, 55 (1975).
119. M. J. Blandamer and J. R. Membrey, *J. Chem. Soc. Chem. Commun.* 514 (1973).
120. D. M. Blow and T. A. Steitz, *Ann. Rev. Biochem.* **39**, 63 (1970).
121. D. M. Blow, C. S. Wright, D. Kukla, A. Ruhlmann, W. Steigemann, and R. Huber, *J. Mol. Biol.* **69**, 137 (1972).
122. T. L. Blundell, G. Dodson, D. C. Hodgkin, and D. Mercola, *Adv. Prot. Chem.* **26**, 279 (1972).
123. C. A. Blyth and J. R. Knowles, *J. Amer. Chem. Soc.* **93**, 3017 (1971).
124. W. Bode and P. Schwager, *FEBS Lett.* **56**, 139 (1975).
125. W. Bode and P. Schwager, *J. Mol. Biol.* **98**, 693 (1975).
126. W. Bode, P. Schwager, and R. Huber, "Proc. 10th FEBS Meeting," p. 3 (1975).
127. W. Bol, G. Gerrits, and C. van Panthaleon van Eck, *J. Appl. Cryst.* **3**, 486 (1970).
128. W. Bol and T. Welzen, *Chem. Phys. Lett.* **49**, 182 (1977).
129. G. Bolis and E. Clementi, *J. Amer. Chem. Soc.* **99**, 5550 (1977).
130. R. Bonacorsi, G. Petrongolo, E. Scrocco, and J. Tomasi, *Theor. Chim. Acta* **20**, 331 (1971).

131. O. D. Bonner, J. M. Bednarek, and R. K. Arisman, *J. Amer. Chem. Soc.* **99**, 2898 (1977).
132. M. Born, *Z. Phys.* **1**, 45 (1920).
133. M. Born and K. Huang, "Dynamical Theory of Crystal Lattices," Oxford University Press (1954).
134. M. J. T. Bowers and R. M. Pitzer, *J. Chem. Phys.* **59**, 163 (1973).
135. L. Bøye and A. Hvidt, *Biopolymer* **11**, 2357 (1972).
136. G. W. Brady, *J. Chem. Phys.* **28**, 464 (1958).
137. G. W. Brady, *J. Chem. Phys.* **29**, 1371 (1958).
138. G. W. Brady, *J. Chem. Phys.* **33**, 1079 (1960).
139. G. W. Brady and J. T. Krause, *J. Chem. Phys.* **27**, 304 (1957).
140. W. L. Bragg and J. West, *Phil. Mag.* **10**, 823 (1930).
141. C.-I. Brändén, H. Eklund, B. Nordström, T. Boiwe, G. Söderland, E. Zeppezauer, I. Ohlsson, and Å. Åkeson, *Proc. Nat. Acad. Sci. USA* **70**, 2439 (1973).
142. J. F. Brandts, *in* "Structure and Stability of Biological Macromolecules" (S. N. Timasheff and G. D. Fasman, eds.), p. 213, Dekker, New York (1969).
143. J. I. Brauman and L. K. Blair, *J. Amer. Chem. Soc.* **90**, 6561 (1968).
144. C. Braun and H. Leidecker, *J. Chem. Phys.* **61**, 3104 (1974).
145. K. J. Breslauer, B. Terrin, and J. M. Sturtevant, *J. Phys. Chem.* **78**, 2363 (1974).
146. L. L. Bright and J. R. Jezorek, *J. Phys. Chem.* **79**, 800 (1975).
147. L. G. S. Brooker, A. C. Craig, D. W. Heseltine, P. W. Jenkins, and L. L. Lincoln, *J. Amer. Chem. Soc.* **87**, 2443 (1965).
148. W. Brostow and Y. Sicotte, *Physica* **A80**, 513 (1975).
149. S. Brownstein, *Can. J. Chem.* **38**, 1590 (1960).
150. J. Brubacher, L. Treindl, and R. E. Robertson, *J. Amer. Chem. Soc.* **90**, 4611 (1968).
151. A. Bruggink, B. Zwanenburg, and J. B. F. N. Engberts, *Tetrahedron* **25**, 5655 (1969).
152. R. G. Bryant, *J. Amer. Chem. Soc.* **96**, 297 (1974).
153. A. D. Buckingham, *Proc. Roy. Soc.* **A238**, 235 (1956).
154. H. B. Bull and K. Breese, *Archiv. Biochem. Biophys.* **137**, 299 (1970).
155. J. Bulla, P. Törmälä, and J. J. Lindberg, *Acta Chem. Scand.* **A29**, 89 (1975).
156. C. A. Bunton, "Nucleophilic Substitution at a Saturated Carbon Atom," Elsevier, Amsterdam (1963).
157. C. A. Bunton, S. K. Huang, and C. H. Paik, *Tetrahedron Lett.* 1445 (1976).
158. R. M. Burnett, G. D. Darling, D. S. Kendall, M. E. LeQuesne, S. G. Mayhew, W. W. Smith, and M. L. Ludwig, *J. Biol. Chem.* **249**, 4383 (1974).
159. N. P. Buslenko, D. I. Golenko, I. M. Sobol, V. G. Srangroich, and Yu. A. Shneider, "The Method of Statistical Imitation," Nauka Publishing House, Moscow (1961).
160. A. R. Butler and V. Gold, *J. Chem. Soc.*, 2305 (1961).
161. S. Cabani, G. Conti, and E. Matteoli, *J. Solution Chem.* **5**, 125 (1976).
162. S. Cabani, G. Conti, and E. Matteoli, *J. Solution Chem.* **5**, 751 (1976).
163. E. F. Caldin and H. P. Bennetto, *J. Solution Chem.* **2**, 217 (1973).
164. G. Calvaruso, F. P. Cavasino, and E. Di Dio, *J. Chem. Soc. Perkin Trans. 2*, 1108 (1974), and papers cited therein.
165. R. Caminiti, G. Licheri, G. Piccaluga, and G. Pinna, *J. Chem. Phys.* **65**, 3134 (1976).
166. R. Caminiti, G. Licheri, G. Piccaluga, and G. Pinna, *J. Chem. Phys.* **19**, 371 (1977).

167. R. Caminiti, G. Licheri, G. Piccaluga, and G. Pinna, *J. Chem. Soc. Faraday Disc.* **64** (1977).

168. B. Capon and S. P. McManus, "Neighboring Group Participation," Vol. 1, p. 101, Plenum Press, New York (1976).

169. B. Capon and R. B. Walker, *J. Chem. Soc. Perkin Trans.* 2, 1600 (1974).

170. P. A. Carapelluci, *J. Amer. Chem. Soc.* **97**, 1278 (1975).

171. C. H. Carlisle, B. A. Gorinsky, D. S. Moss, R. A. Palmer, C. de Rango, Y. Mauguen, and G. Tsoucaris, *Acta Cryst.* **A31** (S3), S21 (1975).

172. C. H. Carlisle, R. A. Palmer, S. K. Mazumdar, B. A. Gorinsky, and D. G. R. Yeates, *J. Mol. Biol.* **85**, 1 (1974).

173. C. W. Carter, J. Kraut, S. T. Freer, and R. A. Alden, *J. Biol. Chem.* **249**, 6339 (1974).

174. C. W. Carter, J. Kraut, S. T. Freer, N. H. Xuong, R. A. Alden, and R. G. Bartsch, *J. Biol. Chem.* **249**, 4212 (1974).

175. D. Y. C. Chan, D. J. Mitchell, B. W. Ninham, and B. A. Pailthorpe, *J. Chem. Soc. Faraday Trans. II* **74**, 2050 (1978); **75**, 556 (1979).

176. D. Y. C. Chan, D. J. Mitchell, B. W. Ninham, and B. A. Pailthorpe, *Mol. Phys.* **35**, 1669 (1978).

177. P. S. Y. Cheung and J. G. Powles, *Mol. Phys.* **30**, 921 (1975).

178. C. H. Chothia, *Nature* **248**, 338 (1974).

179. C. Chothia, *Nature* **254**, 304 (1975).

180. C. Chothia and J. Janin, *Nature*, **256**, 705 (1975).

181. J. J. Christensen, R. M. Izatt, and L. D. Hansen, *J. Amer. Chem. Soc.* **89**, 213 (1967).

182. G. A. Clarke and R. W. Taft, *J. Amer. Chem. Soc.* **84**, 2295 (1962).

183. E. Clementi, *Bull. Soc. Chim. Belg.* **85**, 969 (1976).

184. E. Clementi, "Lecture Notes in Chemistry," Vol. 2, "Determination of Liquid Water Structure, Coordination Numbers for Ions and Solvation for Biological Molecules," Springer-Verlag, Berlin (1976).

185. E. Clementi, R. Brosotti, J. Fromm, and R. O. Watts, *Theor. Chim. Acta* **43**, 101 (1976).

186. E. Clementi, F. Cavallone, and R. Scordamaglia, *J. Amer. Chem. Soc.* **99**, 5531 (1977).

187. E. Clementi and H. Popkie, *J. Chem. Phys.* **57**, 1077 (1972).

188. E. Clementi, G. Ranghino, and R. Scordamaglia, *Chem. Phys. Lett.* **49**, 218 (1977).

189. N. J. Cleve, *Suomen Kemistilehti* **B45**, 235 (1972).

190. N. J. Cleve, *Suomen Kemistilehti*, **B46**, 385 (1973).

191. N. J. Cleve, *Finn. Chem. Lett.* **2**, 78 (1974).

192. J. Clifford, *in* "Water: A Comprehensive Treatise" (F. Franks, ed.), Vol. 5, p. 75, Plenum Press, New York (1975).

193. A. F. Cockerill, *J. Chem. Soc. B* 964 (1967).

194. G. H. A. Cole, "An Introduction to the Statistical Theory of Classical Simple Dense Fluids," Pergamon Press, New York (1967).

195. P. M. Coleman, J. N. Jansonius, and B. W. Matthews, *J. Mol. Biol.* **70**, 701 (1972).

196. P. M. Coleman and B. W. Matthews, *J. Mol. Biol.* **60**, 163 (1971).

197. D. M. Collins, M. D. Brice, T. F. M. La Cour, and M. J. Legg, *in* "Crystallographic Computing Techniques" (F. R. Ahmed, ed.), p. 330, Munksgaard, Copenhagen (1976).

198. D. M. Collins, F. A. Cotton, E. E. Hazen, and M. J. Legg, *in* "Structure and Conformation of Nucleic Acids: Protein–Nucleic Acid Interaction," Proc. 4th Ann. Harry Steenbock Symp. 1974, p. 1317 (1975).
199. R. Collins, *Proc. Phys. Soc.* **86**, 199 (1965).
200. R. Cooke and I. D. Kuntz, *Ann. Rev. Biophys. Bioeng.* **3**, 95 (1974).
201. E. H. Cordes and C. Gitler, *Prog. Bioorg. Chem.* **2**, 1 (1973).
202. F. A. Cotton, J. C. Bier, V. W. Day, E. E. Hazen, and S. Larsen, *Cold Spring Harbor Symp. Quant. Biol.* **36**, 243 (1972).
203. F. A. Cotton and E. E. Hazen, *in* "The Enzymes" (P. D. Boyer, ed.), Vol. 4, p. 153, Academic Press, New York (1971).
204. C. A. Coulson and D. Eisenberg, *Proc. Roy. Soc. (London)* **A291**, 445, 454 (1966).
205. B. G. Cox, *J. Chem. Soc. Perkin Trans.* **2**, 607 (1973).
206. B. G. Cox and A. Gibson, *J. Chem. Soc. Chem. Commun.* **638** (1974).
207. B. G. Cox and A. Gibson, *Faraday Symp. Chem. Soc.* No. 10, 107 (1975).
208. G. Cox and A. Gibson, *J. Chem. Soc. Perkin Trans.* **2**, 1812 (1977).
209. B. G. Cox and P. T. McTigue, *Aust. J. Chem.* **20**, 1815 (1967).
210. B. G. Cox and A. J. Parker, *J. Amer. Chem. Soc.* **95**, 408 (1973).
211. R. D. Cramer, III, *J. Amer. Chem. Soc.* **99**, 5408 (1977).
212. F. H. C. Crick and J. C. Kendrew, *Adv. Prot. Chem.* **12**, 133 (1957).
213. D. M. Crothers and D. J. Ratner, *Biochemistry* **7**, 1823 (1968).
214. P. W. Crowe and D. P. Santry, *Chem. Phys. Lett.* **22**, 52 (1973).
215. G. Cubiotti, G. Maisano, P. Migliardo, and F. Wanderlingh, *J. Phys. Chem.* **10**, 4689 (1977).
216. P. T. Cummings, J. W. Perram, and E. R. Smith, *Mol. Phys.* **31**, 535 (1976).
217. S. Cummings, J. E. Enderby, and R. A. Howe, *J. Phys. C*, in press (1979).
218. L. A. Curtiss and J. A. Pople, *J. Mol. Spectrosc.* **55**, 1 (1975).
219. J. F. Cutfield, E. J. Dodson, G. G. Dodson, D. C. Hodgkin, N. W. Isaacs, K. Sakabe, and N. Sakabe, *Act. Cryst.* **A31** (S3), S21 (1975).
220. M. R. J. Dack, *Aust. J. Chem.* **28**, 1643 (1975).
221. M. R. J. Dack, *Chem. Soc. Rev.* **4**, 211 (1975).
222. D. B. Dahlberg, *J. Phys. Chem.* **76**, 2045 (1972).
223. A. K. Das and K. K. Kundu, *J. Chem. Soc. Faraday Trans. I* **69**, 730 (1973).
224. V. G. Dashevsky and G. N. Sarkisov, *Dokl. Acad. Sci.* **202**, 1356 (1972).
225. V. G. Dashevsky and G. N. Sarkisov, *Mol. Phys.* **27**, 1271 (1974).
226. D. W. Davidson *in* "Water: A Comprehensive Treatise" (F. Franks, ed.), Vol. 2, p. 115, Plenum Press, New York (1973).
227. C. S. Davis and J. B. Hyne, *Can. J. Chem.* **50**, 2270 (1972).
228. C. S. Davis and J. B. Hyne, *Can. J. Chem.* **51**, 1687 (1973).
229. A. Dearing and J. S. Rollett, *in* "Abstr. 4th European Crystallographic Meeting, Oxford," *A*24 (1977).
230. D. J. DeFrees, R. T. McIver, Jr., and W. J. Hehre, *J. Amer. Chem. Soc.* **99**, 3853 (1977).
231. J. Deisenhofer and W. Steigemann, *Acta Cryst.* **B31**, 238 (1975).
232. J. E. Del Bene, *J. Chem. Phys.* **55**, 4633 (1971).
233. J. E. Del Bene, *J. Chem. Phys.* **56**, 4923 (1972).
234. J. E. Del Bene, *J. Chem. Phys.* **57**, 1899 (1972).
235. J. Del Bene and J. A. Pople, *J. Chem. Phys.* **52**, 4858 (1970).
236. J. E. Del Bene and J. A. Pople, *J. Chem. Phys.* **58**, 3605 (1973).

237. C. de Rango, Y. Mauguen, G. Tsoucaris, J. Cutfield, G. Dodson, and N. Isaacs, *Acta Cryst.* **A31** (S3), S21 (1975).

238. B. V. Deryagin (ed.), "Research in Surface Forces," Vol. 2, Plenum Press, New York (1966).

239. J. E. Desnoyers, G. E. Pelletier, and C. Jolicoeur, *Can. J. Chem.* **43**, 3232 (1965).

240. J. E. Desnoyers, G. Perron, L. Avedikian, and J. P. Morel, *J. Solution Chem.* **5**, 631 (1976).

241. D. de Vault and B. Chance, *Biophys. J.* **6**, 825 (1966).

242. D. de Vault and B. Chance, *Nature* **215**, 642 (1967).

243. C. de Visser, P. Del, and G. Somsen, *J. Solution Chem.* **6**, 571 (1977).

243a. C. de Visser, G. Perron, and J. E. Desnoyers, *J. Amer. Chem. Soc.* **99**, 5894 (1977).

244. C. de Visser and G. Somsen, *J. Phys. Chem.* **78**, 1719 (1974).

245. H. DeVoe, *J. Amer. Chem. Soc.* **98**, 1724 (1976).

246. A. L. de Vries, *Science* **172**, 1151 (1972).

247. R. Diamond, *Acta Cryst.* **A27**, 436 (1971).

248. R. Diamond, *in* "Crystallographic Computing Techniques" (F. R. Ahmed, ed.), p. 295, Munksgaard, Copenhagen (1976).

249. G. H. F. Diercksen, *Chem. Phys. Lett.* **4**, 373 (1969).

250. G. H. F. Diercksen, *Theor. Chim. Acta* **21**, 335 (1971).

251. G. H. F. Diercksen, *Theor. Chim. Acta* **23**, 398 (1972).

252. G. H. F. Diercksen, W. P. Kraemer, and B. O. Ross, *Theor. Chim. Acta* **36**, 249 (1975).

253. G. H. F. Diercksen, W. P. Kraemer, and W. van Niessen, *Theor. Chim. Acta* **28**, 67 (1972).

254. J. D. Dill, L. C. Allen, W. C. Topp, and J. A. Pople, *J. Amer. Chem. Soc.* **97**, 7220 (1975).

255. K. Dimroth, C. Reichardt, T. Siepmann, and F. Bohlman, *Justus Liebigs Ann. Chem.* **661**, 1 (1963).

256. O. Dobis, J. M. Pearson, and M. Szwarc, *J. Amer. Chem. Soc.* **90**, 278 (1968).

257. N. Dollet and J. Juillard, *J. Solution Chem.* **5**, 77 (1976).

258. A. K. Dorosh and A. F. Skryshevskii, *J. Struct. Chem.* **5**, 842 (1964).

259. A. K. Dorosh and A. F. Skryshevskii, *J. Struct. Chem.* **8**, 300 (1967).

260. P. Douzou, G. H. B. Hoa, and G. A. Petsko, *J. Mol. Biol.* **96**, 367 (1975).

261. W. Drenth and M. Cocivera, *Can. J. Chem.* **54**, 3944 (1976).

262. J. Drenth, W. Hol, J. N. Jansonius, and R. Koekoek, *Cold Spring Harbor Symp. Quant. Biol.* **36**, 107 (1977).

263. J. Drenth, J. N. Jansonius, R. Koekoek, L. A. A. Sluyterman, and B. G. Wolthers, *Phil. Trans. Roy. Soc.* **B257**, 231 (1970).

264. J. Drenth, J. N. Jansonius, R. Koekoek, and B. G. Wolthers, *Adv. Prot. Chem.* **25**, 79 (1971).

265. J. Drenth, K. H. Kalk, and H. M. Swen, *Biochemistry* **15**, 3731 (1976).

266. B. M. Dunn, C. di Bello, and C. B. Anfinsen, *J. Biol. Chem.* **248**, 4769 (1973).

267. T. R. Dyke, K. M. Mack, and J. S. Muenter, *J. Chem. Phys.* **66**, 498 (1977).

268. J. H. Dymond and B. J. Alder, *J. Chem. Phys.* **54**, 3472 (1971).

269. D. Eagland, *in* "Water: A Comprehensive Treatise" (F. Franks, ed.), Vol. 4, p. 305, Plenum Press, New York (1975).

270. F. G. Edwards, J. E. Enderby, R. A. Howe, and D. I. Page, *J. Phys. C.* **8**, 3483 (1975).

271. P. A. Egelstaff, D. I. Page, and C. R. T. Heard, *J. Phys. C.* **4**, 1453 (1971).
272. M. Eigen and G. G. Hammes, *Adv. Enzymology* **25**, 1 (1963).
273. T. Einwohner and B. J. Alder, *J. Chem. Phys.* **49**, 1458 (1968).
274. D. Eisenberg and W. Kauzmann, "The Structure and Properties of Water," Clarendon, Oxford (1969).
275. P. Eisenberger and B. M. Kinncaid, *Chem. Phys. Lett.* **36**, 134 (1975).
276. H. Eklund, B. Nordström, E. Zeppezauer, A. Söderland, I. Ohlsson, T. Boiwe, and C.-I. Brändén, *FEBS Lett.* **44**, 200 (1974).
277. H. Eklund, B. Nordström, E. Zeppezauer, G. Söderland, I. Ohlsson, T. Boiwe, B.-O. Söderberg, and C. I. Brändén, *Acta Cryst.* **A31** (S3), S29 (1975).
278. H. Eklund, B. Nordström, E. Zeppezauer, G. Söderland, I. Ohlsson, T. Boiwe, B.-O. Söderberg, O. Tapia, C.-I. Brändén, and Å. Åkeson, *J. Mol. Biol.* **102**, 27 (1976).
279. J. E. Enderby, D. M. North, and P. A. Egelstaff, *Phil. Mag.* **14**, 961 (1966).
280. J. F. J. Engbersen, Ph.D. Thesis, Groningen (1976).
281. J. F. J. Engbersen and J. B. F. N. Engberts, *J. Amer. Chem. Soc.* **97**, 1563 (1975).
282. J. B. F. N. Engberts, H. Morssink, and A. Vos, *J. Amer. Chem. Soc.* **100**, 799 (1978).
283. O. Epp, E. E. Lattman, M. Schiffer, R. Huber, and W. Palm, *Biochemistry* **14**, 4043 (1975).
284. E. K. Euranto and N. J. Cleve, *Acta Chem. Scand.* **17**, 1584 (1963).
285. E. K. Euranto and N. J. Cleve, *Acta Chem. Scand.* **27**, 1841 (1973).
286. M. G. Evans and M. Polanyi, *Trans. Faraday Soc.* **31**, 875 (1935).
287. O. Exner, *Nature (London)* **201**, 488 (1964).
288. O. Exner, *Nature (London)* **227**, 366 (1970).
289. O. Exner, *Coll. Czech. Chem. Commun.* **38**, 799 (1973).
290. H. Eyring, *J. Chem. Phys.* **3**, 107 (1935).
291. L. Fabbrizzi, M. Micheloni, P. Paoletti, and G. Schwarzenbach, *J. Amer. Chem. Soc.* **99**, 5574 (1977).
292. A. H. Fainberg and S. Winstein, *J. Amer. Chem. Soc.* **78**, 2770 (1956).
293. M. Falk and O. Knop, in "Water: A Comprehensive Treatise" (F. Franks, ed.), Vol. 2, p. 55, Plenum Press, New York (1973).
294. Y. Fang and J. de la Vega, *Chem. Phys. Lett.* **6**, 117 (1970).
295. I. E. Farquar, "Ergodic Theory in Statistical Mechanics," Interscience, New York (1964).
296. H. Fehlhammer, W. Bode, and R. Huber, *J. Mol. Biol.* **111**, 415 (1977).
297. J. H. Fendler, *Acc. Chem. Res.* **9**, 153 (1976).
298. J. H. Fendler and E. J. Fendler, "Catalysis in Micellar and Macromolecular Systems," Academic Press, New York (1975).
299. J. H. Fendler and L. J. Liu, *J. Amer. Chem. Soc.* **97**, 999 (1975).
300. G. Fermi, *J. Mol. Biol.* **97**, 237 (1975).
301. L. P. Fernandez and L. G. Hepler, *J. Amer. Chem. Soc.* **81**, 1783 (1959).
302. T. H. Fife and D. M. McMahon, *J. Amer. Chem. Soc.* **91**, 7481 (1969).
303. A. L. Fink, *Biochemistry* **12**, 1736 (1973).
304. A. L. Fink, *Biochemistry* **13**, 277 (1974).
305. A. L. Fink and I. A. Ahmed, *Nature* **263**, 294 (1976).
306. J. L. Finney, *J. Mol. Biol.* **96**, 721 (1975).
307. J. L. Finney, *Phil. Trans. Roy. Soc.* **B278**, 3 (1977).
308. J. L. Finney, *J. Mol. Biol.* **119**, 415 (1978).

309. J. L. Finney and P. A. Timmins, *I. L. L. Annual Report* (*1977*), Annex p. 351, Institut Laue-Langevin, Grenoble, France.

310. I. Z. Fisher, *Statistical Theory of Liquids*, Nauka Publishing House, Moscow (1961).

311. P. D. Fleming, III and J. H. Gibbs, *J. Stat. Phys.* **10**, 157, 351 (1974).

312. F. K. Fong, *Acc. Chem. Res.* **9**, 433 (1976).

313. D. W. Fong and E. Grunwald, *J. Phys. Chem.* **73**, 3909 (1969).

314. M. P. Fontana, *Solid State Commun.* **18**, 765 (1976).

315. E. Forslind and A. Jacobsson, *in* "Water: A Comprehensive Treatise" (F. Franks, ed.), Vol. 5, p. 173, Plenum Press, New York (1975).

316. F. W. Fowler, A. R. Katritzky, and R. J. D. Rutherford, *J. Chem. Soc. B* **460** (1971).

317. H. S. Frank, *in* "Chemical Physics of Ionic Solutes," (B. E. Conway and R. G. Barradas, eds.), Wiley, New York (1965).

318. H. S. Frank, *Federation Proc. Suppl. 15*, **24**, 1 (1965).

319. H. S. Frank, *in* "Water: A Comprehensive Treatise," (F. Franks, ed.), Vol. 1, p. 515, Plenum Press, New York (1972).

320. H. S. Frank and M. W. Evans, *J. Chem. Phys.* **13**, 507 (1945).

321. H. S. Frank and F. Franks, *J. Chem. Phys.* **48**, 4746 (1968).

322. H. S. Frank and A. S. Quist, *J. Chem. Phys.* **34**, 604 (1961).

323. H. S. Frank and W. Y. Wen, *Disc. Faraday Soc.* **24**, 133 (1957).

324. L. S. Frankel, T. R. Stengle, and C. H. Langford, *J. Chem. Soc. Chem. Comm.* 393 (1965).

325. F. Franks, *in* "Water: A Comprehensive Treatise" (F. Franks, ed.), Vol. 1, p. 115, Plenum Press, New York (1972).

326. F. Franks, *in* "Water: A Comprehensive Treatise" (F. Franks, ed.), Vol. 2, p. 1, Plenum Press, New York (1973).

327. F. Franks, *in* "Water: A Comprehensive Treatise" (F. Franks, ed.), Vol. 4, p. 1, Plenum Press, New York (1975).

328. F. Franks, *Nature* **270**, 386 (1977).

329. F. Franks, *Phil. Trans. Roy. Soc.* **B278**, 33 (1977).

330. F. Franks and D. Eagland, *Crit. Rev. Biochem.* **3**, 165 (1975).

331. F. Franks and D. E. Morris, *Biochim. Biophys. Acta* **540**, 346 (1978).

332. F. Franks, M. Pedley, and D. S. Reid, *J. Chem. Soc. Faraday Trans. I* **72**, 359 (1976).

333. F. Franks, J. R. Ravenhill, and D. S. Reid, *J. Solution Chem.* **1**, 3 (1972).

334. F. Franks, D. S. Reid, and A. Suggett, *J. Solution Chem.* **2**, 99 (1973).

335. F. Franks and H. T. Smith, *J. Phys. Chem.* **68**, 3581 (1964).

336. S. T. Freer, R. A. Alden, S. A. Levens, and J. Kraut, *in* "Crystallographic Computing Techniques" (F. R. Ahmed, ed.), p. 317, Munksgaard, Copenhagen (1976).

337. H. L. Friedman, *J. Solution Chem.* **1**, 387, 413, 419 (1972).

338. H. L. Friedman and C. V. Krishnan, *in* "Water: A Comprehensive Treatise" (F. Franks, ed.), Vol. 3, p. 1, Plenum Press, New York (1973).

339. M. E. Friedman and H. A. Scheraga, *J. Phys. Chem.* **69**, 3795 (1965).

340. J. A. Frier and M. F. Perutz, *J. Mol. Biol.* **112**, 97 (1977).

341. J. Fromm, E. Clementi, and R. O. Watts, *J. Chem. Phys.* **62**, 1388 (1975).

342. R. Fuchs, C. P. Hagan, and R. F. Rodewald, *J. Phys. Chem.* **78**, 1509 (1974).

343. R. Fuchs and J. R. Jones, *in* "Analytical Calorimetry," Vol. 4, p. 227, Plenum Press, New York (1977).

344. N. Fukuta and B. J. Mason, *J. Phys. Chem. Solids* **24**, 715 (1963).

345. G. W. Gear, *ANL Report* No. ANL-7126 (1966).
346. G. W. Gear, "Numerical Initial Value Problems in Ordinary Differential Equations," Chap. 9, Prentice Hall, Englewood Cliffs, New Jersey (1971).
347. F. W. Getzen, in "Techniques of Chemistry" (M. R. J. Dack, ed.), Vol. VIII, Part II, Wiley, New York (1976).
348. S. J. Gill and L. Noll, *J. Phys. Chem.* **76**, 3065 (1972).
349. D. Gingell and V. A. Parsegian, *J. Theor. Biol.* **36**, 41 (1972).
350. J. A. Glasel, *J. Amer. Chem. Soc.* **92**, 375 (1970).
351. J. P. Glusker, *J. Mol. Biol.* **38**, 149 (1968).
352. W. A. Goddard and W. J. Hunt, *Chem. Phys. Lett.* **24**, 464 (1974).
353. H. S. Golinkin and J. B. Hyne, *Can. J. Chem.* **46**, 125 (1968).
354. H. S. Golinkin, I. Lee, and J. B. Hyne, *J. Amer. Chem. Soc.* **89**, 1307 (1967).
355. H. Goldstein, "Classical Mechanics," Addison-Wesley, Reading, Massachusetts (1953).
356. I. C. Golton, J. D. Nicholas, and J. L. Finney, in "Abstr. 4th European Crystallographic Meeting, Oxford," *A*370 (1977).
357. J. E. Gordon, *J. Amer. Chem. Soc.* **94**, 650 (1972).
358. J. E. Gordon, "The Organic Chemistry of Electrolyte Solutions," Wiley, New York (1975).
359. T. Graafland, J. B. F. N. Engberts, and A. J. Kirby, *J. Org. Chem.* **42**, 2462 (1977).
360. J. Graham and G. F. Walker, *J. Chem. Phys.* **40**, 540 (1964).
361. E. P. Grimstud and P. Kebarle, *J. Amer. Chem. Soc.* **95**, 7939 (1973).
362. E. Grunwald, G. Baughman, and G. Kohnstam, *J. Amer. Chem. Soc.* **82**, 5801 (1960).
363. E. Grunwald and A. Effio, *J. Amer. Chem. Soc.* **96**, 423 (1974).
364. E. Grunwald and D. Eustace, in "Proton-Transfer Reactions" (E. Caldin and V. Gold, eds.), Chap. 4, Chapman and Hall, London (1975).
365. E. Grunwald, R. L. Lipnick, and E. K. Ralph, *J. Amer. Chem. Soc.* **91**, 4333 (1969).
366. E. Grunwald, K.-C. Pan, and A. Effio, *J. Phys. Chem.* **80**, 2937 (1976).
367. E. Grunwald and E. Price, *J. Amer. Chem. Soc.* **86**, 4517 (1964).
368. E. Grunwald and E. K. Ralph, *J. Amer. Chem. Soc.* **89**, 4405 (1967).
369. E. Grunwald and S. Winstein, *J. Amer. Chem. Soc.* **70**, 846 (1948).
370. H. Guilett, L. Avedikian, and J.-P. Morel, *Can. J. Chem.* **53**, 455 (1975).
371. R. W. Gurney, in "Ionic Processes in Solution," p. 251, McGraw-Hill, New York (1953).
372. J. P. Guthrie, *Can. J. Chem.* **51**, 3494 (1973).
373. J. P. Guthrie, *J. C. S. Chem. Commun.* 898 (1973).
374. A. T. Hagler and J. Moult, *Nature* **272**, 222 (1978).
375. G. G. Hall, *Chem. Phys. Lett.* **6**, 501 (1975).
376. J. M. Hammersley and D. C. Handscomb, "Monte Carlo Methods," Wiley, New York (1964).
377. L. P. Hammett, "Physical Organic Chemistry," 2nd ed., McGraw-Hill, New York (1970).
378. D. Hankins, J. W. Moskowitz, and F. H. Stillinger, *Chem. Phys. Lett.* **4**, 527 (1970).
379. D. Hankins, J. W. Moskowitz, and F. H. Stillinger, *J. Chem. Phys.* **53**, 4544 (1970).
380. D. Hankins, J. W. Moskowitz, and F. H. Stillinger, *J. Chem. Phys.* **59**, 995 (1973).
381. J. P. Hansen, and I. R. MacDonald, "Theory of Simple Liquids," Academic Press, New York (1976).

382. H. S. Harned and B. B. Owen, "The Physical Chemistry of Electrolyte Solutions," 3rd ed., Reinhold, New York (1958).
383. H. S. Harned and A. M. Ross, *J. Amer. Chem. Soc.* **63**, 1993 (1941).
384. G. D. Harp and B. J. Berne, *J. Chem. Phys.* **49**, 1249 (1968).
385. G. D. Harp and B. J. Berne, *Phys. Rev.* **2**, A975 (1970).
386. F. E. Harris and B. J. Alder, *J. Chem. Phys.* **21**, 1031 (1953).
387. J. M. Harris, D. J. Raber, W. C. Neal, Jr., and M. D. Dukes, *Tetrahedron Lett.* 2331 (1974).
388. A. Hartkopf and B. L. Karger, *Acc. Chem. Res.* **6**, 209 (1973).
389. J. A. Hartsuck and W. N. Lipscomb, *in* "The Enzymes" (P. D. Boyer, ed.), Vol. 3, p. 1, Academic Press, New York (1971).
390. A. E. V. Haschemeyer, W. Guschlbauer, and A. L. de Vries, *Nature* **269**, 87 (1977),
391. H. Hauser, *in* "Water: A Comprehensive Treatise" (F. Franks, ed.), Vol. 4, p. 209 Plenum Press, New York (1975).
392. K. Heinonen and E. Tommila, *Suomen Kemistilehti* **B38**, 9 (1965).
393. K. Heinzinger and P. C. Vogel, *Z. Naturforsch.* **29a**, 1164 (1974).
394. K. Heinzinger and P. C. Vogel, *Z. Naturforsch.* **31a**, 463 (1976).
395. D. Henderson and P. J. Leonard, *in* "Physical Chemistry: Advanced Treatise" (H. Eyring, ed.), Vol. 8B, Academic Press, New York (1971).
396. R. Henderson, *J. Mol. Biol.* **54**, 341 (1970).
397. R. Henderson, C. S. Wright, G. P. Hess, and D. M. Blow, *Cold Spring Harbor Symp. Quant. Biol.* **36**, 63 (1972).
398. W. G. Henderson, M. Taagepera, D. Holtz, R. T. McIver, Jr., J. L. Beauchamp, and R. W. Taft, *J. Amer. Chem. Soc.* **94**, 4728 (1972).
399. W. A. Hendrickson and W. E. Love, *Nature New Biology* **232**, 197 (1971).
400. J. Hermans and A. Rahman, "Report of CECAM Workshop on Protein Dynamics," p. 153, CECAM, Orsay, France (1976).
401. D. R. Herrick and F. H. Stillinger, *J. Chem. Phys.* **65**, 1345 (1976).
402. T. T. Herskovits and T. M. Kelly, *J. Phys. Chem.* **77**, 381 (1973).
403. F. Hibbert and F. A. Long, *J. Amer. Chem. Soc.* **94**, 7637 (1972).
404. J. H. Hildebrand, J. M. Prausnitz, and R. L. Scott, "Regular and Related Solutions," Van Nostrand Reinhold, New York (1970).
405. B. J. Hiley, J. L. Finney, and T. Burke, *J. Phys. A* **10**, 197 (1977).
406. D. J. T. Hill and C. Malar, *Aust. J. Chem.* **28**, 7 (1975).
407. C. J. Hills and C. A. Viana, in "Hydrogen-Bonded Solvent Systems" (A. K. Covington and P. Jones, eds.), p. 261, Taylor and Francis, London (1968).
408. J. O. Hirschfelder, C. F. Curtis, and R. B. Bird, "Molecular Theory of Gases and Liquids," Wiley, New York (1954).
409. R. W. Hockney, *Meth. Comp. Phys.* **9**, 176 (1970).
410. R. W. Hockney and T. R. Brown, *Technical Report RCS27*, Dept. of Computer Science, University of Reading, Reading RG6 2AX, U.K. (1974).
411. R. W. Hockney, S. P. Goel, and J. W. Eastwood, *Chem. Phys. Lett.* **21**, 589 (1973).
412. R. W. Hockney, S. P. Goel, and J. W. Eastwood, *J. Comp. Phys.* **14**, 148 (1974).
413. F. Hofmeister, *Arch. Exptl. Pathol. Pharmakol.* **24**, 247 (1888).
414. G. Höjer and J. Keller, *J. Amer. Chem. Soc.* **96**, 3746 (1974).
415. H. A. J. Holterman and J. B. F. N. Engberts, *J. Org. Chem.* **42**, 2792 (1977).
416. H. A. J. Holterman and J. B. F. N. Engberts, *J. Phys. Chem.* **83**, 443 (1979).
417. W. G. Hoover and B. J. Alder, *J. Chem. Phys.* **46**, 686 (1967).

418. A. J. Hopfinger, "Conformational Properties of Macromolecules," Academic Press, New York (1973).
419. H. P. Hopkins, Jr., W. C. Duer, and F. J. Millero, *J. Solution Chem.* **5**, 263 (1976).
420. R. A. Horne, *in* "Survey of Progress in Chemistry" (A. F. Scott, ed.), Vol. 4, p. 1, Academic Press, New York (1968).
421. R. A. Horne, "Water and Aqueous Solutions," Chap. 8, Wiley-Interscience, New York (1972).
422. I.-N. Hsu, L. T. J. Delbaere, M. N. G. James, and T. Hofmann, *Nature* **266**, 140 (1977).
423. R. Huber, W. Bode, K. Kukla, U. Kohl, and C. A. Ryan, *Biophys. Struct. Mech.* **1**, 189 (1975).
424. R. F. Hudson, *J. Chem. Soc. B*, 761 (1966).
425. E. D. Hughes and C. K. Ingold, *J. Chem. Soc.*, 244 (1935).
426. E. D. Hughes, C. K. Ingold, and C. S. Patel, *J. Chem. Soc.*, 526 (1935).
427. M. Hunkapillar, S. Smallcomb, D. Whitaker, and J. H. Richards, *Biochem.* **12**, 4732 (1973).
428. R. Huq, *J. Chem. Soc. Faraday Trans. I* **69**, 1195 (1973).
429. P. Huurdeman and J. B. F. N. Engberts, *J. Org. Chem.* **44**, 297 (1979).
430. J. Hylton, R. E. Christoffersen, and G. G. Hall, *Chem. Phys. Lett.* **24**, 501 (1974).
431. J. B. Hyne, *J. Amer. Chem. Soc.* **82**, 5129 (1960).
432. J. B. Hyne, *in* "Hydrogen Bonded Solvent Systems" (A. K. Covington and P. Jones, eds.), p. 99, Taylor and Francis, London (1968).
433. J. B. Hyne, H. S. Golinkin, and W. G. Laidlaw, *J. Amer. Chem. Soc.* **88**, 2104 (1966).
434. J. B. Hyne and R. E. Robertson, *Can. J. Chem.* **34**, 931 (1956).
435. T. Imoto, L. N. Johnson, A. C. T. North, D. C. Phillips, and J. A. Rupley, *in* "The Enzymes" (P. D. Boyer, ed.), Vol. 7, p. 665, Academic Press, New York (1972).
436. C. K. Ingold, "Structure and Mechanism in Organic Chemistry," Bell, London (1953).
437. N. Isaacs, R. James, H. Niall, G. Bryant-Greenwood, G. Dodson, A. Evans, and A. C. T. North, *Nature* **271**, 278 (1978).
438. J. N. Israelachvili, *J. Chem. Soc. Faraday II* **69**, 1729 (1973).
439. J. N. Israelachvili and D. Tabor, *Prog. Surface Membrane Sci.* **7**, 1 (1973).
440. G. Jacucci, I. R. MacDonald, K. Singer, *Phys. Lett.* **50A**, 141 (1974).
441. D. K. Jaiswal and J. R. Jones, *J. Chem. Soc. Perkin Trans. 2*, 102 (1976).
442. M. N. G. James, I.-N. Hsu, and L. T. J. Delbaere, *Nature* **267**, 808 (1977).
443. M. Jaszunski and A. J. Sadlej, *Chem. Phys. Lett.* **15**, 41 (1972).
444. G. A. Jeffrey and R. K. McMullen, *Prog. Inorg. Chem.* **8**, 43 (1967).
445. W. P. Jencks, "Catalysis in Chemistry and Enzymology," McGraw-Hill, New York (1969).
446. W. P. Jencks, *Adv. Enzym.* **43**, 219 (1975).
447. W. P. Jencks and J. Carriuolo, *J. Amer. Chem. Soc.* **83**, 1743 (1961).
448. L. H. Jensen, *Ann. Rev. Biophys. Bioeng.* **3**, 81 (1974).
449. L. H. Jensen, *in* "Crystallographic Computing Techniques" (F. R. Ahmed, ed.), p. 307, Mumksgaard, Copenhagen (1976).
450. A. Johansson and P. A. Kollman, *J. Amer. Chem. Soc.* **94**, 6196 (1972).
451. S. L. Johnson, *Adv. Phys. Org. Chem.* **5**, 237 (1967).

452. C. Jolicoeur, P. Bernier, E. Firkins, and J. K. Saunders, *J. Phys. Chem.* **80**, 1908 (1976).
453. C. Jolicoeur, E. Firkins, and P. Bernier, *J. Phys. Chem.* **78**, 51 (1974).
454. C. Jolicoeur and G. Lacroix, *Can. J. Chem.* **51**, 3051 (1973).
455. J. B. Jones and K. D. Gordon, *Biochemistry* **12**, 71 (1973).
456. J. R. Jones, "The Ionization of Carbon Acids," Academic Press, New York (1973).
457. J. R. Jones and R. Fuchs, *Can. J. Chem.* **55**, 99 (1977).
458. B. L. Kaiser and E. T. Kaiser, *Acc. Chem. Res.* **5**, 219 (1972).
459. E. T. Kaiser and K.-W. Lo, *J. Amer. Chem. Soc.* **91**, 4912 (1969).
460. K. Kalliorinne and E. Tommila, *Acta Chem. Scand.* **23**, 2567 (1969).
461. M. J. Kamlet, J. L. Abboud, and R. W. Taft, *J. Amer. Chem. Soc.* **99**, 6027 (1977) and earlier papers.
462. M. K. Kamlet and R. W. Taft, *J. Amer. Chem. Soc.* **98**, 377, 2886 (1976).
463. A. Kankaanperä and R. Aaltonen, *Suomen Kemistilehti* **B43**, 183 (1970).
464. A. Kankaanperä and R. Aaltonen, *Acta Chem. Scand.* **26**, 1698 (1972).
465. K. K. Kannan, A. Liljas, I. Waara, P.-C. Bergstén, S. Lövgren, B. Strandberg, U. Bengtsson, U. Carlbom, K. Fridborg, L. Järup, and M. Petef, *Cold Spring Harbor Symp. Quant. Biol.* **36**, 221 (1971).
466. K. K. Kannan, B. Notstrand, K. Fridborg, S. Lövgren, A. Ohlsson, and M. Petef, *Proc. Nat. Acad. Sci. USA* **72**, 51 (1975).
467. J. F. Karnicky and C. J. Pings, *Adv. in Chem. Phys. XXXIV*, 157 (1976).
468. W. Karzijn and J. B. F. N. Engberts, *Rec. Trav. Chim.* **96**, 95 (1977).
469. W. Karzijn and J. B. F. N. Engberts, *Tetrahedron Lett.*, 1787 (1978).
470. W. Karzijn and J. B. F. N. Engberts, to be published.
471. J. Kaspi and Z. Rappoport, *Tetrahedron Lett.* 2035 (1977).
472. H. P. Kasserra and K. J. Laidler, *Can. J. Chem.* **48**, 1793 (1970).
473. A. I. Kataygorodsky and K. V. Mirskaya, *Tetrahedron* **9**, 183 (1960).
474. W. Kauzmann, *Adv. Prot. Chem.* **14**, 1 (1959).
475. R. L. Kay and T. L. Broadwater, *J. Solution Chem.* **5**, 57 (1976).
476. D. S. Kemp and K. Paul, *J. Amer. Chem. Soc.* **92**, 2553 (1970).
477. C. W. Kern and M. Karplus, *in* "Water: A Comprehensive Treatise" (F. Franks, ed.), Vol. 1, p. 21, Plenum Press, New York (1972).
478. W. R. Kester and B. W. Matthews, *Biochemistry* **16**, 2506 (1977).
479. R. G. Khalifah, *J. Biol. Chem.* **246**, 2561 (1971).
480. R. G. Khalifah, *Proc. Nat. Acad. Sci. USA* **70**, 1986 (1973).
481. F. Y. Khalil and H. Sadek, *Z. Physik. Chem.* **75**, 308 (1971).
482. K. H. Khoo and C. Chee-Yan, *J. Chem. Soc. Faraday Trans. I*, **71**, 446 (1975).
483. J. V. Kilmartin, J. J. Breen, G. C. K. Roberts, and C. Ho, *Proc. Nat. Acad. Sci. USA* **70**, 1246 (1973).
484. A. Kinchin, "Mathematical Foundations of Statistical Mechanics," Dover, New York, 1949.
485. A. J. Kirby and A. R. Fersht, *Prog. Bioorg. Chem.* **1**, 1 (1971).
486. J. G. Kirkwood, *J. Chem. Phys.* **2**, 351 (1934).
487. J. G. Kirkwood, *J. Chem. Phys.* **7**, 911 (1939).
488. J. G. Kirkwood and F. P. Buff, *J. Chem. Phys.* **19**, 774 (1951).
489. J. G. Kirkwood and F. H. Westheimer, *J. Chem. Phys.* **6**, 506 (1938).
490. H. Kistenmacher, G. C. Lie, H. Popkie, and E. Clementi, *J. Chem. Phys.* **61**, 546 (1974).

491. H. Kistenmacher, H. Popkie, and E. Clementi, *J. Chem. Phys.* **58**, 1689 (1973).
492. H. Kistenmacher, H. Popkie, and E. Clementi, *J. Chem. Phys.* **58**, 5627 (1973).
493. H. Kistenmacher, H. Popkie, and E. Clementi, *J. Chem. Phys.* **59**, 5842 (1973).
494. H. Kistenmacher, H. Popkie, and E. Clementi, *J. Chem. Phys.* **61**, 799 (1974).
495. H. Kistenmacher, H. Popkie, E. Clementi, and R. O. Watts, *J. Chem. Phys.* **60**, 4455 (1974).
496. O. Kiyohara and G. C. Benson, *Can. J. Chem.* **55**, 1354 (1977).
497. M. H. Klapper, *Prog. Bioorg. Chem.* **2**, 55 (1973).
498. I. M. Klotz, *Science* **128**, 815 (1958).
499. I. M. Klotz and J. S. Franzen, *J. Amer. Chem. Soc.* **84**, 3461 (1962).
500. I. M. Klotz, E. C. Stellwagen, and V. H. Stryker, *Biochim. Biophys. Acta*, **86**, 122 (1962).
501. B. R. Knauer and J. J. Napier, *J. Amer. Chem. Soc.* **98**, 4395 (1976).
502. J. Knowles and J. Albery, *Acc. Chem. Res.* **10**, 105 (1977).
503. E. C. F. Ko and R. E. Robertson, *J. Amer. Chem. Soc.* **94**, 573 (1972).
504. S. Kodama, *Bull. Chem. Soc. Japan* **35**, 827 (1962).
505. G. Kohnstam, *Adv. Phys. Org. Chem.* **5**, 121 (1967).
506. P. A. Kollman, *J. Amer. Chem. Soc.* **99**, 4875 (1977).
507. P. A. Kollman and L. C. Allen, *J. Chem. Phys.* **51**, 3286 (1969).
508. P. A. Kollman and L. C. Allen, *J. Amer. Chem. Soc.* **92**, 753 (1970).
509. I. A. Koppel and V. A. Palm, *in* "Advances in Linear Free Energy Relationships" (N. B. Chapman and J. Shorter, eds.), Chap. 5, Plenum Press, New York (1972).
510. R. Korenstein and B. Hess, *Nature* **270**, 184 (1977).
511. K. M. Koshy, K. Mohanty, and R. E. Robertson, *Can. J. Chem.* **55**, 1314 (1977).
512. K. M. Koshy and R. E. Robertson, *Can. J. Chem.* **52**, 2485 (1974).
513. K. M. Koshy, R. E. Robertson, G. S. Dyson, and S. Singh, *Can. J. Chem.* **54**, 3614 (1976).
514. E. M. Kosower, *J. Amer. Chem. Soc.* **80**, 3253 (1958).
515. E. M. Kosower, "An Introduction to Physical Organic Chemistry," Wiley, New York (1968).
516. J. K. Kozak, W. S. Knight, and W. Kauzmann, *J. Chem. Phys.* **48**, 675 (1968).
517. J. W. Kress, E. Clementi, J. J. Kozak, and M. E. Schwartz, *J. Chem. Phys.* **63**, 3907 (1975).
518. R. H. Kretsinger and C. E. Nockolds, *J. Biol. Chem.* **248**, 3313 (1973).
519. M. Krieger, L. M. Kay, and R. M. Stroud, *J. Mol. Biol.* **83**, 209 (1974).
520. J. Kroh and G. Stradowski, *Int. J. Radiat. Phys. Chem.* **5**, 243 (1973).
521. H. Krueger and M. A. Johnson, *Inorg. Chem.* **7**, 679 (1968).
522. R. R. Krug, W. G. Hunter, and R. A. Grieger, *J. Phys. Chem.* **80**, 2335, 2341 (1976).
523. R. F. Kruh, *Chem. Rev.* **62**, 319 (1962).
524. R. F. Kruh and C. L. Standley, *J. Inorg. Chem.* **1**, 941 (1962).
525. I. D. Kuntz, *J. Amer. Chem. Soc.* **94**, 4009 (1972).
526. I. D. Kuntz, *J. Amer. Chem. Soc.* **94**, 8586 (1972).
527. I. D. Kuntz and W. Kauzmann, *Adv. Prot. Chem.* **28**, 239 (1973).
528. J. K. Kurz and G. J. Ehrhardt, *J. Amer. Chem. Soc.* **97**, 2259 (1975).
529. J. L. Kurz and J. M. Farrar, *J. Amer. Chem. Soc.* **97**, 2250 (1975).
530. J. L. Kurz and D. N. Wexler, *J. Amer. Chem. Soc.* **97**, 2255 (1975).
531. A. J. C. Ladd, *Mol. Phys.* **33**, 1039 (1977).

532. R. C. Ladner, E. J. Heidner, and M. F. Perutz, *J. Mol. Biol.* **114**, 385 (1977).

533. K. J. Laidler, *Suomen Kemistilehti* **A33**, 44 (1960).

534. K. J. Laidler and J. C. Polanyi, *Prog. React. Kin.* **3**, 1 (1965).

535. C. H. Langford and J. P. K. Tong, *Acc. Chem. Res.* **10**, 258 (1977).

536. B. Larsen, T. Forland, and K. Singer, *Mol. Phys.* **26**, 1521 (1973).

537. J. W. Larson and L. G. Hepler, *in* "Solute–Solvent Interactions" (J. F. Coetzee and C. D. Ritchie, eds.), Chap. 1, Marcel Dekker, New York (1969).

538. R. M. Lawrence and R. F. Kruh, *J. Chem. Phys.* **47**, 4758 (1967).

539. J. L. Lebowitz and J. S. Rowlinson, *J. Chem. Phys.* **41**, 1331 (1964).

540. B. Lee and F. M. Richards, *J. Mol. Biol.* **55**, 379 (1971).

541. I. Lee and J. B. Hyne, *Can. J. Chem.* **46**, 2333 (1968).

542. J. A. Leffler and E. Grunwald, "Rates and Equilibria of Organic Reactions," Wiley, New York (1963).

543. H. L. Lemberg and F. H. Stillinger, *J. Chem. Phys.* **52**, 1677 (1975).

544. H. L. Lemberg and F. H. Stillinger, *Mol. Phys.* **32**, 353 (1976).

545. D. M. LeNeveu, R. P. Rand, and V. A. Parsegian, *Nature* **259**, 601 (1976).

546. D. M. LeNeveu, R. P. Rand, V. A. Parsegian, and D. Gingell, *Biophys. J.* **18**, 209 (1977).

547. B. R. Lentz and H. A. Scheraga, *J. Chem. Phys.* **58**, 5296 (1973).

548. D. Levesgue and L. Verlet, *Phys. Rev.* **2**, A2514 (1970).

549. M. Levitt, *J. Mol. Biol.* **104**, 59 (1976).

550. M. Levitt and S. Lifson, *J. Mol. Biol.* **46**, 269 (1969).

551. M. Levitt and A. Warshel, *Nature* **253**, 694 (1975).

552. M. Levitt and A. Warshel, *J. Mol. Biol.* **106**, 421 (1976).

553. S. Lewin, "Displacement of Water and Its Control of Biochemical Reactions," Academic Press, New York (1974).

554. J. W. E. Lewis and K. Singer, *J. Chem. Soc. Faraday Trans.* **71**, 41 (1975).

555. J. W. E. Lewis, K. Singer, and L. V. Woodcock, *J. Chem. Soc. Faraday Trans. I* **71**, 301 (1975).

556. G. Licheri, G. Piccaluga, and G. Pinna, *J. Appl. Cryst.* **6**, 392 (1973).

557. G. Licheri, G. Piccaluga, and G. Pinna, *J. Chem. Phys.* **63**, 4412 (1975).

558. G. Licheri, G. Piccaluga, and G. Pinna, *Phys. Lett.* **35**, 119 (1975).

559. G. Licheri, G. Piccaluga, and G. Pinna, *J. Chem. Phys.* **64**, 2437 (1976).

560. G. C. Lie and E. Clementi, *J. Chem. Phys.* **62**, 2195 (1976).

561. A. Liljas, K. K. Kannan, P.-C. Bergstén, I. Waara, K. Fridborg, B. Strandberg, U. Carlbom, L. Järup, S. Lövgren, and M. Petef, *Nature New Biol.* **235**, 131 (1972).

562. S. Linskog and J. E. Coleman, *Proc. Nat. Acad. Sci. USA* **70**, 2505 (1973).

563. C. L. Liotta, A. Abidaud, and H. P. Hopkins, Jr. *J. Amer. Chem. Soc.* **94**, 8624 (1972).

564. W. N. Lipscomb, J. A. Hartsuck, F. A. Quiocho, and G. N. Reeke, Jr., *Proc. Nat. Acad. Sci. USA* **64**, 28 (1969).

565. W. N. Lipscomb, J. A. Hartsuck, G. N. Reeke, F. A. Quiocho, P. H. Bethge, M. L. Ludwig, T. A. Steitz, H. Muirhead, and J. C. Coppola, *Brookhaven Symp. Biol.* **21**, 24 (1968).

566. H. Lipson and C. A. Taylor, "Fourier Transforms and X-Ray Diffraction," Bell, London (1958).

567. G. Livingstone, F. Franks, and L. J. Aspinall, *J. Solution Chem.* **6**, 203 (1977).

568. J. Llor and M. Cortijo, *J. Chem. Soc. Perkin Trans.* **2**, 1111 (1977).

569. J. Long and B. Munson, *J. Amer. Chem. Soc.* **95**, 2427 (1973).
570. F. Lontelme, P. Turq, B. Quentree, and J. Lewis, *Mol. Phys.* **28**, 1537 (1974).
571. M. Losonczy, J. W. Moskowitz, and F. H. Stillinger, *J. Chem. Phys.* **59**, 3264 (1973).
572. R. Lovett and A. Ben-Naim, *J. Chem. Phys.* **51**, 3108 (1969).
573. B. W. Low and F. M. Richards, *J. Amer. Chem. Soc.* **76**, 2511 (1954).
574. T. H. Lowry and K. S. Richardson, "Mechanism and Theory in Organic Chemistry," Harper and Row, New York (1976).
575. M. Lucas, *J. Phys. Chem.* **80**, 359 (1976).
576. M. Lucas and R. Bury, *J. Phys. Chem.* **80**, 999 (1976).
577. W. A. P. Luck, *in* "The Hydrogen Bond" (P. Schuster, G. Zundel, and C. Sandorfy, eds.), Vol. 3, Chap. 28, North Holland Publ. Co. Amsterdam (1976).
578. M. L. Ludwig, R. D. Anderson, P. A. Apgar, R. M. Burnett, M. E. LeQuesne, and S. G. Mayhew, *Cold Spring Harbor Symp. Quant. Biol.* **36**, 389 (1972).
579. R. Lumry, *Adv. Chem. Phys.* **21**, 567 (1971).
580. R. Lumry and S. Rajender, *Biopolymers* **9**, 1125 (1970).
581. I. G. Macara, T. G. Hoy, and P. M. Harrison, *Biochem. J.* **126**, 151 (1972).
582. D. D. MacDonald, B. Dolan, and J. B. Hyne, *J. Solution Chem.* **5**, 405 (1976).
583. D. D. MacDonald and J. B. Hyne, *Can. J. Chem.* **48**, 2494 (1970).
584. D. D. MacDonald, J. B. Hyne, and F. L. Swinton, *J. Amer. Chem. Soc.* **92**, 6355 (1970).
585. G. I. Mackay and D. K. Bohme, *J. Amer. Chem. Soc.* **100**, 327 (1978).
586. M. J. Mackinnon, A. B. Lateef, and J. B. Hyne, *Can. J. Chem.* **48**, 2025 (1970).
587. J. Mahanty and B. W. Ninham, "Dispersion Forces," Academic Press (1976), and references quoted therein.
588. G. A. Mansoori, N. F. Carnahan, K. E. Starling, and T. W. Leland, Jr. *J. Chem. Phys.* **54**, 1523 (1971).
589. S. Marčelja, *Biochim. Biophys. Acta* **455**, 1 (1976).
590. S. Marčelja, *Croatica Chem. Acta* **49**, 347 (1977).
591. S. Marčelja, D. J. Mitchell, R. W. Ninham, and M. J. Sculley, *J. Chem. Soc. Faraday Trans. II* **73**, 630 (1977).
592. S. Marčelja and N. Radic, *Chem. Phys. Lett.* **42**, 129 (1976).
593. N. H. March and M. P. Tosi, *Phys. Lett.* **50A**, 224 (1974).
594. Z. Z. Margolin and F. A. Long, *J. Amer. Chem. Soc.* **95**, 2757 (1973).
595. R. B. Martin, *Nature* **271**, 94 (1978).
596. R. L. Martin and D. A. Shirley, *J. Amer. Chem. Soc.* **96**, 5299 (1974).
597. M. J. Mastroianni, M. J. Pikal, and S. Lindenbaum, *J. Phys. Chem.* **76**, 3050 (1972).
598. O. Matsuoka, E. Clementi, and M. Yoshimine, *J. Chem. Phys.* **64**, 1351 (1976).
599. B. W. Matthews, *Ann. Rev. Phys. Chem.* **27**, 493 (1976).
600. B. W. Matthews, P. M. Colman, J. N. Jansonius, K. Titani, K. A. Walsh, and H. Neurath, *Nature New Biol.* **238**, 41 (1972).
601. B. W. Matthews and L. H. Weaver, *Biochem.* **13**, 1719 (1974).
602. B. W. Matthews, L. H. Weaver, and W. R. Kester, *J. Biol. Chem.* **249**, 8030 (1974).
603. F. S. Mathews, M. Levine, and P. Argos, *Nature New Biol.* **233**, 15 (1971).
604. P. Mazur, *Ann. N.Y. Acad. Sci.* **125**, 658 (1956).
605. P. Mazur, *Science* **168**, 939 (1970).
606. J. H. McCreery, R. E. Christoffersen, and G. G. Hall, *J. Amer. Chem. Soc.* **98**, 7191, 7198 (1976).
607. W. F. McDevit and F. A. Long, *J. Amer. Chem. Soc.* **74**, 1773 (1952).

608. I. R. McDonald, "Report CECAM Workshop, Centre Européen de Calcul Atomique et Moléculaire," 91 Campus D'Orsay, Paris (1972).
609. I. R. McDonald and J. C. Rasaiah, *Chem. Phys. Lett.* **34**, 382 (1975).
610. J. D. McElroy, D. C. Manzerall, and G. Feher, *Biochim. Biophys. Acta* **333**, 261 (1974).
611. R. T. McIver, Jr. and J. S. Miller, *J. Amer. Chem. Soc.* **96**, 4323 (1974).
612. D. A. McQuarrie, "Statistical Mechanics," Harper and Row, New York (1976).
613. P. T. McTigue and A. R. Watson, *Aust. J. Chem.* **25**, 777 (1972).
614. W. R. Melander, *Chem. Phys. Lett.* **28**, 114 (1974).
615. F. M. Menger and K. S. Venkatasubban, *J. Org. Chem.* **41**, 1868 (1976).
616. L. Menninga and J. B. F. N. Engberts, *J. Amer. Chem. Soc.* **98**, 7652 (1976).
617. L. Menninga and J. B. F. N. Engberts, *J. Org. Chem.* **41**, 3101 (1976).
618. L. Menninga and J. B. F. N. Engberts, *J. Org. Chem.* **42**, 2694 (1977).
619. M. Metropolis, A. W. Rosenbluth, M. N. Rosenbluth, A. N. Teller, and E. Teller, *J. Chem. Phys.* **21**, 1087 (1953).
620. A. S. Mildvan, *Ann. Rev. Biochem.* **43**, 357 (1974).
621. J. R. Miller, *Science* **189**, 221 (1975).
622. F. J. Millero, J. C. Ahluwalia, and L. G. Hepler, *J. Chem. Eng. Data* **9**, 319 (1964).
623. D. J. Mitchell, B. W. Ninham, and B. A. Pailthorpe, *Chem. Phys. Lett.* **51**, 257 (1977).
624. D. J. Mitchell, B. W. Ninham, and B. A. Pailthorpe, *J. Chem. Soc. Faraday Trans. II* **74**, 1098, 1116 (1978).
625. D. J. Mitchell, B. W. Ninham, and B. A. Pailthorpe, *J. Colloid Interface Sci.* **64**, 194 (1978).
626. E. A. Moelwyn-Hughes, "Physical Chemistry," 2nd ed., p. 389, Pergamon Press, Oxford (1961).
627. P. C. Moews and R. H. Kretsinger, *J. Mol. Biol.* **91**, 201 (1975).
628. R. K. Mohanty and R. E. Robertson, *Can. J. Chem.* **55**, 1319 (1977).
629. R. K. Mohanty, S. Sunder, and J. C. Ahluwalia, *J. Phys. Chem.* **76**, 2577 (1972).
630. W. J. Moore, "Physical Chemistry," 5th ed., p. 705, Longman, London (1972).
631. K. Morokuma and L. Pedersen, *J. Chem. Phys.* **48**, 3275 (1968).
632. K. Morokuma and J. R. Winick, *J. Chem. Phys.* **52**, 1301 (1970).
633. C. Moser (ed.), "Report of CECAM Workshop on Long Time Scale Events in Molecular Dynamics," CECAM, Orsay, France (1974).
634. C. Moser (ed.), "Report of CECAM Workshop on Protein Refinement," CECAM, Orsay, France (1974).
635. J. Moult, A. Yonath, W. Traub, A. Smilansky, A. Podjarny, D. Rabinovich, and A. Saya, *J. Mol. Biol.* **100**, 179 (1976).
636. P. Mukerjee, P. Kapauan, and H. G. Meyer, *J. Phys. Chem.* **70**, 783 (1966).
637. P. Mukerjee and A. Ray, *J. Phys. Chem.* **70**, 2138, 2144, 2150 (1966).
638. J. Murto, *Suomen Kemistilehti* **B34**, 92 (1961).
639. J. Murto and A. M. Hirro, *Suomen Kemistilehti* **B37**, 177 (1964).
640. A. H. Narten, *J. Phys. Chem.* **74**, 765 (1970).
641. A. H. Narten, OWRL Department, No. ONRL-4578 (1970).
642. A. H. Narten, *J. Chem. Phys.* **56**, 5681 (1972).
643. A. H. Narten, M. D. Danford, and H. A. Levy, *Disc. Faraday Soc.* **43**, 97 (1967).
644. A. H. Narten and H. A. Levy, *J. Chem. Phys.* **55**, 2263 (1971).

645. A. H. Narten and H. A. Levy, *in* "Water: A Comprehensive Treatise" (F. Franks, ed.), Vol. 1, p. 311, Plenum Press, New York (1972).

646. A. H. Narten, F. Vaslow, and H. A. Levy, *J. Chem. Phys.* **58**, 5017 (1973).

647. T. W. Nee and R. W. Zwanzig, *J. Chem. Phys.* **52**, 6353 (1970).

648. G. W. Neilson, "Workshop on Structure and Dynamics of Liquids with Ionic Interaction at I.L.L." (Grenoble) (1976).

649. G. W. Neilson and J. E. Enderby, *J. Phys. C.* **11**, L625 (1978).

650. G. W. Neilson, J. E. Enderby, and R. A. Howe, "I.L.L. Annual Report" (1976).

651. G. W. Neilson, R. A. Howe, and J. E. Enderby, *Chem. Phys. Lett.* **33**, 284 (1975).

652. G. Nemethy and H. A. Scheraga, *J. Chem. Phys.* **36**, 3382 (1962).

653. H. M. Neumann, *J. Solution Chem.* **6**, 33 (1977).

654. M. D. Newton and S. Ehrenson, *J. Amer. Chem. Soc.* **93**, 4971 (1971).

655. J. D. Nicholas, P. Barnes, and J. L. Finney, "Abstr. 4th European Crystallographic Meeting, Oxford," A375 (1977).

656. A. Nickon, J. J. Frank, D. F. Covey, and Y.-i. Lin, *J. Amer. Chem. Soc.* **96**, 7574 (1974).

657. B. W. Ninham and V. A. Parsegian, *Biophys. J.* **10**, 646 (1970).

658. N. Nodelman and J. C. Martin, *J. Amer. Chem. Soc.* **98**, 6597 (1976), and references cited therein.

659. C. E. Nockolds and R. H. Kretsinger, *Proc. Nat. Acad. Sci. USA* **69**, 581 (1972).

660. D. M. North, J. E. Enderby, and P. A. Egelstaff, *J. Phys. C.* Ser. 22, 1075 (1968).

661. B. Notstrand, I. Vaara, and K. K. Kannan, *Isozymes* **1**, 575 (1975).

662. T. Novak, A. S. Mildvan, and G. L. Kenyon, *Biochemistry* **12**, 1690 (1973).

663. D. G. Oakenfull, *J. Chem. Soc. Perkin Trans. 2*, 1006 (1973).

664. D. G. Oakenfull, *Aust. J. Chem.* **27**, 1423 (1974).

665. D. G. Oakenfull and D. E. Fenwick, *Aust. J. Chem.* **27**, 2149 (1974).

666. D. G. Oakenfull and D. E. Fenwick, *J. Phys. Chem.* **78**, 1759 (1974).

667. D. G. Oakenfull and D. E. Fenwick, *Aust. J. Chem.* **38**, 715 (1975).

668. R. A. Ogg, Jr. and O. K. Rice, *J. Chem. Phys.* **5**, 140 (1937).

669. I. Olovsson and P.-G. Jönsson, *in* "The Hydrogen Bond" (P. Schuster, G. Zundel, and C. Sandorfy, eds.), Vol. 3, p. 393, North-Holland, Amsterdam (1976).

670. A. R. Olson and R. J. Miller, *J. Amer. Chem. Soc.* **60**, 2687 (1938).

671. A. R. Olson and L. K. J. Tong, *J. Amer. Chem. Soc.* **66**, 1555 (1944).

672. L. Onsager, *J. Amer. Chem. Soc.* **58**, 1486 (1936).

673. L. Onsager, *J. Amer. Chem. Soc.* **43**, 189 (1939).

674. L. Onsager and M. Dupuis, *in* "Electrolytes" (B. Pesce, ed.), p. 27, Pergamon Press, New York (1967).

675. C. G. Overberger, R. C. Glowaky, and P. H. Vandewijer, *J. Amer. Chem. Soc.* **95**, 6008, 6014 (1973).

676. J. C. Owicki and H. A. Scheraga, *J. Amer. Chem. Soc.* **99**, 7403, 7413 (1977).

677. J. C. Owicki, L. L. Shipman, and H. A. Scheraga, *J. Phys. Chem.* **79**, 1794 (1975).

678. K. J. Packer, *Phil. Trans. Roy. Soc. B (London)* **278**, 59 (1977).

679. D. I. Page, *in* "Water: A Comprehensive Treatise" (F. Franks, ed.), Vol. 1, p. 333, Plenum Press, New York (1972).

680. M. I. Page, *Biochem. Biophys. Res. Commun.* **72**, 456 (1976).

681. U. K. Pandit and F. R. Mas Cabre, *J. Chem. Soc. Chem. Commun.* 552 (1971).

682. A. J. Parker, *Adv. Phys. Org. Chem.* **5**, 173 (1967).

683. A. J. Parker, *Chem. Rev.* **69**, 1 (1969).

684. V. A. Parsegian and B. W. Ninham, *Biophys. J.* **10**, 664 (1970).

685. V. A. Parsegian and B. W. Ninham, *J. Chem. Phys.* **52**, 4578 (1970).

686. G. H. Parsons and C. H. Rochester, *J. Chem. Soc. Faraday Trans. I*, **71**, 1069 (1975).

687. S. Patai and Y. Israeli, *J. Chem. Soc.*, 2020 (1960).

688. S. Patai and J. Zabicky, *J. Chem. Soc.*, 2030 (1960).

689. G. N. Patey and J. P. Valleau, *J. Chem. Phys.* **63**, 2334 (1975).

690. D. Patterson and M. Barbe, *J. Phys. Chem.* **80**, 2435 (1976).

691. Peking Insulin Structure Research Group, *Sci. Sin.* **17**, 752 (1974).

692. J. K. Percus and G. J. Yevick, *Phys. Rev.* **110**, 1 (1958).

693. G. Perdoncin and G. Scorrano, *J. Amer. Chem. Soc.* **99**, 6983 (1977).

694. B. Perlmutter-Hayman and R. Shinar, *Int. J. Chem. Kin.* **9**, 1 (1977).

695. B. Perlmutter-Hayman and R. Shinar, *Int. J. Chem. Kin.* **7**, 453, 798 (1975).

696. M. F. Perutz, *Trans. Faraday Soc.* **42B**, 187 (1946).

697. M. F. Perutz, *Nature* **228**, 726 (1970).

698. M. F. Perutz, H. Muirhead, L. Mazzarella, P. A. Crowther, J. Greer, and J. V. Kilmartin, *Nature* **222**, 1240 (1969).

699. M. F. Perutz and H. Raidt, *Nature* **255**, 256 (1975).

700. M. F. Perutz and L. F. Ten Eyck, *Cold Spring Harbor Symp. Quant. Biol.* **36**, 295 (1972).

701. R. C. Petersen, *J. Org. Chem.* **29**, 3133 (1964).

702. G. A. Petsko, *J. Mol. Biol.* **96**, 381 (1975).

703. D. C. Phillips, *Sci. Amer.* **215** (11), 78 (1966).

704. D. C. Phillips, *Harvey Lect.* **66**, 135 (1971).

705. D. C. Phillips and D. Grace, "Abstr. 4th European Crystallographic Meeting, Oxford," A352 (1977).

706. R. A. Pierotti, *J. Phys. Chem.* **67**, 1840 (1963).

707. R. A. Pierotti, *J. Phys. Chem.* **69**, 281 (1965).

708. R. A. Pierotti, *Chem. Rev.* **76**, 717 (1976).

709. G. Placzek, *Phys. Rev.* **86**, 377 (1952).

710. Y. Pointud, J. Juillard, L. Avedikian, J.-P. Morel, and M. Ducros, *Thermodynamic Acta* **8**, 423 (1974).

711. R. J. Poljak, L. M. Amzel, H. P. Avey, B. L. Chen, R. P. Phizackerley, and F. Saul, *Proc. Nat. Acad. Sci. USA* **70**, 3305 (1973).

712. H. Popkie, H. Kistenmacher, and E. Clementi, *J. Chem. Phys.* **59**, 1325 (1973).

713. J. A. Pople, *Proc. Roy. Soc. A* **205**, 163 (1951).

714. G. N. J. Port and A. Pullman, *FEBS Lett.* **31**, 70 (1973).

715. G. N. J. Port and A. Pullman, *Theor. Chim. Acta* **31**, 231 (1973).

716. G. N. J. Port and A. Pullman, *Int. J. Quantum Chem., Quantum Biol. Symp.* **1**, 21 (1974).

717. K. T. Potts, *Chem. Rev.* **61**, 87 (1961).

718. J. G. Powles, *Adv. in Physics* **22**, 1 (1973).

719. K. P. Prasad and J. C. Ahluwalia, *J. Solution Chem.* **5**, 491 (1976).

720. L. R. Pratt and D. Chandler, *J. Chem. Phys.* **67**, 3683 (1977).

721. P. L. Privalov and N. N. Khechinashvili, *J. Mol. Biol.* **86**, 665 (1974).

722. A. Pullman, G. Alagona, and J. Tomasi, *Theor. Chim. Acta* **33**, 87 (1974).

723. A. Pullman and A. M. Armbrusta, *Int. J. Quantum Chem. Symposium* **1S**, 169 (1974).

724. A. Pullman and B. Pullman, *Quart. Rev. Biophys.* **7**, 505 (1975).

725. B. Pullman, *in* "Environmental Effects on Molecular Structure and Properties," (B. Pullman, ed.), pp. 55–80, Reidel Publishing Co., Dordrecht, Holland (1975).
726. F. A. Quiocho and W. N. Lipscomb, *Adv. Prot. Chem.* **25**, 1 (1971).
727. N. Quirke and A. K. Soper, *J. Phys. C.* **10**, 1802 (1977).
728. I. V. Radchenko and A. I. Ryss, *J. Struct. Chem.* **6**, 171 (1965).
729. A. Rahman, *Phys. Rev.* **136**, A405 (1964).
730. A. Rahman, *J. Chem. Phys.* **45**, 2585 (1966).
731. A. Rahman, "Molecular Dynamics Workshop, CECAM Report," Orsay, 1974.
732. A. Rahman and F. H. Stillinger, *J. Chem. Phys.* **55**, 3336 (1971).
733. A. Rahman and F. H. Stillinger, *J. Amer. Chem. Soc.* **95**, 7943 (1973).
734. A. Rahman, F. H. Stillinger, and H. L. Lemberg, *J. Chem. Phys.* **63**, 5223 (1975).
735. C. N. R. Rao, *in* "Water: A Comprehensive Treatise" (F. Franks, ed.), Vol. 1, p. 93, Plenum Press, New York (1972).
736. Z. Rappoport and J. Kaspi, *J. Amer. Chem. Soc.* **96**, 586 (1974).
737. J. C. Rasaiah and H. L. Friedman, *J. Chem. Phys.* **50**, 3965 (1969).
738. J. A. Raymond, Y. Lin, and A. L. de Vries, *J. Ex. Zool.* **193**, 125 (1975).
739. C. Reichardt, *Angew. Chem.* **77**, 30 (1965).
740. C. Reichardt and K. Dimroth, *Forschr. Chem. Forsch.* **11**, 1 (1968).
741. H. Reiss, *Adv. Chem. Phys.* **9**, 1 (1965).
742. J. Requena, D. E. Brooks, and D. A. Haydon, *Colloid Interface Sci.* **1**, 27 (1976).
743. W. L. Reynolds and R. W. Lumry, *J. Chem. Phys.* **23**, 2460 (1959).
744. S. A. Rice and P. Gray, "The Statistical Mechanics of Simple Liquids," Interscience, New York (1965).
745. F. M. Richards, *J. Mol. Biol.* **82**, 1 (1974).
746. F. M. Richards, *Ann. Rev. Biophys. Bioeng.* **6**, 151 (1977).
747. F. M. Richards and H. W. Wyckoff, in "The Enzymes" (P. D. Boyer, ed.), Vol. 4, p. 647, Academic Press, New York (1970).
748. W. G. Richards, "Quantum Pharmacology," Butterworths, London (1977).
749. W. G. Richards, T. E. H. Walker, and R. K. Hinkley, "Bibliography of *Ab Initio* Molecular Wave Functions," Clarendon Press, Oxford (1970), Supplement for 1970–73 (1974), Supplement for 1974–77 (1978).
750. M. Richardson and F. G. Soper, *J. Chem. Soc.* 1873 (1929).
751. C. D. Ritchie, *Acc. Chem. Res.* **5**, 348 (1972).
752. C. D. Ritchie, *J. Amer. Chem. Soc.* **97**, 1170 (1975) and papers cited therein.
753. R. E. Robertson, *Can. J. Chem.* **33**, 1536 (1955).
754. R. E. Robertson, *Can. J. Chem.* **42**, 1707 (1964).
755. R. E. Robertson, *Prog. Phys. Org. Chem.* **4**, 213 (1969).
756. R. E. Robertson and B. Rossall, *Can. J. Chem.* **49**, 1441 (1971).
757. R. E. Robertson, B. Rossall, and W. A. Redmond, *Can. J. Chem.* **49**, 3665 (1971).
758. R. E. Robertson and S. Singh, *Can. J. Chem.* **55**, 2582 (1976).
759. R. E. Robertson and S. E. Sugamori, *J. Amer. Chem. Soc.* **91**, 7254 (1969).
760. R. E. Robertson and S. E. Sugamori, *Can. J. Chem.* **50**, 1353 (1972).
761. D. R. Robinson and W. P. Jencks, *J. Amer. Chem. Soc.* **87**, 2462 (1965).
762. R. A. Robinson and A. Peiperl, *J. Phys. Chem.* **67**, 1723 (1963).
763. C. H. Rochester, "Acidity Functions," Chap. 6, Academic Press, New York (1970).
764. G. A. Rogers and T. Bruice, *J. Amer. Chem. Soc.* **96**, 2473 (1974).
765. D. Ronis, L. Martina, and J. M. Deutch, *Chem. Phys. Lett.* **46**, 53 (1977).

766. M. Roseman and W. P. Jencks, *J. Amer. Chem. Soc.* **97**, 631 (1975).
767. M. P. Rosi, and F. G. Fumi, *J. Phys. Chem. Solids* **25**, 45 (1964).
768. B. Rossall and R. E. Robertson, *Can. J. Chem.* **53**, 869 (1975).
769. J. S. Rowlinson, *Trans. Faraday Soc.* **45**, 974 (1949).
770. J. S. Rowlinson, *Trans. Faraday Soc.* **47**, 120 (1951).
771. J. S. Rowlinson, *J. Chem. Phys.* **19**, 827 (1951).
772. J. S. Rowlinson, "Liquids and Liquid Mixtures," Butterworth, London (1969).
773. A. Rühlmann, D. Kukla, P. Schwager, K. Bartels, and R. Huber, *J. Mol. Biol.* **77**, 417 (1973).
774. J. A. Rupley, *in* "Structure and Stability of Biological Macromolecules" (S. N. Timasheff and G. D. Fasman, eds.), Dekker, New York (1969).
775. A. I. Ryss and I. V. Radchenko, *J. Struct. Chem.* **5**, 489 (1964).
776. J. R. Sabin, R. E. Harris, T. W. Archibald, P. A. Kollman, and L. C. Allen, *Theor. Chim. Acta* **18**, 235 (1970).
777. H. Sadek and F. Y. Khalil, *Z. Physik. Chem.* **61**, 63 (1968).
778. P. P. S. Saluja, T. M. Young, R. F. Rodewald, F. H. Fuchs, D. Kohli, and R. Fuchs, *J. Amer. Chem. Soc.* **99**, 2949 (1977).
779. D. R. Sandstrom and H. W. Dodgen, *J. Chem. Phys.* **67**, 473 (1977).
780. G. N. Sarkisov, V. G. Dashevsky, and G. G. Malenkov, *Mol. Phys.* **27**, 1249 (1974).
781. K. Sasaki, S. Dockerill, D. A. Adamiak, I. J. Tickle, and T. L. Blundell, *Nature* **257**, 751 (1975).
782. J. J. Savage and R. H. Wood, *J. Solution Chem.* **5**, 733 (1976).
783. D. Sayre, *Acta Cryst.* **A28**, 210 (1972).
784. D. Sayre, *Acta Cryst.* **A30**, 180 (1974).
785. D. Sayre, *in* "Crystallographic Computing Techniques" (F. R. Ahmed, ed.), p. 322, Munksgaard, Copenhagen (1976).
786. W. J. Scanlon and D. Eisenberg, *J. Mol. Biol.* **98**, 485 (1975).
787. F. L. Schadt, T. W. Bentley, and P. v. R. Schleyer, *J. Amer. Chem. Soc.* **98**, 7667 (1976).
788. F. L. Schadt, III, C. J. Lancelot, and P. v. R. Schleyer, *J. Amer. Chem. Soc.* **100**, 228 (1978).
789. L. L. Schaleger and F. A. Long, *Adv. Phys. Org. Chem.* **1**, 1 (1963).
790. L. L. Schaleger and C. N. Richards, *J. Amer. Chem. Soc.* **92**, 5565 (1970).
791. M. F. Schmid and J. R. Herriott, *J. Mol. Biol.* **103**, 175 (1976).
792. B. P. Schoenborn, *Nature* **224**, 143 (1969).
793. B. P. Schoenborn, *Cold Spring Harbor Symp. Quant. Biol.* **36**, 569 (1971).
794. B. P. Schoenborn and J. C. Norvell, *Acta Cryst.* **A31** (S3), S32 (1975).
795. G. E. Schultz, M. Elzinga, F. Marx, and R. H. Schirmer, *Nature* **250**, 120 (1974).
796. R. Scordamaglia, F. Cavallone, and E. Clementi, *J. Amer. Chem. Soc.* **99**, 5545 (1977).
797. J. M. W. Scott and R. E. Robertson, *Can. J. Chem.* **50**, 167 (1972).
798. N. C. Seeman, R. O. Day, and A. Rich, *Nature* **253**, 324 (1975).
799. I. M. Shapovalov and I. V. Radchenko, *J. Struct. Chem.* **10**, 804 (1969).
800. I. M. Shapovalov, I. V. Radchenko, and M. K. Lisovitskaya, *Zh. Strukt. Khim.* **4**, 10 (1963).
801. I. M. Shapovalov, I. V. Radchenko, and M. K. Lisovitskaya, *J. Struct. Chem.* **13**, 121 (1972).
802. C. Shin, I. Worsley, and C. M. Criss, *J. Solution Chem.* **5**, 876 (1976).

803. K. Shinoda, *J. Phys. Chem.* **81**, 1300 (1977).

804. L. L. Shipman, *Chem. Phys. Lett.* **31**, 361 (1975).

805. L. L. Shipman, J. C. Owicki, and H. A. Scheraga, *J. Phys. Chem.* **78**, 2055 (1974).

806. L. L. Shipman and H. A. Scheraga, *J. Chem. Phys.* **78**, 909 (1974).

807. J. D. Shore, H. Gutfreund, R. L. Brooks, D. Santiago, and P. Santiago, *Biochem.* **13**, 4185 (1974).

808. A. Shrake and J. M. Rupley, *J. Mol. Biol.* **79**, 351 (1973).

809. O. Sinanoglu, "Mol. Ass. Biol. Proc. Int. Symp. 1967," p. 427 (1968).

810. O. Sinanoglu, *in* "The World of Quantum Chemistry," (R. Daudel and B. Pullman, eds.), pp. 265–71, Reidel Publishing Co., Dordrecht, Holland (1974).

811. K. Singer, A. Taylor, and J. V. L. Singer, *Mol. Phys.* **33**, 1757 (1977).

812. S. J. Singer, *Adv. Prot. Chem.* **17**, 1 (1962).

813. S. Singh and R. E. Robertson, *Can. J. Chem.* **54**, 1246 (1976).

814. T. D. Singh and R. W. Taft, *J. Amer. Chem. Soc.* **97**, 3867 (1975).

815. S. G. Smith, A. H. Fainberg, and S. Winstein, *J. Amer. Chem. Soc.* **83**, 618 (1961).

816. R. A. Sneen, *Acc. Chem. Res.* **6**, 46 (1973).

817. A. K. Soper, Ph.D. thesis (Univ. of Leicester), 1977.

818. A. K. Soper, G. W. Neilson, J. E. Enderby, and R. A. Howe, *J. Phys. C.* **10**, 1793 (1977).

819. P. E. Sørensen, *Acta Chem. Scand.* **A30**, 673 (1976).

820. H. A. Staab, *Chem. Ber.* **89**, 1927 (1956).

821. C. L. Standley and R. F. Kruh, *J. Chem. Phys.* **34**, 1450 (1961).

822. L. J. Stangeland, L. Senatore, and E. Ciuffarin, *J. Chem. Soc., Perkin Trans. 2*, 852 (1972).

823. W. Steigemann, J. Deisenhofer, and R. Huber, *in* "Crystallographic Computing Techniques" (F. R. Ahmed, ed.), p. 302, Munksgaard, Copenhagen (1976).

824. W. Steigemann and E. Weber, "Abstr. 4th European Crystallographic Meeting, Oxford," A315 (1977).

825. T. A. Steitz, R. Henderson, and D. M. Blow, *J. Mol. Biol.* **46**, 337 (1969).

826. F. H. Stillinger, *J. Chem. Phys.* **74**, 3677 (1970).

827. F. H. Stillinger, *J. Solution Chem.* **2**, 141 (1973).

828. F. H. Stillinger, *Adv. Chem. Phys.* **31**, 1 (1975).

829. F. H. Stillinger, *Israel J. Chem.* **14**, 130 (1975).

830. F. H. Stillinger and A. Rahman, *J. Chem. Phys.* **57**, 1281 (1972).

831. F. H. Stillinger and A. Rahman, *J. Chem. Phys.* **60**, 1545 (1974).

832. F. H. Stillinger and A. Rahman, *J. Chem. Phys.* **61**, 4973 (1974).

833. G. W. Stockton and J. S. Martin, *J. Amer. Chem. Soc.* **94**, 6921 (1972).

834. G. Stout and J. B. F. N. Engberts, *J. Org. Chem.* **39**, 3800 (1974).

835. A. Streitwieser, "Solvolytic Displacement Reactions," McGraw-Hill, New York (1962).

836. R. M. Stroud, L. M. Kay, and R. E. Dickerson, *J. Mol. Biol.* **82**, 185 (1974).

837. E. J. R. Sudhölter and J. B. F. N. Engberts, *Rec. Trav. Chim.* **96**, 85 (1977).

838. A. Suggett, *in* "Water: A Comprehensive Treatise" (F. Franks, ed.), Vol. 4, p. 519, Plenum Press, New York (1975).

839. J. Suurkuusk, *Acta Chem. Scand.* **B28**, 409 (1974).

840. C. G. Swain and D. C. Dittmer, *J. Amer. Chem. Soc.* **75**, 4627 (1953).

841. C. G. Swain, D. C. Dittmer, and L. E. Kaiser, *J. Amer. Chem. Soc.* **77**, 3737 (1955).

842. C. G. Swain, R. B. Mosely, and D. E. Brown, *J. Amer. Chem. Soc.* **77**, 3731 (1955).

843. K. M. Swamy and K. L. Narayana, *Z. Phys. Chem.* **256**, 1 (1975).
844. R. M. Sweet, H. J. Wright, J. Janin, C. H. Chothia, and D. M. Blow, *Biochem.* **13**, 4212 (1974).
845. C. A. Swenson and L. Koob, *J. Phys. Chem.* **74**, 3376 (1963).
846. M. C. R. Symons, *J. Chem. Res. (S)*, 140 (1978).
847. R. W. Taft and L. S. Levitt, *J. Org. Chem.* **42**, 916 (1977).
848. A. D. Tait, and G. G. Hall, *Theoret. Chim. Acta* **31**, 311 (1973).
849. M. J. Tait and F. Franks, *Nature* **230**, 91 (1971).
850. M. J. Tait, A. Suggett, F. Franks, S. Ablett, and P. Quickenden, *J. Solution Chem.* **1**, 131 (1972).
851. T. Takano, *J. Mol. Biol.* **110**, 537 (1977).
852. T. Takano, *J. Mol. Biol.* **110**, 560 (1977).
853. N. Tanaka and E. R. Thornton, *J. Amer. Chem. Soc.* **98**, 1617 (1976).
854. N. Tanaka and E. R. Thornton, *J. Amer. Chem. Soc.* **99**, 7300 (1977).
855. C. Tanford, "The Hydrophobic Effect," Wiley-Interscience, New York (1973).
856. H. N. V. Temperley, J. S. Rowlinson, and G. S. Rushbrooke, "Physics of Simple Liquids," North Holland Publishing Co., Amsterdam (1968).
857. D. S. Terekhova and I. V. Radchenko, *J. Struct. Chem.* **10**, 980 (1969).
858. D. S. Terekhova, A. I. Ryss and I. V. Radchenko, *J. Struct. Chem.* **10**, 807 (1969).
859. A. Thomson, *J. Chem. Soc. B*, 1798 (1970).
860. E. Tommila and P. J. Antikainen, *Acta Chem. Scand.* **9**, 825 (1955).
861. E. Tommila and S. Heitala, *Acta Chem. Scand.* **18**, 257 (1964).
862. E. Tommila and A. Hella, *Ann. Acad. Sci. fennicae* **AII**, 53 (1954).
863. E. Tommila, A. Koivisto, J. P. Lyyra, K. Antell, and S. Heino, *Ann. Acad. Sci. fennicae* **AII**, 47 (1952).
864. E. Tommila and M. L. Murto, *Acta Chem. Scand.* **17**, 1957 (1963).
865. E. Tommila, E. Paakala, U. K. Virtanen, A. Erva, and S. Varila, *Ann. Acad. Sci. fennicae* **AII**, 91 (1959).
866. E. Tommila and I. P. Pitkanen, *Acta Chem. Scand.* **9**, 825 (1955).
867. E. Tommila and I. P. Pitkanen, *Acta Chem. Scand.* **20**, 937 (1966).
868. M. L. Tonnett and A. N. Hambly, *Aust. J. Chem.* **24**, 703 (1971).
869. A. Tran, *Tetrahedron* **32**, 1903 (1976).
870. C. Treiner, *Can. J. Chem.* **55**, 682 (1977).
871. R. Triolo and A. H. Narten, *J. Chem. Phys.* **63**, 3624 (1975).
872. Y.-C. Tse and M. D. Newton, *J. Amer. Chem. Soc.* **99**, 611 (1977).
873. G. Tsoucaris, *in* "Crystallographic Computing Techniques" (F. R. Ahmed, ed.), p. 328, Munksgaard, Copenhagen (1976).
874. A. Tulinsky, R. L. Vandlen, C. N. Morimoto, N. V. Mani, and L. H. Wright, *Biochemistry* **12**, 4185 (1973).
875. H. Uedaira, *Bull. Chem. Soc. Japan* **45**, 3068 (1972).
876. R. Ueoka, M. Kato, and K. Ohkubo, *Tetrahedron Lett.*, 2163 (1977).
877. P. van Eikeren and D. L. Grier, *J. Amer. Chem. Soc.* **98**, 4655 (1976).
878. J. M. van Leeuwen, J. Groenveld, and J. de Boer, *Physica* **25**, 792 (1959).
879. L. Verlet, *Phys. Rev.* **159**, 98 (1967).
880. L. Verlet, *Phys. Rev.* **165**, 201 (1967).
881. C. A. Vernon, *Proc. Roy. Soc.* **B167**, 389 (1967).
882. A. Vesala, *Acta Chem. Scand.* **A28**, 839, 851 (1974).
883. P. C. Vogel and K. Heinzinger, *Z. Naturforsch.* **30a**, 789 (1975).

884. P. C. Vogel and K. Heinzinger, *Z. Naturforsch.* **31a**, 476 (1976).

885. K. Wada and D. Arnon, *Proc. Nat. Acad. Sci. USA* **68**, 3064 (1971).

886. A. Wagenaar, A. J. Kirby, and J. B. F. N. Engberts, *Tetrahedron Lett.* 489 (1976).

887. T. Graafland, A. Wagenaar, A. J. Kirby, and J. B. F. N. Engberts, submitted for publication.

888. T. E. Wainwright and B. J. Alder, *Nuovo Cimento*, **9**, *Suppl. Sec.* **10**, 116 (1958).

889. C. Walling and D. D. Tanner, *J. Amer. Chem. Soc.* **85**, 612 (1963).

890. D. T. Warner, *Ann. N.Y. Acad. Sci.* **125**, 605 (1965).

891. B. E. Warren, "X-Ray Diffraction," Addison-Wesley, Reading, Massachusetts (1969).

891a. K. D. Watenpaugh, T. N. Margulis, L. C. Sieker, and L. H. Jensen, *J. Mol. Biol.* **122**, 175 (1978).

892. K. D. Watenpaugh, L. C. Sieker, J. R. Herriott, and L. H. Jensen, *Acta Cryst.* **B29**, 943 (1973).

893. K. D. Watenpaugh, L. C. Sieker, and L. H. Jensen, *Proc. Nat. Acad. Sci. USA* **70**, 3857 (1973).

894. K. D. Watenpaugh, L. C. Sieker, and L. H. Jensen, ACA Meeting, Abstract EA7 (1977).

895. H. E. Watson, *Prog. Stereochem.* **4**, 255 (1969).

896. R. O. Watts, *Specialist Periodical Report: Statistical Mechanics* (K. Singer, ed.), Vol. 1, p. 1, Chem. Soc., London (1973).

897. R. O. Watts, E. Clementi, and J. Fromm, *J. Chem. Phys.* **61**, 2550 (1974).

898. R. O. Watts and I. J. McGee, "Liquid State Chemical Physics," Wiley-Interscience, New York (1976).

899. D. F. Waugh, *Adv. Protein Chem.* **9**, 325 (1954).

900. L. H. Weaver, W. R. Kester, and B. W. Matthews, *J. Mol. Biol.* **114**, 119 (1977).

901. J. L. Webb, "Enzyme and Metabolic Inhibitors," Vol. I, Academic Press, New York (1963).

902. J. D. Weeks, D. Chandler, and H. C. Andersen, *J. Chem. Phys.* **54**, 5237 (1971).

903. M. Weissmann and N. V. Cohan, *J. Chem. Phys.* **43**, 119 (1965).

904. O. Weres and S. A. Rice, *J. Amer. Chem. Soc.* **94**, 8983 (1972).

905. D. L. Wertz and J. R. Bell, *J. Inorg. Nucl. Chem.* **35**, 137 (1973).

906. D. L. Wertz and J. R. Bell, *J. Inorg. Nucl. Chem.* **35**, 861 (1973).

907. D. L. Wertz and R. F. Kruh, *J. Chem. Phys.* **50**, 4313 (1969).

908. D. L. Wertz, R. M. Lawrence, and R. F. Kruh, *J. Chem. Phys.* **43**, 2163 (1965).

909. A. V. Westerman, "Abstr. 4th European Crystallographic Meeting, Oxford," A372 (1977).

910. F. H. Westheimer, *Chem. Rev.* **61**, 265 (1961).

911. E. Whalley, *Adv. Phys. Org. Chem.* **2**, 93 (1964).

912. E. P. Wigner, *Phys. Rev.* **46**, 1002 (1934).

913. J. Winkelman, *Z. Phys. Chem.* **255**, 1109 (1974).

914. S. Winstein and A. H. Fainberg, *J. Amer. Chem. Soc.* **79**, 5937 (1957).

915. S. Winstein, A. H. Fainberg, and E. Grunwald, *J. Amer. Chem. Soc.* **79**, 1608, 4146 (1957).

916. S. Winstein, E. Grunwald, and H. W. Jones, *J. Amer. Chem. Soc.* **73**, 2700 (1951).

917. J. G. Winter and J. M. W. Scott, *Can. J. Chem.* **46**, 2887 (1968).

918. S. Wold and O. Exner, *Chemica Scripta* **3**, 5 (1973).

919. W. W. Wood, "Physics of Simple Liquids" (H. M. V. Temperley, G. S. Rush-

brooke, and J. S. Rowlinson, eds.), North Holland Publishing Co., Amsterdam (1968).

920. L. V. Woodcock and K. Singer, *Trans. Faraday Soc.* **67**, 12 (1971).

921. H. J. Wright, *J. Mol. Biol.* **79**, 1 (1973).

922. Y. C. Wu, *in* "Structure of Water and Aqueous Solutions" (W. A. Luck, ed.), Verlag Chemie und Physik, Verlag, Weinheim, B.R.D. (1974).

923. H. W. Wyckoff, D. Tsernoglou, A. W. Hansen, J. R. Knox, B. Lee, and F. M. Richards, *J. Biol. Chem.* **245**, 305 (1970).

924. M. Yaacobi and A. Ben-Naim, *J. Solution Chem.* 2, 425 (1973).

925. M. Yaacobi and A. Ben-Naim, *J. Phys. Chem.* 78, 175 (1974).

926. M. Ya Fishkis and T. B. Sobdeva, *J. Struct. Chem.* 15, 175 (1974).

927. M. Ya Fishkis and V. A. Zhmak, *J. Struct. Chem.* 15, 1 (1974).

928. P. L. Yeagle, C. H. Lochmuller, and R. W. Henkens, *Proc. Nat. Acad. Sci. USA* **72**, 454 (1972).

929. M. D. Zeidler, *in* "Structure of Water and Aqueous Solutions" (W. A. P. Luck, ed.), Chap. VII, Verlag Chemie, Weinheim (1974).

Author Index

Subject Index

Acid–base catalysis, 86, 210–220
Activation parameters, 172-176
  isochoric, 175
Acyl-enzyme intermediate, 101, 102, 107, 116
Amino acid substitution, 96, 98, 117
Antifreeze glycoprotein, 57

Bohr effect, 89
Bragg's law, 63

Cage structures, solvent effects, 236
Calorimetry, 59, 71, 74
Central force model, 327, 332
Charge ordering, 38, 41, 42
Charge shielding, 90
Charge transfer, 57
Chemical potential, 242
Clathrate hydrate, 56, 98, 116, 117, 252-256
Cohesive energy density, 152-154, 237
Compensation temperature, 179
Compressibility, 175, 176, 286
Concentration fluctuation, 145
Condensation reaction, 211
Correlation length, 272, 273
Crystal hydrate, 91
Crystal structure analysis, 54
Cybotactic region, 147

Decarboxylation, 139
Density maximum, 255
Deprotonation, 211, 213, 215
Detritiation, 140, 199, 211
Dielectric constant, 58, 115, 118, 121, 147, 286, 370, 371
  local, 86, 90

Differential scattering cross section, 21
Diffusional separation of caged radical pairs, 235
Dipole-dipole interactions, 297
Dipole moment, 150-152
  of water molecule, 283
Dispersion forces, 166
Distribution function, 242, 244; *see also* Pair correlation function

Electron density, 65, 114
  map, 64, 65, 66, 67, 109
Electron spin resonance, 149
Electron transfer, 48, 107, 117
Electron tunneling, 58, 117
Electrostatic interactions, 167-170
β-Elimination, 213
Endostatic conditions, 175
Enthalpy-entropy compensation, 172, 177, 179, 180, 185, 191, 193, 205, 210, 221, 222, 228, 229
  substrate dependence of extrema, 194
  compensation temperature, 179
Enthalpy of transfer, deuterium isotope effect, 187
Enzyme activation, 93
Enzyme activity, 82, 85, 86
Enzyme mechanism, 58, 100, 101, 102, 103, 105, 107
Enzyme-substrate interactions, 80, 86
Euler angles, 281, 302, 303, 333
Expansivity, 176

Fluctuation theory, 5
Fourier space, 63
Fourier transform, 63
Free energy of transfer, 185, 275

Compound Index